Connective Tissue in Health and Disease

Editor

Marcos Rojkind
Professor
Departments of Medicine and Pathology
Liver Research Center
Albert Einstein College of Medicine
Bronx, New York

CRC Press
Taylor & Francis Group
Boca Raton London New York

CRC Press is an imprint of the
Taylor & Francis Group, an **informa** business

First published 1990 by CRC Press
Taylor & Francis Group
6000 Broken Sound Parkway NW, Suite 300
Boca Raton, FL 33487-2742

Reissued 2018 by CRC Press

ISBN 13: 978-1-138-50678-7 (hbk)

DOI: 10.1201/9781315150512

Visit the Taylor & Francis Web site at http://www.taylorandfrancis.com and the
CRC Press Web site at http://www.crcpress.com

PREFACE

Frequently consider the connection of all things in the universe and their relation to one another. For in a manner all things are implicated with one another, and all in this way are friendly to one another; for one thing comes in order after another, and this is by virtue of the active movement and mutual conspiration and the unity of the substance.

The Meditations of Marcus Aurelius

The connective tissues are composed of cells and a rich extracellular matrix. The latter is complex and contains collagens, noncollagenous proteins and glycosaminoglycans. Different connective tissues contain different types of cells and variable quantities of extracellular matrix components. The compostion of the connective tissue as well as the tridimensional organization of its components is made to fit the structure of the tissue and, accordingly, there are significant qualitative and quantitative difrferences in the various tissues.

For many years it was considered that the connective tissue was a kind of glue made by mesenchymal cells, and whose function was to hold together the different cellular elements of a given tissue. Now, our concept of the connective tissue, and in particluar of the extracellular matrix, has changed drastically.

We have learned that the extracellular matrix forms a radar system that maintains the cells informed of the changes occurring in the microenvironment. The molecules that conform the extracellular matrix are in communication with each other and with similar molecules that conform the extracellular matrix are in communication with each other and with similar molecules that are associated with the cell surface or that are integral components of the plasma membrane. By this extensive mesh of multiple interactions, the extracellular matrix establishes contact with the cytoskeleton. The extracellular matrix is made by the different cells that compose the tissue. Its results from a complex network of signals that act by autocrine and paracrine mechanisms and indicate each cells the amount and type of matrix components to be produced. Accordingly changes in the types or amounts of cells, or in the cytokines that regulate the expression of the cells will result in abnormal deposition of matrix components. Moreover, the abnormal matrix will strongly influence the phenotypic expression of the cells in the surroundings.

The complexity of the matrix components, of their effects on gene expression, and of the cytokines that provide the signals that regulate extracellular matrix production are the topics of the series on *Connective Tissue in Health and Disease*. We shall try to cover the topics as they develop, but more importantly, to cover the gap between the basic apsects of the connective tissue structure and function and its implications in disease. We shall attempt to analyze animal models of human connective tissue diseases, describe recently developed methodologies that may prove to be useful for the aficionados. Experts in the field will review the composition and orginization of the extracellular matrix of specific organs. They will describe the biological activities of the different connective tissue cells and will analyze in great detail the cytokines and growth factors that modify cell proliferation and extracellular matrix production.

We are aware of the limitations of books, mainly the short half-life of the printed information. Therefore, when appropriate, the same topic may be discussed again in future issues. The editor and the members of the advisory board will attempt to maintain the proper balance of contents of this series. To publish issues that contain topics of scientific merit and of clinical relevance.

Marcos Rojkind
July 1989

THE EDITOR

Marcos Rojkind, M.D., Ph.D. is a professor of medicine and pathology and a member of the Liver Research Center at Albert Einstein College of Medicine in New York.

In 1960 he received an M.D. from the School of Medicine of the National University of Mexico. He completed his training in pathology in the Pathology Unit of the General Hospital under Professor Ruy Perez-Tamayo. He received a Helen Hay Whitney Foundation Fellowship and from 1962 to 1965 he received training in biochemistry, in the Department of Biochemistry, Albert Einstein College of Medicine, under Paul M. Gallop. In 1971 he received a Ph.D. degree in Biochemistry from the Centro de Investigacion y Estudios Avanzados in Mexico City.

From 1966 to 1967 he was a senior investigator of the Department of Biochemistry, at the Instituto Nacional de la Nutricion in Mexico. From 1967 to 1970 he was the chariman of the Department. From 1970 to 1975 he was Professor of the Department of Cell Biology of the Centro de Investigacion y de Estudios Avanzados in Mexico. From 1975 to 1976 he was visiting professor of medicine and biochemistry at Albert Einstein College of Medicine, and from 1976 to 1978 he was associate professor of medicine and biochemistry at the above mentioned institution. He returned to Mexico and from 1978 to 1986 he was professor of biochemistry at the Centro de Investigacion y de Estudios Avanzados del I.P.N.

Dr. Rojkind has over 100 publications in journals and books, and has participated in many national and international meetings and has trained many scientists. He was the president of the Mexican Association for the Study of the Liver from 1985 to 1986. In collaboration with Dr. Joan Rodes and Roberto Groszmann, Dr. Rojkind founded and organized the First Symposium International Hispanoparlante De Hepatologia in 1982. They are currently organizing the fourth Symposium.

His major interest is the connective tissue of the liver, including the mechanisms leading to cirrhosis and the mechanism of action of colchicine as an antifibrogenic agent. He is also interested in the mechanisms by which pathogenic strains of Entamoeba histolytica invade tissues.

ADVISORY BOARD

CONTRIBUTORS

Mark J. Czaja, M.D.
Assistant Professor
Department of Medicine
Albert Einstein College of Medicine
Bronx, New York

Lynetta J. Freeman, D.V.M., M.S.
Senior Scientist-Veterinary Surgeon
Ethicon Research Foundation
Ethicon Inc.
Sommerville, NJ

Jean-Alexis Grimaud, M.D. Ph.D.
Director
Laboratoire de Pathologie Cellulaire
Institut Pasteur
Lyon Cedex, France

Gerald A. Hegreberg, D.V.M., Ph.D.
Associate Professor
Department of Veterinary Microbiology and Pathology
Washington State University
Pullman, Washington

Simone Peyrol
Laboratoire de Pathologie Cellulaire
Institut Pasteur
Lyon Cedex, France

Juha Risteli, M.D., Ph.D.
Collagen Research Unit
Biocenter and Department of Medical Biochemistry
University of Oulu
Oulu, Finland

Leila Risteli, M.D., Ph.D.
Collagen Research Unit
Biocenter and Department of Medical Biochemisty
University of Oulu
Oulu, Finland

Moises Selman Lama, M.D.
Head
Clinical Research Division
National Institute of Respiratory Diseases
Mexico City, Mexico

Francis R. Weiner, M.D.
Assistant Professor
Department of Medicine
Albert Einstein College of Medicine
Bronx, New York

Mark A. Zern, M.D.
Director
Division of Gastroenterology
Roger Williams General Hospital
Associate Professor of Medicine
Brown University
Providence, Rhode Island

TABLE OF CONTENTS

Chapter 1

IMMUNOLOCALIZATION OF CONNECTIVE MATRIX COMPONENTS: METHODOLOGICAL APPROACH, LIMITS, AND SIGNIFICANCE IN THE LIVER

Jean-Alexis Grimaud and Simone Peyrol

TABLE OF CONTENTS

I. INTRODUCTION

During the last decade, fascinating progress has been made in the research of the structure and function of connective tissue.[22] The classic mechanical properties of the major collagen-made organs have been explained and reexamined using recent data on the properties of the macromolecular complex edifice which associate collagen, glycoproteins, and proteoglycans. As one of the best examples, the bidimensional parallel fiber collagen bundles, which regulate the tensile forces between muscle to skeleton, were mechanically compared to the tridimensional network of thin fibers associating water-binding proteoglycans adapted to regulate the multidirectional pressure which acts on articular cartilage. Thus, the appropriate biological property of a connective tissue is the result of a precise macromolecular organization of the cellular and extracellular components. Although there are no data directly pertaining to the biological and functional significance of the observed collagen polymorphism, wide variations in the relative distribution of collagen isotypes in various tissue and organs suggest that collagen polymorphism may play an important role in modulating the properties of the tissues in various physiological and pathological states. Besides structural and mechanical diversity of connective tissue, a large chapter of biological functions of connective matrix components has been opened in the last decade: gene expression, proliferation, maintenance of cell phenotype, and differentiation are regulated by extracellular matrix components.[23] In other words, data showing interactions between cells and collagen and/or associated glycoproteins are most important to understand the general mechanism which regulates tissular growth regeneration and repair in embryonic adult or pathological conditions.

Actually, it is evident that the considerable advance in knowledge of the molecular diversity and the ultrastructural polymorphism of the extracellular matrix components[8] has not been followed by significant progress in the meaning of such polymorphism at the cellular or physiopathological level. It seems time to note that the precise significance of a collagen isotype codistribution, the presence of minor collagen, and the rate of intermolecular cross-link pattern of fibrillar organization have to be understood and interpreted. Immunocytochemical techniques have been successfully applied to connective tissue research.[1,3,6,10,13,19,28] These studies have been possible due to established and precise procedures for the isolation of specific antibody preparation directed against a chosen antigenic determinant in the genetically distinct collagens and procollagens.[8,11,19]

The aim of this chapter is to focus on the interest of immunolocalization using the light and electron microscope with the following objectives: (1) to design the best methodological strategy for optimal morphological evidence of the matrix components and (2) to open perspectives on the mediation or interactive functions of connective matrix which is revealing its leading part in cell-cell dialogue and cell population interaction.

II. TECHNIQUES FOR IMMUNOLOCALIZATION OF HEPATIC CONNECTIVE MATRIX COMPONENTS: METHODOLOGICAL STRATEGY

Review of the literature concerning the various methods for immunolocalization of matrix proteins shows a large diversity of successful procedures. A comparative analysis reveals a common strategy based on four essential points:

1. Immunochemical properties of the antibodies (purity and specificity) and applicability to histological visualization
2. Antigenic preservation in tissues whose morphological aspects allow a precise ultrastructural localization (tissue fixation)
3. Access to the target antigen present in extra- and intracellular compartments (permeabilization)

4. Quality of the detection of the immune complex (choice of the tracer, choice of the method of detection, and choice of specific controls)

As a matter of fact, these four essential points proved to be sufficient to get satisfactory results, even if the different proposed processing exhibited minor changes.

A. THE PERFORMANCE OF THE ANTIBODIES

Several reviews are dedicated to the immunochemical properties of the anticollagen antibodies.[8,11,19] More precisely for the morphologist, the specific antibody has to be representative of its target antigen. In this sense, polyclonal antibodies very often offer a wider range of recognitive determinants in their arrangement, characteristic of the protein deposited in the tissue. For precise recognition of special configuration of the protein (degradative aspect—intermolecular cross-links, precursors) monoclonal antibodies may be used. Oligoclonal antibodies could help to understand the complexity of the target antigen at the ultrastructural level.

B. ANTIGENIC PRESERVATION AND TISSUE FIXATION

For immunolocalization of collagens, fixation appears as the critical step during the procedure. Two combined objectives are wanted: (1) retain the antigen *in situ* and (2) preserve the tissue structure. The main disadvantage is that fixation introduces intra- and intermolecular cross-links which dramatically decrease the antigenicity masking the original epitope recognized specifically by the antibodies. With preembedding methods, an additional problem occurs: the penetration of the antibodies is partly diminished by the fixative-induced cross-linking.

Before starting an electron microscopy (EM) study of immunolabeling, it is strongly recommended that the compatibility of the antigen and the specific antibody under investigation be screened using light-microscopic immunohistochemistry.

In Tables 1 and 2, various technical approaches indicate some critical stages in the immunolabeling techniques. The instability of collagen antigenicity under chemical fixative treatments has led workers in the field[5,10,13,18,24] to choose a method which is a compromise that takes care of the protein antigenicity and of the ultrastructural preservation of the tissue. Currently, matrix protein immunofluorescent staining is performed on fresh-frozen tissue sections, and EM labeling, on paraformaldehyde-fixed liver. Concentrations of glutaraldehyde (2%) and osmium tetroxide (1%) usually used in EM must be avoided. Fixation by perfusion through the portal vein of rat liver or through the largest vascular orifice of surgical human liver biopsies[5] allows a high frequency of intracellular labeling of collagen and glycoprotein precursors. Such results have also been obtained after fixation by immersion of small delicate samples such as chick embryonic liver 2 to 6 d old[2] (Figure 5) and human needle biopsies. In this latter case, fixation has to be carried out under controlled vacuum depression. It could be suggested that the above conditions allow a rapid fixation, and prevent artefactual diffusion of the precursors, outside their intracellular compartments. The duration of fixation does not seem to be a limiting factor since (1) intracellular precursors of matrix proteins have been visualized in chick embryonic hepatoblasts using preembedding immunoperoxidase labeling after 12-h PLP immersion fixation;[2] and (2) small blocks of perfused rat livers, stored after perfusion in fixative at 4°C until cryoprotection and ultracryotomy, are suitable for immunogold detection of procollagens in hepatocytes.[9] On the contrary, a heavy processing of cryopreservation necessary for subsequent thin sectioning[17] and a long-time storage of the specimen in the frozen state (−80°C) appeared unfavorable to such a visualization as could have been ascertained in retrospective studies of human liver biopsies.[13] This feature might be related to the denaturing effect of low temperature on collagen molecules, all the more as they are present in their precursor form unstabilized by

TABLE 1
Diversity of the Methods for Tissular Collagen Immunolabeling in the Liver by Light Microscopy

Histological Preimmunolabeling Tissue Treatments				Labeling of Hepatic Matrix Proteins					
Fixation	Embedding	Sectioning	Tracer	Species	Interstitial collagens	Basement membrane collagens	Glycoprotein	Proteoglycans	Refs.
Unfixed	Frozen	4—6 μm Air dried	Indirect IF	Calf	I, III + +				19
	Frozen	4—6 μm Air dried	Indirect IF	Rat/CCl_4	I, III, p-III + + +				15
	Frozen	4—6 μm Air dried	Indirect IF	Human	I, III, p-III + + +	IV +	FN, LAM + +		13
	Frozen	4—6 μm Air dried	Indirect IF	Human fibrotic		IV +	FN, LAM + +		14
	Frozen	4—6 μm Air dried	Indirect IF	Human fibrotic	I, III + +	AB, B + +			4, 27
	Same in *n*-hexane cooled with dry ice acetone	Same + acetone 10 min	Indirect IF	Bovine-human	I, III + +	IV +			16
	Frozen	4—6 μm Air dried	Indirect IF			IV and 7S + V, SC_1(VI) + +			28
Unfixed or perfusion with paraformaldehyde (PF) or PF-glutaraldehyde <0.2% in phosphate buffer	Same in liquid Freon	4—6 μm Air dried	Indirect IF	Rat	I, III + +	IV, V, VI + + +	FN +	LAM +	10
Unfixed	Frozen	4—6 μm Air dried	Indirect IF	Human				PDS[a]	26
Formalin/Gendre's or Bouin's fixative	Paraffin	3.5 μm and trypsin digestion	PAP	Human	I, III + +	IV +			3
Paraformaldehyde in cacodylate buffer	EPON	Ultramicrotome 1—2μm	Indirect IF or PAP after elimination of embedding medium	Chick embryo	I, III + +	IV +		FN, LAM + +	12

[a] PDS: Proteo dermatan sulfate.

TABLE 2
Diversity of the Methods for Tissular Collagen Immunolabeling in the Liver by Electron Microscopy

Histological Preimmunolabeling Tissue Treatments					Labeling of Hepatic Matrix Proteins					
Fixation	Embedding	Sectioning	Tracer	EM embedding	Species	Interstitial collagens	Basement membrane collagens	Glycoprotein	Proteoglycans	Refs.
Paraformaldehyde 4% in cacodylate buffer (immersion 5 h)	Cryopreservation with glycerol and freezing	Cryostat 10 μm	Indirect immunoperoxidase with hyaluronidase pretreatment	Epon	Human normal fibrotic	I III + +	IV +	FN +		13, 21
PLP[a] fixative	Same as above	4 μm	Same without pretreatment		Bovine	I III + +	IV +			24
Paraformaldehyde 4% in phosphate buffer (perfusion 30 min + immersion 3 h)	Cryopreservation with glycerol and freezing in methylbutane pre-cooled in liquid nitrogen	Cryostat 8 μm	Fab as primary antibodies + PAP	Med cast araldite	Rat, normal	I +	IVFN + +	LAM +		18
Extra- and intracellular										
Paraformaldehyde 4% in cacodylate buffer (perfusion 15 min)	Cryopreservation and freezing in isopentane cooled with liquid nitrogen	Vibratome 10—15 μm Cryostat 8 μm	Indirect immunoperoxidase with saponin treatment	Epon	Rat, human normal and fibrotic	I III pIII + +	IVFN + +			5, 6, 13
Paraformaldehyde 4% in cacodylate buffer PLP fixative (17) (immersion)	—	—	Indirect immunoperoxidase on entire embryos or hepatic rudiment	Epon	Chick embryos 2—6 d	I III pIII + + +	IVFN + +	LAM +		2
Paraformaldehyde 2.5% in phosphate buffer	Araldite	Ultramicrotome	PAP after elimination of embedding medium	—	Human CALD[b]		FN +			1
Extra- and intracellular										
Paraformaldehyde 2—4% in phosphate buffer (perfusion 10 min + immersion)	Cryopreservation sucrose	Ultracryotome 100—200 nm	Indirect protein A-gold complex 5—8 nm		Rat	pIII +	FN +		HSPG[c] +	9

[a] PLP: Periodate-lysine-paraformaldehyde
[b] CALD: Chronic active liver disease
[c] HSPG: Heparan sulfate proteoglycan

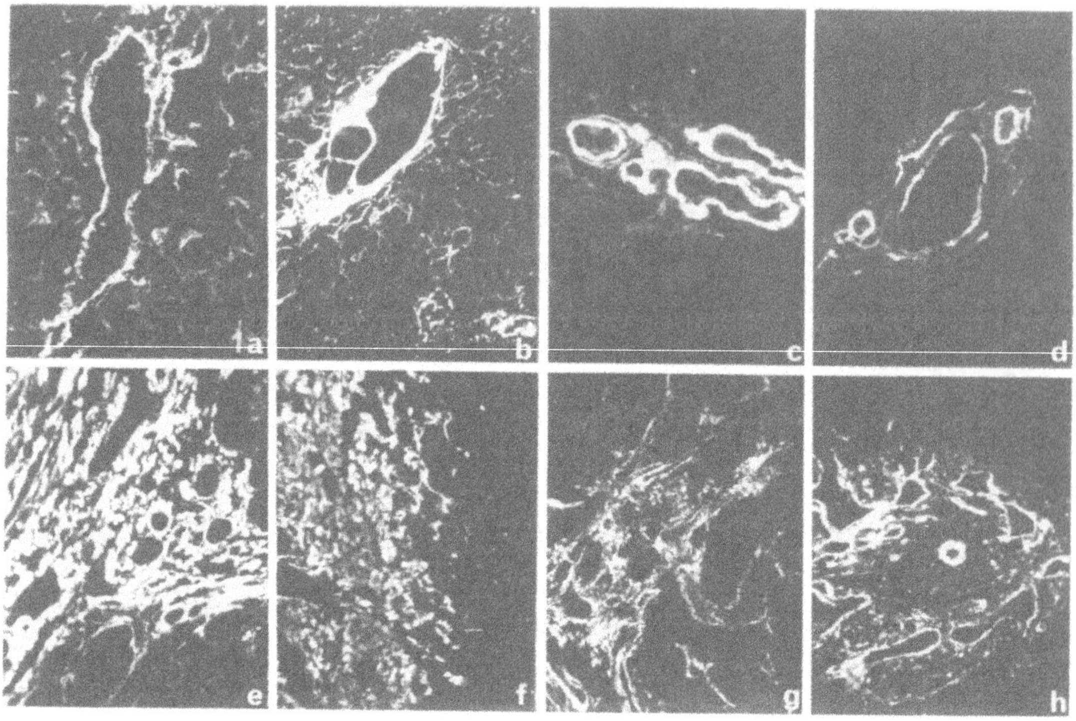

FIGURE 1. Immunofluorescence staining of human liver. (a, b, c, and d): normal; (e, f, g, and h): fibrotic. a, e: type I collagen; b, f: type III collagen; c, g: type IV collagen; a, h: laminin. (Magnification × 400.)

polymerization. This hypothesis has to be reexamined with the development of low temperature embedding procedure for EM using LOWICRYL 4 KM medium which has already given successful results for immunolabeling of extracellular collagens and laminin in skeletal muscle.[25]

C. DEMASKING THE ANTIGEN

The possibility that in multicomponent structures the reactive antigenic determinant may be covered or masked by one or more components has to be considered. Enzyme digestion of tissue sections is commonly used in immunohistochemical practice.[7] So, the second limiting factor for matrix protein immunolabeling is the difficulty of the specific immunoglobulins to reach the antigen in the tissue. First, matrix protein precursors are synthesized, elaborated, and secreted by the intracellular cisternal compartment of producing cells. Second, in the extracellular compartment, most of these proteins are involved in multimolecular edifices either to form large polymers stabilized by intra- and intermolecular cross-links or to build, with other proteins, functional connective units such as basement membranes. Furthermore, large polymers may associate to form larger structures such as bundles of fibers, stabilized, too, by chemical linkages. This situation led to the use of technical procedures, either to present directly by ultrathin sectioning antibodies to tissular antigens or to permeabilize the cytomembranes or the connective matrix which had trapped the proteins to be labeled.

Cryotomy of unfixed liver (3 to 5 μm) followed by immunofluorescent staining is the most commonly used method to examine the histological distribution of the different matrix proteins (Figure 1). Ultracryotomy of adequately fixed liver tissue followed by simple or multiple immunogold labeling represents the optimal way for EM detection of extra- and intracellular proteins. Recent evidence of the fine resolution of the method applied to rat liver[9] promises rapid future morphological demonstration of the codistribution of the matrix deposits in Disse space. Until now, immunoperoxidase staining of presectioned (4 to 20 μm) fixed liver tissue[18,24], combined with hyaluronidase treatment in order to permeabilize collagen-proteoglycan edifice[13] or combined with saponin treatment in order to permeabilize cytomembranes,[6] was practiced with good reproducibility. This group of preembedding immunolabeling techniques allowed (1) identification and localization of extracellular collagens and glycoproteins (fibronectin and laminin) and (2) detection of them in their site of intracellular biosynthesis.

Attempts for collagen immunotyping on embedded material have been developed in order to investigate both retrospective series of pathological material and small size specimens such as embryonic organ rudiments which have to be explored by serial semithin (1 to 2 μm) sectioning. Histological sections (3 to 5 μm) of routinely prepared biopsies (fixation with formalin or Bouin's solution and paraffin embedding) as well as semithin (1 to 2 μm) sections of paraformaldehyde-fixed, epon- or araldite-embedded specimens are suitable for subsequent light microscopic immunofluorescent or immunoperoxidase (PAP) staining of matrix components under the following conditions:

1. Sections have been stuck to glass slides.
2. Embedding medium has been carefully removed.
3. Tissue sections have been submitted to limited trypsin digestion, prior to immunolabeling.[3]

Under these conditions, interstitial collagens, fibronectin, elastin, and basement membrane protein (type IV collagen and laminin) have been detected. Postembedding immunolabeling techniques have been extended to EM. Collagens, fibronectin, and laminin can be visualized after LOWICRYL embedding, as previously noted,[25] and GMA embedding (Figure 3).

D. VISUALIZING THE IMMUNE COMPLEX WITH SPECIFICITY

Several performant tracers have been used with success: ferritin, horseradish peroxidase, and colloidal gold. When endogenous forms of the tracer (ferritin) exist, either a precise investigation of its tissular distribution or a previous efficient inhibition of its activity (peroxidase) must be performed. Sometimes the detection method requires several sequential steps (immunological linkings or affinity adsorption) which may introduce nonspecific labeling by tissular cross-reactivity. Each step has to be accurately controlled. Combined labeling using simultaneously multiple tracers is now necessary to demonstrate the intimate codistribution of proteins in the matrix macromolecular edifice.

III. SIGNIFICANCE OF CONNECTIVE MATRIX IMMUNOLABELING

A. MATRIX PROTEIN DISTRIBUTION AND TISSULAR COMPARTMENTS

In the normal liver, immunofluorescence staining of matrix proteins clearly shows two distinct domains of hepatic vascularization. The first one was interstitial collagens (types I, III) (Figures 1a and b) and basement membrane proteins (type IV collagen and laminin) (Figures 1c and d). They contribute together to the building of the supportive framework of the arborescent vascular and biliary tree.

The second one is poorly shown by light microscopy. However, the codistribution of types I and III collagen with fibronectin and type IV collagen was clearly demonstrated forming the perisinusoidal Disse space connective matrix. The similarity of this labeled structure with the silver-stained reticulin network is obvious, although the silver staining of

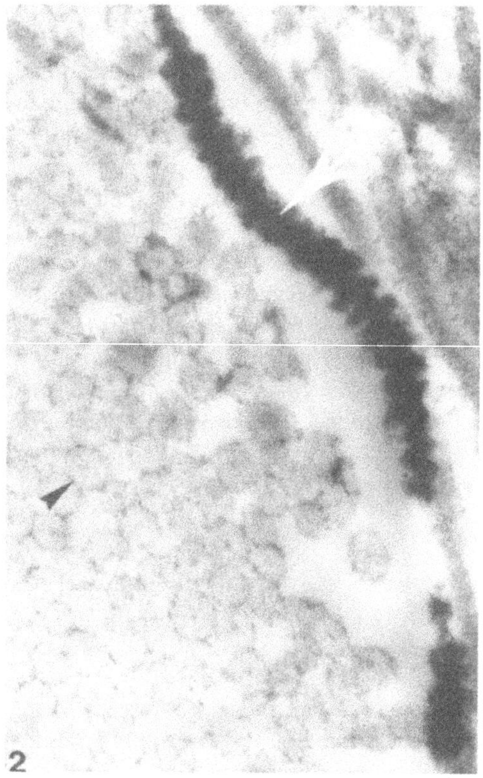

2

FIGURE 2. EM of immunoperoxidase labeling of type I collagen fibers in human cornea. White arrow: longitudinal section; black arrow: transverse section. (Magnification × 30,000.)

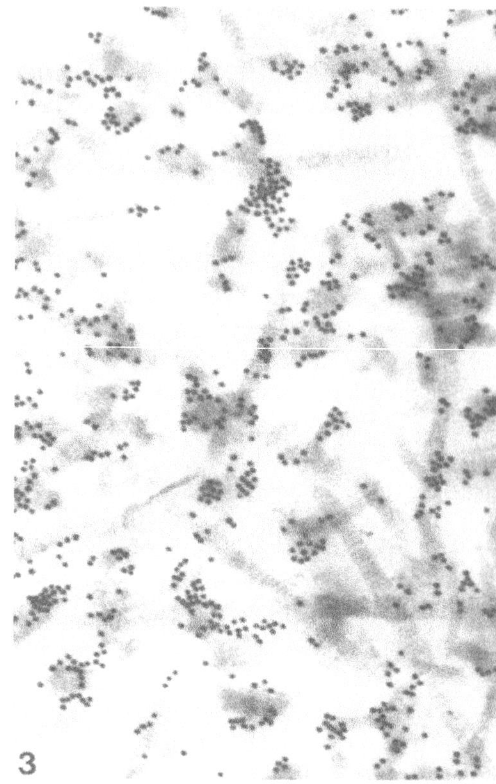

3

FIGURE 3. EM of immunogold labeling of type I collagen fibers in human dental pulp. (Magnification × 15,000.)

reticulin does not show specifically any ECM components. Its particular interest is to give a tridimensional and synthetic aspect of the perisinusoidal blood-hepatocyte connective tissue interface.

In fibrotic liver, complex deposits of polymorphic ECM components are present not only associated with the preexisting portal spaces, but also as neoformed matrix deposit forming interlobular fibrotic septa, perineoductular fibrosis, and periportal Disse space fibrosis. All the ECM components (interstitial and basement membranes) take part in the establishment of a new fibrotic state in the liver (Figures 1e, f, g, and h). Both mechanical and exchange functions are impaired.

B. MATRIX PROTEIN POLYMORPHISM: ELECTRON MICROSCOPIC VISUALIZATION IN EXTRA- AND INTRACELLULAR COMPARTMENTS

EM immunolabeling allows identification of collagen isotypes in their tissular polymeric form. The fibrillar form of interstitial collagens is organized in bundles, and the amorphous basement lamina is found around epithelial ducts and vascular endothelium (Figures 2 and 4). Other methods of detection give the same results (Figure 3). Intracellular processing of collagen precursors could have been demonstrated in the situation of active hepatic fibrogenesis (Figure 5). These data have pointed out the abilities of different cell types, mesenchymal or not, for synthesizing connective matrix.

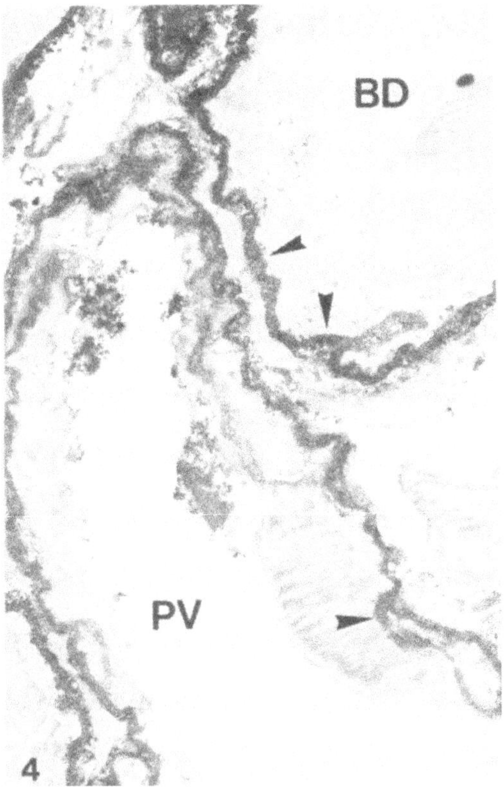

FIGURE 4. EM of immunoperoxidase labeling of type IV collagen in human fibrotic liver. Black arrow: type IV collagen of the basal lamina; (PV): portal vein; (BD): biliary duct. (Portal space magnification × 5000.)

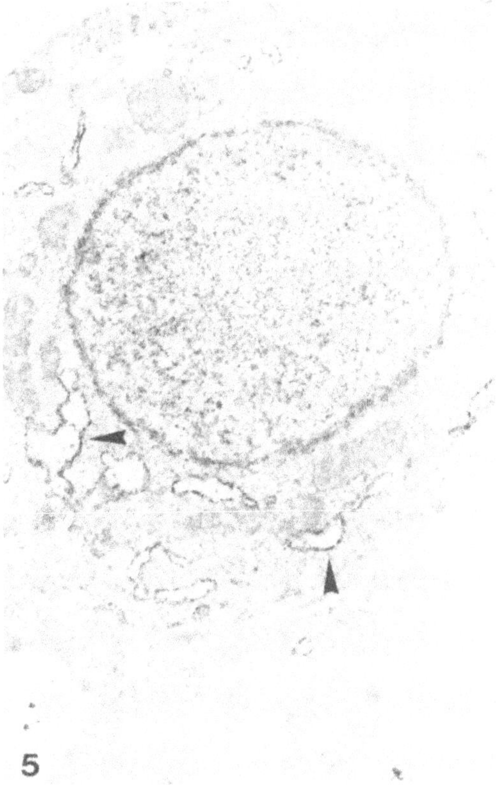

FIGURE 5. EM of immunoperoxidase labeling of type III collagen. Black arrow indicates labeling of intracellular precursors in ergastoplasmic cisternae of a hepatoblast in 4-d-old chick embryonic liver. (Magnification × 12,000.)

C. FIBRONECTIN — UBIQUITOUS AND PERMANENT MOLECULE IN THE MATRIX EDIFICE

In the initial stage of the granuloma formation, fibronectin acts as an intercellular adhesive molecule between the macrophages (Figure 6). Later, persistent deposits of fibronectin are present in the pericellular connective matrix as well as inside the matrix deposit as a macromolecular bridge between collagen fibers and bundles (Figure 7).[20] Out of the granuloma, fibronectin is always detected, associating perivascular or periepithelial connective areas to the dense fibrillar matrix framework.

D. CODISTRIBUTION OF INTERSTITIAL COLLAGENS AND PATTERNS OF CONNECTIVE MATRIX ORGANIZATION

As a quite general feature, type I and type III collagens appear codistributed. EM immunolocalization differentiates types I and III, respectively, as long polymers, periodic fibers with a large diameter and thin short fibrils. They may appear separately, forming dense fiber bundles (type I—Figure 13) or a loose fibrillar network (type III—Figure 12). More often they are associated in mixed fibrillar bundles. In normal conditions where type I and type III appear codistributed, the ratio type I/type III is variable. Type I predominance in well-ordered fiber bundles suggests a supportive function of such connective matrix organization as in lung alveolar septa (Figure 8). In contrast, predominance of type III found in loose perivascular connective matrix traduces a high rate of permeability and molecular diffusion (Figure 9).

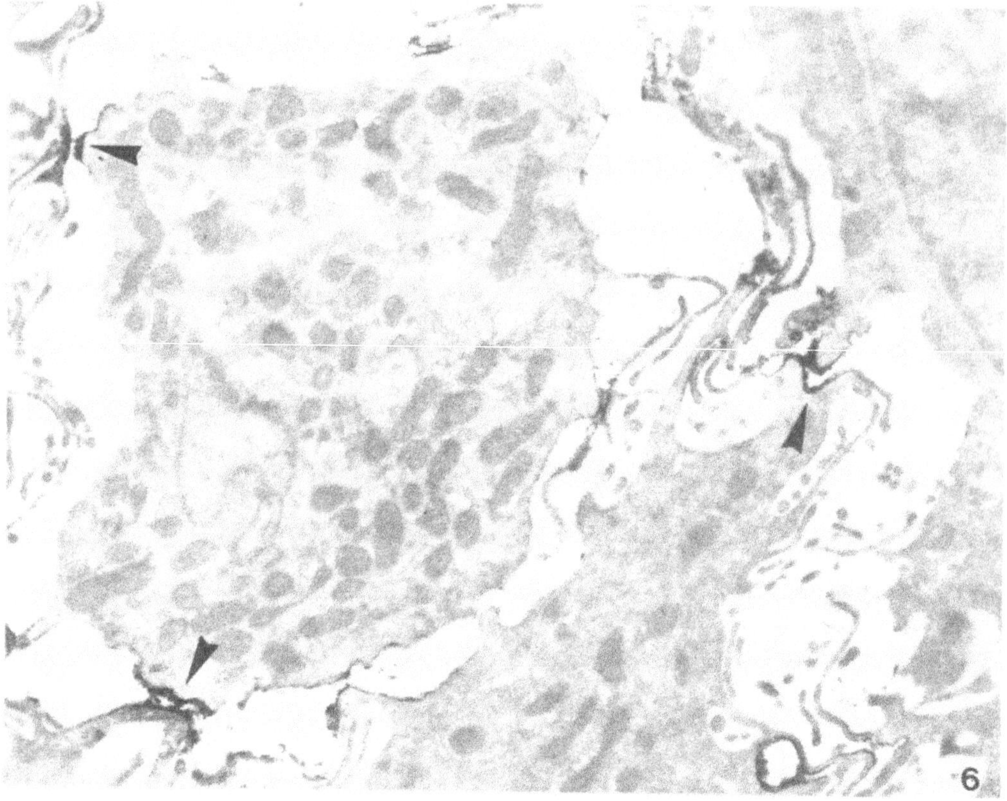

FIGURE 6. EM of immunoperoxidase labeling of fibronectin in human lung. Sarcoid granuloma. Black arrows indicate pericellular connective matrix, cell to cell contact. (Magnification × 13,600.)

In pathological conditions, such patterns are found in the neoformed fibrotic deposit. In some areas (perivascular and peribiliary), a loose connective matrix organization (LCMO, Figure 10) is abundant. A mixed association of fibrillar and basement membrane-like material with lamellar cytoplasmic cell processes characterizes LCMO (Figure 11).[13] The loose pattern observed in areas of active cellular transit and vascular exchange suggests permeability properties leading to a facilitated process of cell-cell recognition and activation.

In close contact to these zones (Figure 10) a pattern of dense connective matrix organization (DCMO) is generally observed. Dense bundles of monodirectional collagen fibers are themselves associated in larger tissular dense complex. They are characterized by a high ratio of type I/type III collagen. Such organization probably results in a weak accessibility of remodeling enzymatic attacks and a great stability during the fibrotic process.

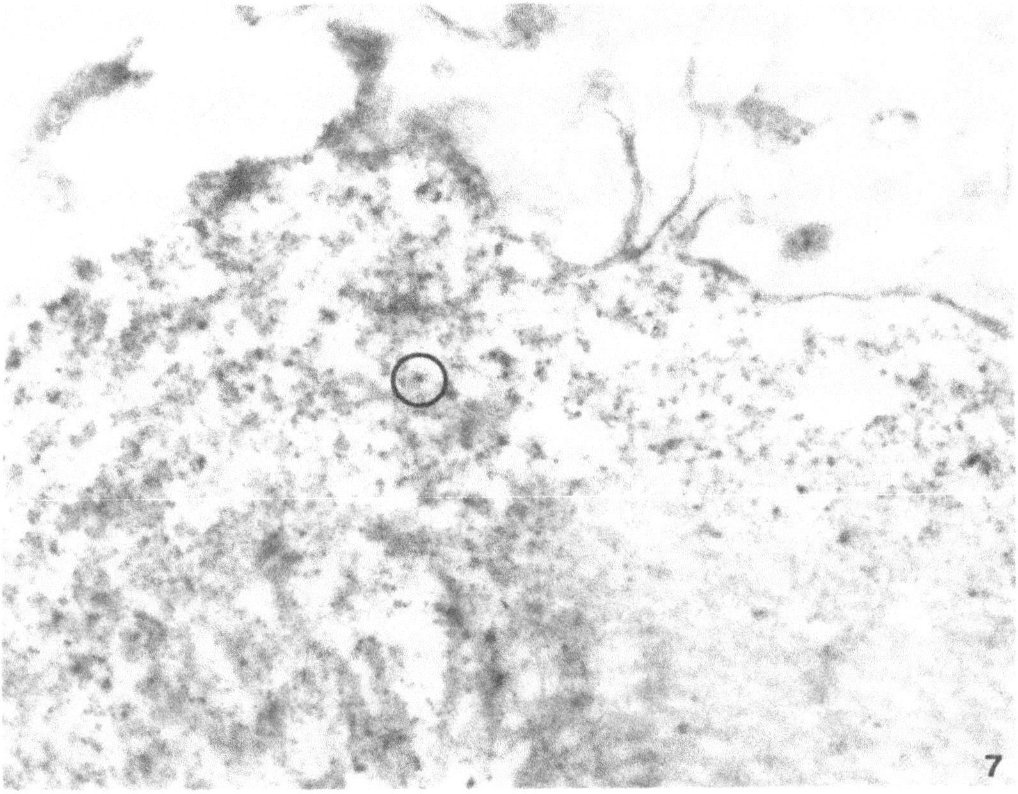

FIGURE 7. Same as Figure 6 but open circle indicates matrix deposit. (Magnification × 25,000.)

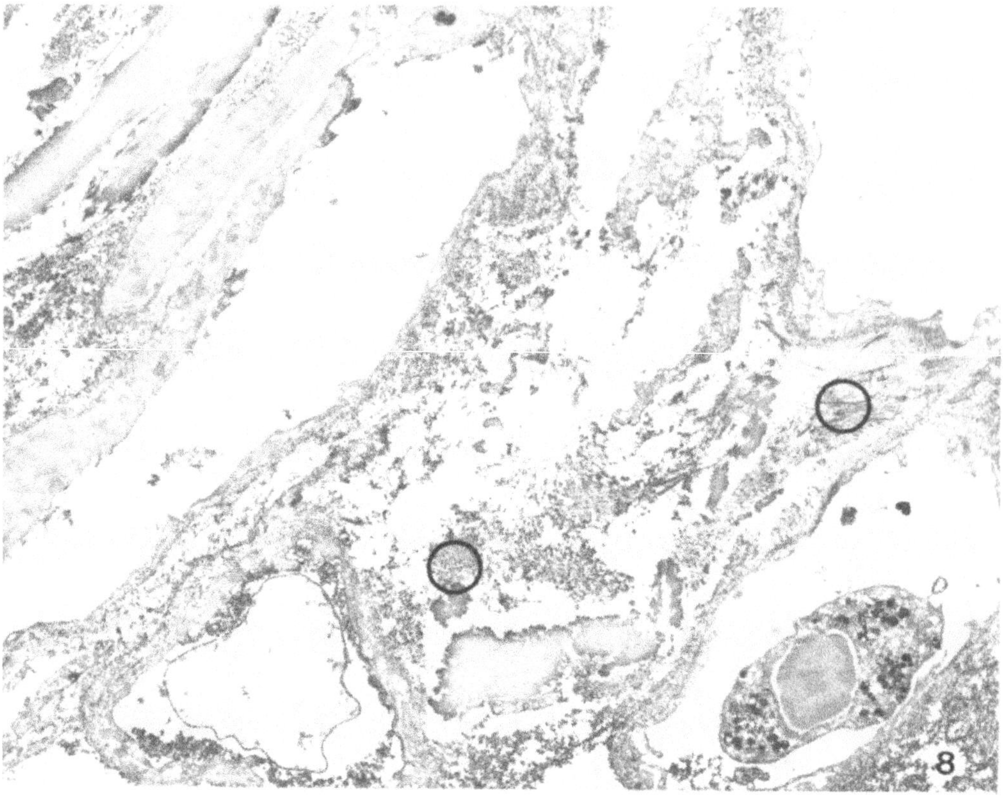

FIGURE 8. EM of immunoperoxidase labeling of type I collagen in human fibrotic lung. ○ indicates transverse and longitudinal sections. (Magnification × 5300.)

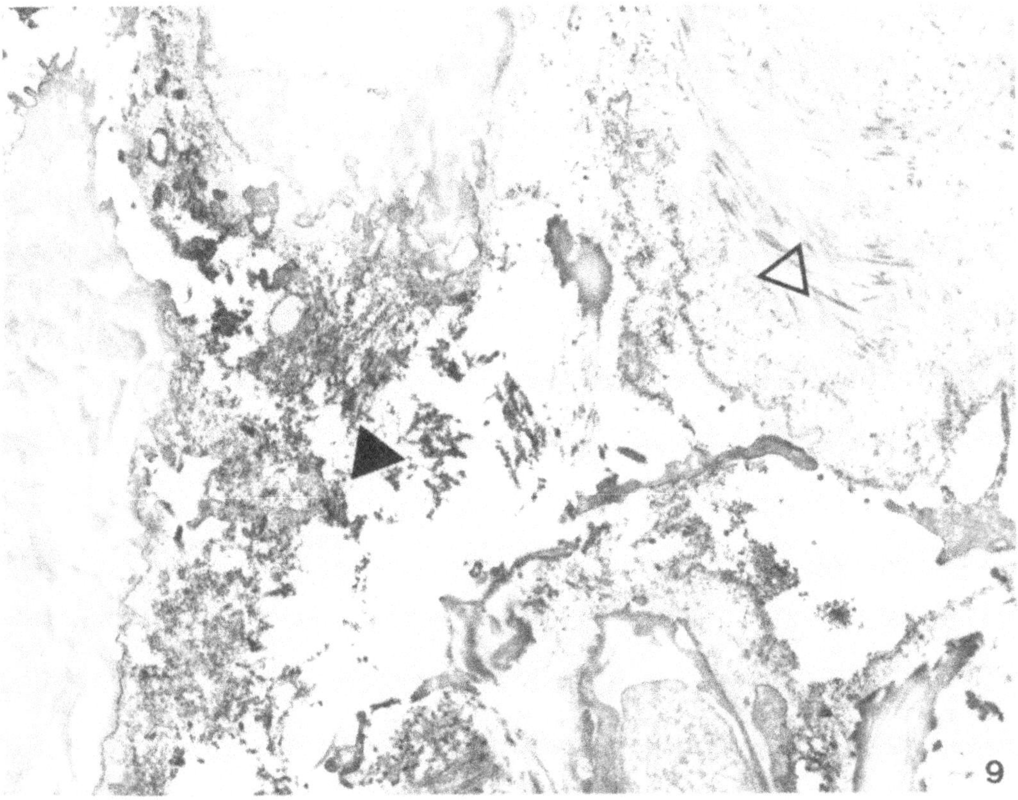

FIGURE 9. EM of immunoperoxidase labeling of type III collagen in human fibrotic lung. ▲ shows perivascular network of type III collagen; △ notes the absence of labeling on type I collagen deposits. (Magnification × 8000.)

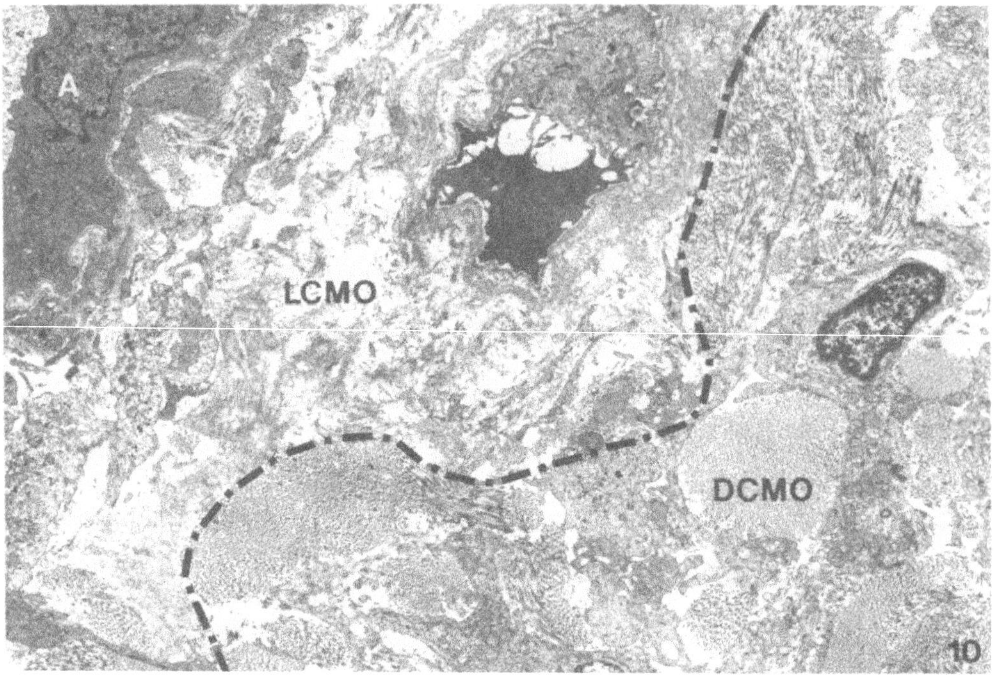

FIGURE 10. Standard EM of portal space in fibrotic liver. Periarteriolar (A). Loose connective matrix (LCMO). Adjacent dense connective matrix organization (DCMO). (Magnification × 8000.)

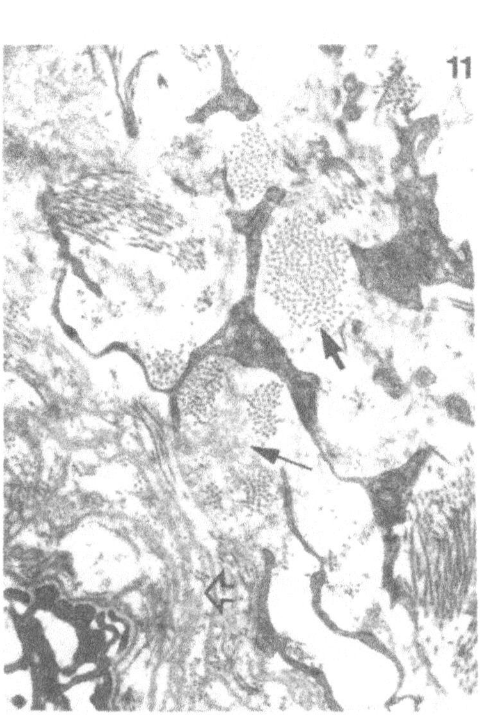

FIGURE 11. Higher magnification of LCMO (arrows thin fibrillar structures and basement membranes). (Magnification × 8000.)

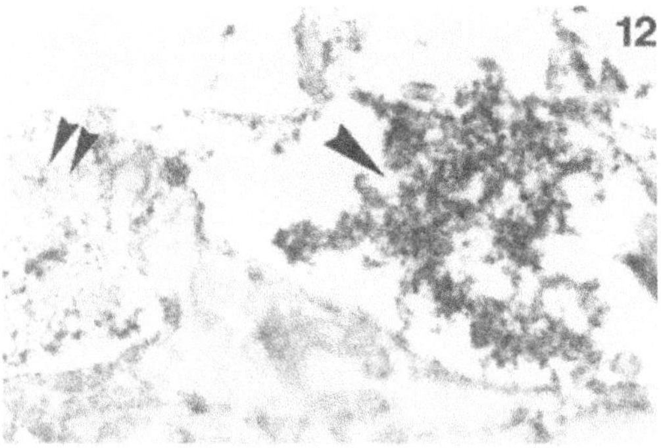

FIGURE 12. EM immunoperoxidase labeling of type III collagen. (➤) isolated network. (≽) mixed bundle of type I and type III collagen. (Magnification × 25,000.)

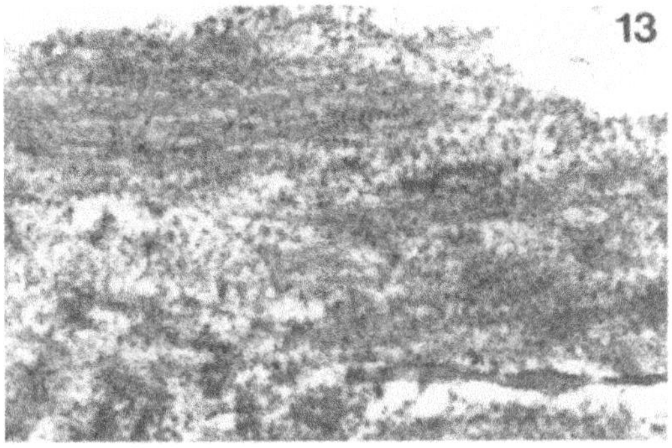

FIGURE 13. EM immunoperoxidase labelling of type I collagen. Dense bundle. (Magnification × 18,700.)

REFERENCES

1. **Badiali de Giorgi, L., Busachi, C. A., Baigini, G., Cenacchi, G., Ballardini, G., Bianchi, F. B., and Del Rosso, M.,** A post-embedding method: demonstration of fibronectin on human liver by PAP technique, *Basic Appl. Histochem.*, 27, 291, 1983.
2. **Barioz, C. M., Sengel, P., and Grimaud, J. A.,** Immunolocalization of matricial components during the early stages of chick embryogenic liver development, Biological Structures and Morphogenesis, 1, 3, 1988.
3. **Bedossa, P., Bacci, J., Martin, E., and Lemaigre, G.,** Immunotyping of collagens in paraffin embedded liver tissue: influence of fixation, enzyme digestion and pattern of staining in alcoholic liver disease, *Histochemistry*, in press.
4. **Biempica, L., Morecki, R., Wu, C. H., Giambrone, M. A., and Rojkind, M.,** Immunocytochemical localization of type B collagen, *Am. J. Pathol.*, 98, 591, 1980.
5. **Clément, B., Emonard, H., Rissel, M., Druguet, M., Grimaud, J. A., Herbage, D., Bourel, M., and Guillouzo, A.,** Cellular origin of collagen and fibronectin in the liver, *Cell. Mol. Biol.*, 30(5), 489, 1984.
6. **Clément, B., Rissel, M., Peyrol, S., Mazurier, Y., Grimaud, J. A., and Guillouzo, A.,** A procedure for light and electron microscopic intracellular immunolocalization of collagen and fibronectin in rat liver, *J. Histochem. Cytochem.*, 33, 407, 1985.
7. **Finley, J. and Petruz, P.,** The use of proteolytic enzymes for improved localization of tissue antigens with immunocytochemistry, in *Techniques in Immunochemistry*, Bullock, G. and Petruz, P., Eds., Academic Press, New York, 1982, 239.
8. **Furthmayr, H. and Von der Mark, K.,** The use of antibodies to connective tissue proteins in studies on their localization in tissues, in *Immunochemistry of the Extracellular Matrix*, Vol. 2, Futhmayr, H., Ed., CRC Press, Boca Raton, FL, 1982, 89.
9. **Geerts, A., Geuze, H. J., Slot, J. W., Voss, B., Schuppan, D., Schellinck, P., and Wisse, E.,** Immunogold localization of procollagen III, Fibronectin and heparan sulfate proteoglycan on ultrathin frozen sections of the normal rat liver, *Histochemistry*, 84, 335, 1986.
10. **Geerts, A., Voss, B., Rauterberg, J., and Wisse, E.,** On the distribution of collagen, fibronectin and laminin in the normal rat liver, in *Sinusoidal Liver Cells*, Knook, D. L. and Wisse, E., Eds., Elsevier Biomedical Press, Amsterdam, 1982, 209.
11. **Grimaud, J. A., Druguet, M., Peyrol, S., and Guerret, S.,** Collagen, in *Methods of Enzymatic Analysis*, Vol. 9, Bergmeyer, H. U., Ed., VCH Publishers, Weinheim, West Germany, 1986, 186.
12. **Grimaud, J. A., Druguet, M., Emonard, H., Peyrol, S., Barioz, C. M., and Guerret, S.,** Collagens immunotyping, in *Biology of Invertebrate and Lower Vertebrate Collagens*, Bairati, A. and Garrone, R., Eds., Plenum Press, New York, 1985, 487.
13. **Grimaud, J. A., Druguet, M., Peyrol, S., Chevalier, O., Herbage, D., and El Badrawy, N.,** Collagen immunotyping in human liver: light and electron microscope study, *J. Histochem. Cytochem.*, 28, 1145, 1980.
14. **Hahn, E., Wick, G., Pencev, D., and Timpl, R.,** Distribution of basement membrane proteins in normal and fibrotic human liver: collagen type IV, laminin and fibronectin, *Gut*, 21, 63, 1980.

15. **Kent, G., Gay, S., Inouye, T., Bahu, R., Minick, O. T., and Popper, H.,** Vitamin A-containing lipocytes and formation of type III collagen in liver injury, *Proc. Natl. Acad. Sci. U.S.A.*, 73, 3719, 1976.
16. **Konomi, H., Sano, J., and Nagai, Y.,** Immunohistochemical localization of type I, III and IV (basement membrane) collagens in the liver, *Acta Pathol. Jpn.*, 31, 973, 1981.
17. **MacLean, I. I. W. and Nakane, P. K.,** Periodate-lysine-paraformaldehyde fixative. A new fixative for immunoelectron microscopy, *J. Histochem. Cytochem.*, 22, 1077, 1974.
18. **Martinez-Hernandez, A.,** The hepatic extracellular matrix. I. Electron immunohistochemical studies in normal rat liver, *Lab. Invest.*, 51, 57, 1984.
19. **Nowack, H., Gay, S., Wick, G., Becker, U., and Timpl, R.,** Preparation and use in immunohistology of antibodies specific for type I and type III collagen and procollagen, *J. Immunol. Methods*, 12, 117, 1976.
20. **Peyrol, S., Takiya, C., Cordier, J. F., and Grimaud, J. A.,** Organization of the connective matrix of the sarcoid granuloma. Evolution and cell-matrix interactions, *N.Y. Acad. Sci.*, 465, 268, 1986.
21. **Peyrol, S., Stocker, S., Guillame, M., and Grimaud, J. A.,** Fine immunolocalization of fibronectin in the Disse space of normal and fibrotic human liver: comparison of polyclonal and monoclonal antibodies as markers, 9th Meet. Fed. Eur. Connective Tissue Societies, Abstr., Budapest, July 1984, 147.
22. **Rojkind, M. and Ponce-Noyola, F.,** The extracellular matrix of the liver, *Collagen Relat. Res.*, 2, 151, 1982.
23. **Rojkind, M. and Kershenobish, D.,** The extracellular matrix fibrosis and cirrhosis, in *The Liver Annual*, Vol. 6, Elsevier, Amsterdam, 1987, 302.
24. **Sano, J., Sato, S., Ishizaki, M., Yajima, G., Konomi, H., Fujiwara, S., and Nagai, Y.,** Types I, III and IV (basement membrane) collagens in the bovine liver parenchyma: electron microscopic localization by the peroxidase-labeled antibody method, *Biomed Res.*, 2, 546, 1981.
25. **Stephens, H., Bendayan, M., and Silver, M.,** Immunocytochemical localization of collagen types and laminin in skeletal muscle with the protein A-gold technique, *Biol. Cell*, 44, 81, 1982.
26. **Voss, B., Glössl, J., and Kresse, H.,** Localization of proteodermatan sulfate in human liver, in *Cells of the Hepatic Sinusoid*, Vol. 1, Kirn, A., Knook, D. L., and Wisse, E., Eds., Kupffer Cell Foundation, 1986, 265.
27. **Voss, B., Rauterberg, J., Allam, S., and Pott, G.,** Distribution of collagen type I and type III and of two collagenous components of basement membranes in the human liver, *Pathol. Res. Pract.*, 170, 50, 1980.
28. **Voss, B., Rauterberg, J., Sander, R., and Herrenpoth, B.,** Immunohistologic localization of several basement membrane associated collagens in the human liver in connective tissue of the normal and fibrotic human liver, Gerlag, V. Pott, G., Rauterberg, J., and Voss, B., Eds., Georg Thieme Verlag, Stuttgart, 1982, 41.

Chapter 2

ANIMAL MODELS OF CONNECTIVE TISSUE DISEASES

Gerald A. Hegreberg and Lynetta J. Freeman

TABLE OF CONTENTS

I. INTRODUCTION

Connective tissue, the macromolecular matrix which forms the scaffold for organized structure and growth in the body, consists of fibers, an interfibrillar matrix or ground substance, and the cellular components responsible for synthesis of the fibrillar and ground substance components. Connective tissue is ubiquitous in the body in its various forms, including the supporting and connecting fabric in soft tissues, the biomechanically important structures in the form of ligaments, tendons, cartilage, and bones, and the vascular system conduits necessary for transport of the blood constituents. This great diversity of structure and function lends biochemical complexity even though many of the structural differences in these connective tissues represent more or less variations of similar molecular species in slightly different forms or in different amounts.

Over the past 20 to 30 years, we have witnessed the progression of understanding of many of the connective tissue diseases of both animals and people from a clinical identification to a deciphering of their molecular pathologies.[1-6] With rare exception, the characterization of the human condition has preceded the study in animals. There are several reasons for the time lag between the identification of the human and animal conditions. Paramount among these is an awareness of the human condition and the acumen to recognize it in animals, usually based on clinical and pathological criteria. An increased awareness has been heightened and nurtured by biomedical educators who have encouraged students and faculty to study and compare rare animal diseases, by the interaction and collaboration of researchers studying both animal and human conditions, and by the availability of some moneys for these comparative studies.

In order to recognize rare animal disorders, the research diagnostician optimally should have a large population of animals for clinical and pathological evaluation. The human physician witnesses virtually every human birth, while the veterinarian is in attendance for only a small portion of the total number of animals born. Furthermore, newborn animals affected with a potentially heritable defect are often shunned by their mothers and die of neglect and malnourishment early in life before the disease is clinically manifested or recognized. Also, for economic reasons and because of genetic protectionism, some owners dispose of the bodies of animals with potential genetic disorders without disclosure of the problem and without clinical or pathological examinations.

Some connective tissue diseases of both animals and people are more readily identified because of the tissues involved or the severity of the involvement. For example, the skin fragility associated with Ehlers-Danlos syndrome (EDS) is an important visible clinical feature which is very useful to the clinican to form the basis for a diagnosis. The massive hemorrhage associated with ruptured aneurysms, a sequela to the weakened vascular connective tissue found in Marfan's syndrome, is readily recognized by the pathologist and assists in making a diagnosis. Conversely, other involvements of connective tissue have been more subtle and perplexing in their clinical and pathological presentations. These include those less well-defined conditions, some of which have significant overlap with other diseases, such as the autoimmune diseases. Still other conditions which have potent temporally associated components are primarily manifested in older people and animals, such as osteoporosis and atherosclerosis. In animals, some of these naturally occurring age-related connective tissue changes have not been defined as readily because the animals often have not been studied during a similar age period corresponding to that studied in the human condition. This is in part a result of the animals not being allowed to live out their normal or maximum life span and because of our parochial approach to comparative studies of chronic human diseases.

The validity of a particular animal model is usually an issue to be addressed, especially whether the animal disorder represents a homologous example of the human disease or whether the similarities can only be drawn at the level of an analogy. Some animal models have identical enzyme deficits to their human counterpart, but even in some of those cases, molecular homology cannot be assumed. Often the study of animal and human disorders have been restricted to clinical and pathological comparisons, especially when the primary enzymatic or structural protein defects in the similar human conditions are unknown. On the other hand, we can point out studies in which the enzymatic defect associated with particular animal genetic models have been identified, and these models have been used to pioneer transplantation and therapeutic reconstitution of the affected animals to a more functional state.

We can expect to witness even greater advances in our understanding of the molecular basis of connective tissue diseases as we employ molecular genetic techniques to uncover molecular derangements which previously defied understanding because of their poorly defined pathological and biochemical bases. Now, the use of gene probes to identify and compare similar disorders of animals and people should usher in new and refreshing approaches to determining the appropriate "fit" or molecular relationship of the animal and human disorders. Even as this chapter is being written, state-of-the-art technologies are further defining our base of understanding of this diverse and complicated group of diseases.

The importance of an animal model of a human disease obviously transcends its recognition and comparison to human diseases, and their true usefulness resides in their use to decipher the still remaining mysteries of disease processes in their simplest molecular form. The study of disorders transmitted in a simple Mendelian manner where the disease process results from a single enzyme or a structural protein defect is a prime example. These experiments in nature provide a most valuable biological tool to evaluate and study the biochemical basis of comparable human disorders and to determine the series of aberrant developmental events which result in disease.

Several previous reports have discussed the comparative aspects of connective tissue diseases.[6-8] It is the intent of this chapter to present both established and emerging animal models of connective tissue diseases. This chapter has been structured to include both inherited and induced disorders of connective tissue. Major headings and subheadings include disease processes at different levels of biological organization, i.e., collagen diseases vs. arteriosclerosis.

II. PRIMARY COLLAGEN DISORDERS

A. EHLERS-DANLOS SYNDROMES

1. General

The EDSs are classics among the group of inherited collagen disorders. They are perhaps the oldest of the reported collagen diseases, and McKusick has suggested that a form of the EDS group was described by Hippocrates when he told of the Scythians and their inability to draw their bows because of their connective tissue laxity.[9] In the early 1900s, Ehlers and Danlos independently recognized a syndrome in people characterized by a triad of cardinal clinical signs including skin fragility, skin hyperextensibility, and joint hyperextensibility.[10,11] Skin fragility was considered to be the most distinctive clinical manifestation of this syndrome.[12] In many cases, the collagen defect appeared to be restricted to the skin; however, some EDS patients exhibited connective tissue defects in other organ systems, including the ocular, cardiovascular, respiratory, gastrointestinal, and central and peripheral nervous systems.[4,9] Ehlers and Danlos recognized a familial pattern of transmission of this syndrome, and the mode of inheritance of the classic form(s) of the disease was later shown to be autosomal dominant.[9,13]

For many years, three similar but distinct forms of EDS were recognized in people and were designated as types I, II, and III. At this time, there are at least ten forms of EDS which have been identified in people and clinical, pathological, and biochemical changes have been described.[3,4,14-16]

2. EDS Types I, II, and III

a. People

Type I EDS, also referred to as the gravis form, is accompanied by the most severe connective tissue changes including very fragile and hyperextensible skin; thin, poorly contracted scars; and prominent joint hypermobility. In the gravis form, the skin is soft and velvety, and minor trauma often results in extensive lacerations which gape. These wounds heal leaving papyraceous scars of the characteristic "cigarette paper" appearance. Hernias, varicose veins, prolapse of mitral valves, and vascular fragility are other clinical manifestations resulting from the connective tissue weakness.

The clinical features of type II EDS, also termed the mitis form, are similar to those observed in type I EDS, but differ in that they are less severe. The skin bruises easily and is moderately hyperextensible. While skin fragility, joint hypermobility, and skin hyperextensibility are present in type II EDS, they are less pronounced than the changes found in type I EDS.

Type III EDS, termed benign familial hypermobility, is characterized by minimal skin changes but marked joint hypermobility, especially affecting the hips and the fingers. Joint dislocations may occur. Although the skin of type III EDS is often soft, it does not have increased compliance or increased fragility.

b. Animals

EDS was one of the first heritable collagen diseases recognized in animals. Although ten forms of EDS have been identified in people, not all forms have been found in animals. Forms of EDS inherited as autosomal dominant traits have been recognized in dogs, cats, and mink.[17-20] The disease in dogs and mink most closely resemble the human type I EDS. Hegreberg has observed several bovine cases which resemble the condition in dogs and mink on pathological and biochemical bases; however, it was not possible to determine the genetic basis. Skin fragility, skin laxity, and skin hyperextensibility form the distinguishing clinical characteristics of the condition in these species. In dogs and mink, skin fragility is the most pronounced and consistent clinical feature.[18,21] In affected dogs, skin lacerations are a common clinical problem, and the skin heals leaving broad, shallow scars. The tensile strength of the skin of affected dogs was only 4% that of nonaffected dogs of the same age, sex, and breed. In young affected mink, small abrasions and lacerations are formed about the nape of the neck as a result of the mother carrying the young kits in her mouth. The tensile strength of the affected mink skin is reduced to approximately 8% that of nonaffected mink skin, and this reduced tensile strength is usually noticed during the pelting process when the mink's skin tears during its removal from the carcass or during the defleshing process. The rate of skin wound healing in the EDS in dogs and mink does not differ appreciably from nonaffected animals, providing that adequate approximation of the wound edges are maintained to facilitate primary healing.[8,22]

Hyperextensibility of the skin has been noted in the affected dogs, cats, and mink.[18-21] Skin laxity varied in severity among individual affected animals, but in dogs became more pronounced with age, especially in regions of the head, legs, and rump. Other clinical abnormalities occasionally associated with the disease include umbilical and inguinal hernias, luxating patellas, and rectal prolapses. Complications experienced by affected dams during parturition were minimal.

The autosomal dominant form of the EDS of dogs, mink, and cats appears to be compatible with a normal life span. Affected animals have survived to ages of 8 years or greater without serious complications from the connective tissue defect, providing that their environment and interaction with other animals is carefully monitored to reduce chances of skin lacerations.

The autosomal dominant disorder of dogs and mink[17] is transmitted with complete penetrance. Studies have not been performed to determine if there is a difference between the heterozygous form and the homozygous dominant genotypes. Some workers have speculated that the condition in the homozygous form may not be compatible with life, based on the small litter size observed from affected-to-affected matings.[17]

The histologic appearance of the condition varies from one affected animal to another. Identification of the histologic changes in affected skin often requires experience in examining tissues from EDS cases and the availability of appropriate control tissues for comparison. Histologically, dermal collagen bundles from affected dogs, mink, and cats are fragmented and frayed compared to the homogeneous and uniformly sized collagen bundles of the normal dermis.[19,20] Collagen bundles in the papillary dermis appear sparse compared to the nonaffected dermis, while collagen bundles of the reticular dermis are haphazardly arranged and often lack the syncytoid interlacing arrangement seen in the normal reticular dermis. In the affected skin, the collagen bundles varied in diameter; some were small in diameter (3 to 4 μm) and appeared as fibrillar strands while other collagen bundles were larger than normal, measuring up to 40 μm in diameter. Collagen bundles occasionally were whorled around small blood vessels of the dermis.

Ultrastructurally, the collagen fibrils from affected dogs and mink demonstrate a variation in fiber diameters and an irregular, branched, and unraveled appearance in longitudinal profile.[23] These alterations were observed using both transmission and scanning electron microscopic methods (TEM and SEM), but the unraveled appearance was best observed using SEM methods. Collagen bundles from the dermis of affected dogs were small and lacked the tight, interwoven appearance of the collagen bundles from normal dogs,[22] TEM changes in the dominant feline EDS have included formation of large, irregular fibrils with indistinct banding patterns in the fiber centers and a disorganized packing of fibrils. A population of larger fibrils was noted in an affected cat dermis, giving a bimodal distribution to the fibril diameter size.[20]

Skin from dogs and mink with EDS revealed that dermal collagen from affected animals was more soluble in 0.5 *M* acetic acid than was nonaffected dermal collagen.[24] No differences in the solubility of collagen in 1 *M* NaCl were found, and the alpha-to-beta ratio of solubilized collagen from affected dogs' skin was not significantly different from that of nonaffected dogs' skin.[24] The aldehyde content of purified collagen from affected dogs' skin was similar to that of purified nonaffected dogs' skin collagen. These results suggested a defect in the formation or stabilization of intermolecular collagen cross-links.[24] Prolyl hydroxylase and lysyl oxidase, measured in the dermis of affected mink, were not significantly different from those measured in the dermis of nonaffected mink.[25]

3. Type V EDS

A sex-linked recessive form of EDS, with clinical signs similar to type II EDS, was initially believed to result from a defect in lysyl oxidase activity, but subsequent studies indicated that the levels of lysyl oxidase in these human patients were normal.[4,14,26] The biochemical status of this disease is unclear at this time. Another form of EDS, type IX, has been identified in which the levels of activity of lysyl oxidase are depressed and will be discussed below along with several naturally occurring and induced disorders of animals in which the activity levels of lysyl oxidase are decreased by either primary or secondary causes. The animal disorders include the mottled mouse, mutants, copper deficiency, lathyrism, and penicillamine-induced changes. These animals conditions are discussed with type IX EDS.

4. Type VI EDS

a. People

Clinically, type VI EDS displays severe scoliosis, hyperextensibility of the skin and joints with recurrent joint dislocations, moderate skin scarring and bruising, and keratoconus with fragility of the ocular globe.[3,4,14,27,28] The condition is inherited as an autosomal recessive trait. The paramount biochemical findings include decreased hydroxylation of lysyl residues in dermal collagen and decreased activity of lysyl hydroxylase in skin fibroblasts.[27,28] Lysyl hydroxylase, along with two species of prolyl hydroxylase, requires ascorbic acid, iron, alpha ketoglutarate, and oxygen as cofactors.[4,15,16] Studies have indicated that the defective hydroxylation of lysyl residues occurs only in type I and III collagens and that the hydroxylation of this amino acid is normal in type II, IV, and V collagens.[14,27,29] It is not clear whether these differential effects of lysyl hydroxylase on different types of collagen are a result of multiple isoenzymes, tissue-specific effects, or different hydroxylation kinetics for the different collagens.[14]

b. Animals

Deficiencies of some cofactors necessary for lysyl hydroxylase activity, especially ascorbic acid, may result in connective tissue alterations. However, their similarity to EDS type VI is only partial. Ascorbic acid influences a number of biochemical steps in collagen synthesis and maturation, including hydroxylation of proline and lysine, collagen secretion, and maturation of fibroblasts.[1,4,29] Ascorbic acid is also involved in many other processes including neurotransmitter formation, steroid synthesis, and proper functioning of the immune system.[29]

Most animal species are able to synthesize ascorbic acid and, therefore, have minimal exogenous requirements under normal health conditions. There are several species which are unable to synthesize ascorbic acid, including primates, guinea pigs, Indian fruit bats, certain passeriform bird species, fish, and shrimp.[1,30-32] Ascorbic acid deficiency in these species usually presents with clinical signs of vascular fragility and poor wound healing. Since ascorbic acid affects a number of metabolic pathways and its effect is not limited to lysyl hydroxylase, its role as an experimental model to study EDS type VI has not been clear cut. Furthermore, EDS type VI is present at birth, and many of the clinical changes in this inherited disease represent developmental alterations which cannot be reproduced in mature scorbutic animals.

5. Type VII EDS

a. People

EDS type VII in people is associated with extreme joint hypermobility, short stature, and joint dislocation.[4,14] The biochemical change associated with this form of EDS involves the inability of a precursor form of type I collagen, termed type I procollagen, to be converted to collagen, resulting in the accumulation of procollagen in the tissues.[14,34] Initially, it was believed that the accumulation of procollagen resulted from decreased activity of N-terminus procollagen peptidase, an enzyme involved in the conversion of procollagen to collagen. Subsequent studies have indicated that the defect responsible for this condition involves an amino acid substitution in the alpha-2 procollagen molecule at the peptidase cleavage site.[35]

b. Animals

A condition termed dermatosparaxia, which resembles type VII EDS in many respects, has been identified in cattle, sheep, cats, and dogs.[36-43] Rather than a defect in the primary structure of the procollagen molecule as occurs in the human type VII EDS, the animal disorder involves a decreased activity of the N-terminus procollagen peptidase, one of the enzymes responsible for the conversion of procollagen to collagen. In all of the above species, the disorder is accompanied by extreme skin fragility; however, an additional milder form of the condition in sheep is generally not recognized until they reach adulthood and are shorn.[44] The condition is inherited as an autosomal recessive trait. In cattle, sheep, and cats, the biochemical changes include an accumulation of the procollagen I precursor chains for both the alpha-1 and alpha-2 polypeptide chains and a decreased activity of the enzyme N-terminus procollagen peptidase.[37,40,41]

The pathologic changes found in affected animals have been primarily associated with the skin. Histologically, there is a decrease in dermal thickness with variation in the size of the collagen bundles. Bundles in the papillary dermis appear fine and sparse, while bundles in the reticular dermis are varied in size and consist of both abnormally small and abnormally large bundles. Collagen bundles of the dermis lacked the typical interwoven appearance seen in the dermis of normal animals.[36,39,42,43] The most characteristic changes have been found in ultrastructural examination of the dermal collagen. When viewed with the scanning electron microscope, the collagen bundles from the dermis of affected animals appeared tangled and loosely arranged as opposed to the parallel and tightly woven appearance of the collagen bundles from comparable regions of the dermis of nonaffected animals.[43] Transmission electron micrographs disclosed fibrils which had bizarre shapes, often referred to as the "cauliflower" or "hieroglyphic" shapes, and fibrils which varied in size.[37,39,42-45] These unusually shaped collagen fibrils appear to be unique to this type of EDS.

6. Type IX EDS and Menkes' Disease

a. People

Type IX EDS is a sex-linked recessive trait characterized by soft skin, moderate to severe skin bruisability, moderate joint hypermobility, varicose veins and hernias.[3,14,46-48] A form of this disease has been reported in which lax skin is also an important clinical feature.[46] Although similar to Menkes' disease in many respects, especially the alteration of collagen, type IX EDS lacks the severe vascular (elastic tissue changes) and neurologic involvement seen in Menkes' disease.[14,46]

Menkes' disease (kinky hair disease, steely-hair syndrome, trichopoliodystrophy) is an inherited multiple-systems disorder characterized by abnormalities of the hair, arteries, bones, and central nervous system.[49,50] Hypothermia, unusual facies, slow growth, and the presence of grayish-to-ivory-colored fragile hair are several of the early clinical changes observed in affected male infants. At several months of age, progressive cerebral degeneration with myelin deficiency and neuronal loss, especially in the cerebellum, are noted. Vascular complications which develop include vascular fragility with subdural hematomas, aneurysm formation, and thrombosis. The bone changes include osteoporosis with pathological fractures and widening of the metaphyses. Affected children usually survive from 3 months to 3 years after birth. The disorder is inherited as a sex-linked trait.

Type IX EDS and Menkes' disease share a close association in their pathologic and biochemical changes, especially involving the connective tissue. Both conditions share several biochemical changes including decreased intestinal copper absorption, depressed levels of serum copper and serum ceruloplasmin, elevated intracellular concentrations of copper, and abnormally low levels of lysyl oxidase activity.[14,48] It has been suggested that there may be abnormal allelic forms involving a primary defect in copper metabolism, especially involving a protein involved with copper transport, and that the connective tissue changes are secondary.[14]

b. Animals

An inherited connective tissue disorder which resembles Menkes' disease and, to some extent, type IX EDS has been recognized in the mouse.[51-54] This sex-linked recessive condition in the mouse is not a single entity, but rather several similar mutant forms involving allelic sites on the X chromosome.[54] Because it is a sex-linked trait, affected offspring are male except in those forms where affected females can survive to adulthood and produce offspring. The most severe of these mutants are the mottled (Mo) and the tortoise-shell (Mo,^{to}) forms, both of which die *in utero* or shortly after birth. Mice affected with the brindled form (Mo,^{br}) die at approximately 2 weeks of age and demonstrate neurologic manifestations including ataxia. The dappled (Mo,^{dp}) mutant is another form in which affected males survive for 2 to 3 weeks. The viable brindled (Mo,^{vbr}) and the blotchy (Mo,^{blo}) mutants can survive to adulthood and are fertile.

The resemblance of the mouse mutants to Menkes' disease is based on the clinical appearance, involving the connective tissue and nervous systems, and the biochemical changes.[53,54] The connective tissue changes include decreased formation of mature collagen in the skin and bone resulting in fragility of these structures. A defect in elastin maturation results in vascular fragility, aneurysm formation, and rupture. These changes are especially seen in those mutants (especially the brindled [Mo,^{vbr}] and the blotchy [Mo,^{blo}] forms) whose longevity permits the study of these tissues. Pulmonary changes resembling panacinar emphysema have been described in the blotchy mouse.[55]

Mice with this trait have a marked increase in the extractability of skin collagen with an elevation in the collagen alpha-to-beta chain ratio.[52] The defect in elastic fibers is manifested as an increase in the solubility of elastin and a decrease in the elastin cross-linkage with a decrease in desmosine and isodesmosine. These molecular changes in collagen and elastin, which include an increase in solubility and a decrease in the lysine-derived aldehyde content of purified skin collagen and aortic elastic tissue, result from decreased activity of lysyl oxidase. The levels of lysyl oxidase activity in the skin[52] and fibroblasts and lung tissues[56] of the mottled mouse were found to be significantly reduced.

Several phenocopies of type IX EDS and Menkes' disease can be experimentally produced by dietary manipulations including the restriction of copper from the diet and the administration of lathyrogens and penicillamine. A well-documented condition which resembles type IX EDS and Menkes' disease is produced by feeding a copper-deficient diet to young growing animals.[57-60] The connective tissue changes consist of osteoporosis, spontaneous fractures, unstable joints, and aneurysms with sudden rupture of large vessels, including the aorta. Since lysyl oxidase is a copper-dependent enzyme, it appears that copper deficiency, in part, has its effect by preventing the proper functioning of this enzyme, resulting in deficient elastin and collagen cross-links.[61,62]

Osteolathyrism is a condition caused by the administration of toxic factors native to certain species of peas, especially *Lathyrus odoratus*.[63] B-aminoproprionitrile has been identified as the major toxic principle. When this compound is administered to animals during their active growth phase, connective tissue weaknesses are produced, including skeletal deformities and aneurysms of major blood vessels. B-aminoproprionitrile inhibits the enzyme lysyl oxidase, resulting in the improper cross-linking of collagen and elastin.[64] Thus, this experimental disease represents a phenocopy of the inherited connective tissue disorders including type IX EDS and Menkes' disease.

Penicillamine is a drug which causes similar changes to osteolathyrism, including changes in the connective tissue of the skin and vascular system.[65-67] Penicillamine produces its effect by complexing with the aldehyde moieties of the forming cross-links of both collagen and elastin, resulting in the inability to form mature intermolecular cross-links.[68] Penicillamine also inhibits the activity of the enzyme, lysyl oxidase, apparently by chelating copper.[69]

B. WILSON'S DISEASE

1. People

Wilson's disease, also referred to as hepatolenticular degeneration, is an inherited disorder which is clinically accompanied by liver disease with acute, chronic, or recurrent episodes, jaundice, edema, acute hemolytic crises, connective tissue changes involving both bones and the soft tissues of blood vessels and joints, renal stones, renal tubular acidosis, and several neurologic manifestations.[70-74] The condition is transmitted as an autosomal recessive gene carried on chromosome 13.[72]

Clinically, an important and pathognomonic sign in Wilson's disease is the deposition of a copper-containing yellow-brown pigment on the Descemet's membrane at the limbus of the cornea, termed the Kayser-Fleischer rings.[70,71] Significant clinical alterations involving the connective tissue include joint laxity, osteoporosis, osteomalacia, reduction in the joint spaces of the limbs and spine, and enlarged joints.[70]

Severe fibrosis of the liver accompanies the chronic stages of the disease.[70] The primary pathological changes in the liver include degeneration and cirrhosis. The ultrastructural changes include abnormal shape and size of mitochondria with the formation of paracrystalline inclusions and the formation of multivesiculated intracytoplasmic bodies.[74]

Two neurological forms of the disease are recognized.[70,73] One form is associated with lenticular degeneration and accompanied by spasticity, rigidity, tremor, muscle contracture, dysarthria, and dysphagia. Another form, termed pseudosclerosis, is accompanied by flapping tremors of the wrists and shoulders, rigidity, and spasticity. Disturbances of intellectual function and behavior have also been noted.

The lesions in the central nervous system are distributed to the basal ganglia (including the lenticular nucleus), the cerebral hemispheres, and the cerebellum (especially the dentate nucleus). A very characteristic microscopic change is the proliferation of a type of protoplasmic astrocytes, termed Alzheimer type II cells, which are especially prominent in the basal ganglia. The Alzheimer type II cells have a large, irregular- to lobulated-shaped nucleus and poorly defined cytoplasm. An additional characteristic cell, called the Opalski cell, is found in the gray matter bordering degenerating areas. Opalski cells have an abundant granular, periodic acid-Schiff (PAS)-positive cytoplasm and an eccentric, pyknotic nucleus, and they are believed to represent degenerating astrocytes. Neuronal degeneration and necrosis contribute to the generalized cortical and cerebellar atrophy and laminar cortical necrosis.[73]

Copper accumulates in the liver, kidneys, brain, and cornea, resulting in damage to the tissues.[70,71] This disease represents a disturbance of copper metabolism including a severe reduction in the rate of incorporation of copper into ceruloplasmin and a reduction in excretion of copper in bile.[71] The connective tissue change may be secondary to the abnormal accumulation of copper or to the deficiency of a normal unidentified protective mechanism.[70] Wilson's disease, along with type IX EDS and Menke's disease are members of a closely related family of human and animal disorders involving altered copper metabolism.

2. Animals

An inherited copper toxicosis which resembles Wilson's disease has been described in several breeds of dogs including Bedlington terriers and West Highland white terriers.[75-79] Doberman pinschers may also have a similar copper-storage disease.[79] The mode of inheritance has been established as autosomal recessive for the Bedlington terrier condition.[80] The mode of inheritance for the similar conditions in other breeds has not been determined.[78]

The condition in the Bedlington terriers is progressive and is accompanied by a chronic hepatic degeneration, and the hepatic changes have been grouped depending on the severity of the pathologic changes.[75] Clinical signs are often not observed in group I stage and may not be observed until group III. All dogs in group V have widespread acute hepatic necrosis and postnecrotic cirrhosis. Signs noted include anorexia, vomiting, depression, diarrhea, weight loss, abdominal pain, hemolytic anemia, and jaundice.[75]

In the affected West Highland white terriers, the severe and massive hepatic necrosis may cause death.[78] Kayser-Fleischer rings have not been observed in the condition in the affected dogs. Serious central nervous system involvement seen in human Wilson's disease has not been reported in the canine disorders, but behavioral changes have been noted in older affected dogs.[77]

Clinical laboratory changes include defective biliary excretion of copper and an increased accumulation of copper in the liver.[76,78] Normal dogs have liver copper concentrations of less than 400 ppm (dry weight), and affected Bedlington terriers have levels 30 to 100 times higher than normal dogs.[76] West Highland white terriers accumulate hepatic copper up to 8 to 10 times that of normal, but do not accumulate the massive amounts of hepatic copper found in Bedlington terriers.[79] Copper levels in the central nervous system are also increased in some cases.[81] Ceruloplasmin levels are slightly increased in Bedlington terriers[77] rather than decreased as in the human condition.

The pathological change in the liver begins with the accumulation of hepatic copper but with no clinical signs or disruption of the hepatic architecture. Changes are noted in the hepatic architecture when the copper concentrations approach 2000 ppm (on a dry weight basis). Multifocal hepatitis, located predominantly in centrilobular regions, may progress to diffuse hepatic necrosis as the hepatic copper levels rise to levels above 2000 ppm. If the animal survives, postnecrotic cirrhosis occurs.[75,78,79]

III. OSTEOCHONDRODYSTROPHIES

A. OSTEOGENESIS IMPERFECTA

1. People

Osteogenesis imperfecta (OI) represents a group of closely related connective tissue disorders primarily involving bone, but also affecting tendons, ligaments, fasciae, sclera, and dentin.[3,4] There are four major recognized forms of OI in people, and these forms have varied modes of inheritance (both autosomal recessive and autosomal dominant) and some variation in clinical features.[3,4,9,82] Of special note is that bone fragility is more severe in some forms, especially type II which results in death either *in utero* or shortly after birth.

Clinically, in OI the bones are osteoporotic and result in increased fragility.[4,9] Fractures often result from minimal trauma including weight bearing. The repeated fractures result in bone deformities including bowing of the long bones of the upper and lower extremities and kyphoscoliosis as a result of the vertebral abnormalities. The deformities of the bones of the head include frontal and parietal bossing with a prominence of the occiput, producing a helmet head or mushroom-shaped deformity. Otosclerosis may develop leading to deafness.

Changes which occur in the soft connective tissues include joint laxity, rupture of tendons, and the development of hernias.[4] The skin is thin, and wound sites may have either atrophic or hypertrophic scars. The connective tissue in the sclera is thin, and scleras appear blue as a result of the choroid pigment showing through the thinned sclera. Several changes in the cardiovascular system are recognized in cases of OI including improper closure of valve cusps, rupture of cordae tendineae, dilated aortic and mitral valve rings, and aortic cystic medionecrosis.[83]

In bone, the pathological changes include rarification of the cortex with a decrease in the number of trabeculae and abnormal production and remodeling of Haversian systems.[4,84] Ultrastructural studies of scleral collagen revealed reduced diameters of the collagen fibers and disrupted lamellar architecture.[84] The skin is thin, and there is a decrease in the skin thickness and a decrease in the number and size of dermal collagen fibrils.[85] Studies of OI type I revealed that fibrils of the reticular dermis had extremely large diameters with a lobulated appearance and that the bundle size was small.[86]

The biochemical defects in several forms of OI appear to involve principally alterations in the primary structure of the collagen type I precursor molecules (alpha-1 and alpha-2 procollagens).[3,5,16,86-88] In the lethal type of OI (type II), a deletion of about 500 bases has been identified for the pro-alpha-1(I) collagen gene.[89] This gene deletion resulted in the defective formation of half of the pro-alpha-1(I) chains, leading to defective formation and secretion of procollagen molecules. Modifications of this theme have been identified in type I OI in which amino acid deletions were found near the N terminus of the pro-alpha-2(I) chain.[86] This defect prevented cleavage of the chain by procollagen *N*-proteinase, resulting in the formation and persistence of an unstable type I procollagen. Another mutant included a case involving alterations in the C terminus of the pro-alpha-2(I) chain which resulted in the inability of the pro-alpha-2 chain to become incorporated into the collagen type I molecule.[90] An additional case involved formation of a propeptide with an altered amino acid sequence. This change resulted in decreased collagen solubility and an increased mannose content.[91] Several other variants have been found, including a case in which half of the pro-alpha-1(I) chains contained cysteine, an amino acid exceptional to the collagen molecule.[92,93] Prockop and Kivirikko[3] have pointed out that several biochemical changes often noted in OI may represent overmodifications of type I procollagen (increased hydroxylysine and glycosylated hydroxylysine content) resulting from a delayed helix formation. The decrease in the ratio of type I to type III collagen reported by some in OI may result from either a decreased rate of synthesis of type I procollagen or the synthesis of a type I collagen which is structurally imperfect and which is degraded *in vivo*.

2. Animals

A lethal form of OI was identified in the Holstein-Friesian calves in Australia.[94] The condition is transmitted as an autosomal dominant trait and is accompanied by bone fragility, joint laxity, ruptured tendons, domed skulls, and thin, soft skin. The syndrome was lethal, and affected calves survived an average of 33 d. The bone fragility resulted in fractures, most of which were located in the ribs and long bones of the extremities. Joint laxity was severe in the affected calves and resulted in walking impairment. Bowing of the long bones occurred during the postnatal period and was especially pronounced in the radius, ulna, tibia, and metacarpal and metatarsal bones. Fibroblasts and odontoblasts had dilation of the rough endoplasmic reticulum and an intracisternal accumulation of nonfibrillar material. Collagen fibrils had significantly smaller diameters than fibrils from age-matched controls.[95]

A similar condition has been identified in calves of the Holstein-Friesian breed from Texas.[96] This variety of OI is also transmitted as an autosomal dominant trait. Although the clinical appearances are apparently similar in both the Australian and Texas varieties, the Texas variety is reported to have more severe osteopenia than the Australian variety. Blue scleras and fragile teeth with translucent pink discoloration have been noted in the affected calves. Biochemical studies performed on calves affected with the Texas variety showed decreased electrophoretic mobility of both alpha-1(I) and alpha-2(I) chains.[96] Collagen isolated from the affected animals had an increased hydroxylysine content and a depletion of osteonectin, a mineral-promoting protein, and phosphophoryn, a dentin-specific protein.[96]

A severe form of OI was identified in the Charollais breed of cattle.[97] Affected calves either died or were sacrificed within days after birth. Fractures of the long bones occurred both *in utero* and shortly after birth. The cortex and spongiosa of the bones were reduced in amount. Pedigree analysis suggested an autosomal recessive mode of inheritance.

A form of OI has been reported in lambs from both New Zealand and England.[98,99] The New Zealand lambs were born dead and had fragile teeth. The bones were extremely fragile, and osteopenia was noted in the diaphysis.[98]

An autosomal recessive form of OI termed fragilitas ossium (fro) has been recognized in the mouse.[100] Deformities have been observed in all limbs and are characterized by curving of the long bones. When compared to nonaffected mice, the diaphyses were wider and translucent with small, irregular denser areas. The cortices were thin, irregular, and had discontinuities.

Familial OI was reported in poodle dogs.[101] The affected puppies were reluctant to stand on their rear limbs, apparently because of the associated pain. The joints of the extremities were hypermobile. The scleras did not appear blue. Radiographic changes consisted of cortical thinning, a decrease in the number of fine intramedullary trabeculae, and irregular radiolucent zones in the metaphysis which were believed to be fractures. In many instances, the epiphyses were partially separated. The condition was first apparent in affected puppies at 6 to 8 weeks of age, and the clinical and radiographic osseous changes began to resolve when the dogs reached 6 months of age.

B. CHONDRODYSTROPHIES

1. People

The chondrodystrophies are generalized forms of inherited skeletal dysplasias which are accompanied by alterations in endochondral ossification.[15,102-105] This is a heterogeneous *group* of disorders, and their classification in the past has been confusing.[15,102] Investigators have classified the conditions on the basis of their mode of inheritance; their clinical appearance, including the age of onset and their anatomical distribution; and the radiographic features of the diseases. Although this classification has been most useful, the morphological and biochemical bases of these disorders are presently being studied and will provide more meaningful criteria for their classification.[102-104] The acid muscopolysaccharide (MPS) storage disorders often are included in this group of disorders, but they are discussed in a separate portion of this chapter. No attempt will be made to further subdivide or compare the various types of chondrodystrophies of human beings or animals because many of the critical morphological and biochemical comparisons cannot be made at this time.

2. Animals

A number of chondrodystrophies have been described in animals, and in 1963 Gruneberg[106] provided a classic description of the systemic disorders of cartilage of animals described up to that time. Since then, several inherited forms of chondrodysplasia have been identified in animals, including the brachymorphic mouse (bm/bm),[107] murine cartilage matrix deficiency (cmd/cmd),[108] nanomelia in chicks,[109] chondrodysplasia in the Alaskan mala mute,[110,111] miniature poodles,[112] beagles,[113] and English pointer dogs.[114] Several inherited types of chondrodysplasia have been identified in cattle, including the Dexter type,[113,115] the Telemark type,[113,116] and the snorter type.[113,117,118] Inherited chondrodysplasias have also been reported in sheep.[113,119,120]

Dyschondroplasia, or osteochondrosis, of fast-growing chickens and turkeys is accompanied by leg weakness with retention of the nonvascularized cartilage, especially of the tibia.[113,121,122] The condition is observed most frequently in birds fed high-energy diets or diets high in chloride or with an altered calcium/phosphorous ratio. Although the underlying cause of the dyschondroplasia is not known, it is believed that the failure of endochondral ossification results from improper vascular invasion of the cartilage because of a disruption of normal cartilage matrix development.[121]

A generalized cartilage defect called osteochondrosis has been reported in rapidly growing dogs,[123] horses,[124] and pigs.[125,126] This defect in animals has similarities to metaphyseal chondroplasia of people, which is a defect in endochondral ossification.

IV. GLYCOSAMINOGLYCAN DISORDERS

A. GENERAL

Glycosaminoglycans, also termed acid mucopolysaccharides, (MPSs), are large complexes of negatively charged polysaccharide chains which are often associated with a small amount of protein. These compounds bind with water, producing a gel matrix which forms the basis for the ground substance of the connective tissue. There are six major classes of

MPSs which include the chondroitin sulfates (chondroitin 4- and 6-sulfate), keratin sulfate, dermatan sulfate, hyaluronic acid, heparan sulfate, and heparin. All except heparin are found in the extracellular ground substance matrix in the body. After ingestion by the cell, MPSs are degraded within lysosomes by lysosomal enzymes.

There are a group of hereditary disorders which are characterized by a deficiency or malfunction of the lysosomal enzymes responsible for the degradation of acid MPSs. As a result of the enzyme deficiency, the acid MPSs accumulate in the tissues and are excreted in excess in the urine. Seven distinct forms of the MPS storage disorders have been identified in people[127,128] and involve enzyme deficiencies associated with the degradation of one or more of three MPSs, namely, dermatan sulfate, heparan sulfate, and keratan sulfate. Each acid MPS storage disease has a characteristic clinical and pathological form. In animals, the two forms of MPS storage disease which have been identified are type I and type VI.

B. ACID MUCOPOLYSACCHARIDOSES

1. Type I Mucopolysaccharidosis

a. People

Several clinical forms of type I mucopolysaccharidosis have been identified, including a severe form (Hurler's disease), a least severe form (the Scheie syndrome), and a form which has characteristics of both forms (the Hurler-Scheie syndrome). Hurler's disease is clinically accompanied by progressive clouding of the cornea; skeletal abnormalities resulting in dwarf stature and an altered shape of the head and vertebrae; joint stiffness; mental deterioration; an accumulation of the polysaccharide material in viscera and connective tissue, including the cardiovascular system; and early death. In the Scheie syndrome, corneas are severely clouded, deformities are observed in the hands, and cardiac abnormalities include aortic stenosis or regurgitation. Disorders of stature and intelligence are not severe clinical manifestations of the disease as seen in the Hurler's syndrome. These forms of type I mucopolysaccharidosis are inherited as autosomal recessive traits.

Patients with type I mucopolysaccharidosis have a reduced alpha-L-iduronidase activity, accumulation of both dermatan sulfate and heparan sulfate in tissues, and an increased urinary excretion of these substances. The acid MPSs accumulate in lysosomes causing their enlargement. Acid MPSs accumulate in lysosomes of blood lymphocytes, and these structures are termed Adler-Reilly bodies.[127]

b. Animals

A form of type I mucopolysaccharidosis similar to the Hurler-Scheie phenotype was identified by Haskins et al.[129] in the domestic short-haired cat. The condition is transmitted as an autosomal recessive trait. Clinically, affected cats were identified by their broad, short maxilla and depressed nasal bridge, small ears, and corneal clouding. The propositus was lame, apparently as a result of the bilateral coxofemoral subluxation associated with the disease. Cervical vertebrae were wide with some vertebral fusion. No serious neurologic deficits were observed.

Clinical laboratory examination of the affected cats revealed an excessive urinary excretion of both heparan sulfate and dermatan sulfate. Alpha-L-iduronidase activity of blood polymorphonuclear leukocytes and cultured fibroblasts from the affected cats was less than 1% that of normal cats. Heterozygotes displayed a level of activity intermediate between the affected cats and the controls.[129,130] Adler-Reilly inclusion bodies were not observed in the peripheral white blood cells.

Pathological changes involved enlargement of the liver and spleen and thickening of the mitral valves. Membrane-bound inclusions were present in neurons, hepatocytes, smooth muscle cells, and fibroblasts of the liver, kidney, and mitral valve. The membrane-bound inclusions of the neurons contained zebra-bodies in some cells and granular material and membranous whorls in other cells.[130]

A type I mucopolysaccharidosis of the Hurler-Scheie variety was described in three Plott Hound dogs.[131-132] The clinical features of the disease involved corneal clouding, insufficiency of the heart valves, bone abnormalities especially involving the face and limbs, and stunted growth.[131] The condition appeared to be inherited as an autosomal recessive trait.[132] Similar to the feline condition, cytoplasmic inclusions were observed in neurons, glial cells, fibroblasts throughout the body, and in cells in the skeletal muscle and visceral and endocrine organs.[131] The stored lysosomal material was not observed as Adler-Reilly inclusion bodies in the white blood cells.

The level of activity of alpha-L-iduronidase in fibroblasts and white blood cells was markedly decreased in the affected dogs.[132,133] Urinary dermatan sulfate and heparan sulfate excretion were increased up to 25-fold over control values. Glycosaminoglycans were markedly elevated in neural tissue and in the cerebrospinal fluid of affected dogs.[132,133]

2. Type VI Mucopolysaccharidosis

a. People

Type VI mucopolysaccharidosis, the Maroteaux-Lamy syndrome, is recognized as an autosomal recessive disorder which is clinically similar to the Hurler syndrome. The form of MPS which accumulates and is excreted in excess is dermatan sulfate,[127,128] resulting from a deficiency of the enzyme arylsulfatase B.[128] This disorder in people is accompanied by severe osseous changes including growth retardation; joint changes, especially involving the stiffness of the hands and degeneration of the femoral heads; corneal cloudiness; thickening of the heart valves and aortic stenosis. Intelligence does not appear to be impaired. Growth retardation and short stature distinguishes this condition from Sheie's disease.

Lysosomal inclusions which are metachromatic are found in leukocytes (Adler-Reilly bodies). Lysosomal inclusions are also found in Kupffer cells, hepatocytes, and cells of the cornea, conjunctiva, and skin.[128] Other diagnostic changes include the excessive urinary excretion of dermatan sulfate.

b. Animals

Mucopolysaccharidosis type VI has been recognized in the Siamese cat breed.[130,134-138] The clinical changes in the affected cats involve primarily the skeletal and ocular systems.[134,135] The affected cats are small in size and have a flattened face, depressed nasal bridge and small ears. The osseous changes, which are especially pronounced in older affected cats, include fusion of the spine and proliferative changes in the vertebrae, shallow pelvic acetabulae, flattened femoral heads, and severe epiphyseal dysplasia of the long bones. The severe vertebral lesions with resulting overgrowth of bone into the spinal canal result in compression of the spinal cord and paraparesis. Corneal cloudiness is a prominent clinical feature. The condition is transmitted as an autosomal recessive trait.[134]

Pathological changes include thickening of the mitral and tricuspid heart valves and bony epiphyseal proliferations.[136] Cytoplasmic vacuoles have been observed in the fibroblasts of the heart valves, the eye, and the skin. In addition, cytoplasmic vacuoles have been observed in the smooth muscle, bone marrow granulocytes, hepatocytes, and keratocytes. Metachromatic intracytoplasmic lysosomal inclusions were present in many polymorphonuclear leukocytes. Ultrastructurally, these cytoplasmic inclusions were membrane-bound structures resembling lysosomes.

Marked depression in the level of activity of arylsulfatase B was detected in fibroblasts and peripheral blood leukocytes of the affected cats.[134,137] Intermediate levels of enzyme activity were detected in cats heterozygous for this condition. The Berry spot test for urinary MPSs was positive. Dermatan sulfate was excreted in excess in the urine.[134,137]

Transplantation studies, grafting bone marrow from normal donor cats to affected cats, appeared to correct the enzyme defect and to reverse some of the ocular and osseous clinical changes.[138] A drop in urinary glycosaminoglycan excretion and an increased leukocyte arylsulfatase B activity persisted for over 200 d after bone marrow transplantation.

C. CUTANEOUS MUCINOSIS

1. People

The primary cutaneous mucinoses are a heterogeneous group of diseases in which a mucinous material accumulates as solitary or multiple, elevated (papular), cutaneous lesions located on the face, trunk, or extremities.[139,140] Several clinical variations of the cutaneous mucinoses have been identified, including both generalized and localized forms. The generalized forms include generalized myxedema and a circumscribed or pretibial myxedema form. Both of these generalized forms are commonly associated with altered thyroid metabolism. The myxedema associated with hypothyroidism usually results in a diffuse edema of the skin. Circumscribed or pretibial myxedema is almost always associated with exophthalmic goiter with an associated thyrotoxicosis. Papular mucinosis is a third form of generalized cutaneous mucinosis. The cause of papular mucinosis is unknown, and there is no evidence of an endocrine imbalance, i.e., thyroid disease, in this form. Investigators have reported a monoclonal immunoglobulin in the serum of affected patients, and there is evidence that a serum factor stimulates mucin production in papular mucinosis.[141]

Focal forms of mucinosis include follicular mucinosis, cutaneous focal mucinosis, and myxoid cysts. Cutaneous focal mucinosis is a focal form in which asymptomatic nodules, which are usually solitary, are found on the face, neck, trunk and extremities.[142]

Secondary mucinoses with abnormal mucin deposition have been reported in several connective tissue diseases which are associated with an inflammatory and/or immunologic components. Some of these diseases are systemic lupus erythematosus (SLE), dermatomyositis, and scleroderma.

Microscopically, the skin lesions in the primary mucinoses, including cutaneous focal mucinosis, are accompanied by a dermal accumulation of mucin with the presence of many spindle-shaped fibroblasts.[140] The numerous fibroblasts are believed to produce the increased amounts of mucin. The accumulated mucin is distributed throughout the dermis and separates the collagen bundles into irregular arrangements or into individual collagen fibers. The predominant dermal mucin is hyaluronic acid.

2. Animals

Focal or localized mucinosis accompanied by asymptomatic nodules, papules, or plaques on the skin or oral mucosa has been described in dogs.[143] The canine lesions were usually solitary and rarely multiple. Microscopically, the dermal lesion was characterized by the accumulation of a mucinous substance (mostly hyaluronic acid) which caused a disruption and separation of the dermal collagen bundles. There was a mild to extensive proliferation of fibroblasts and a mild, diffuse infiltration of mononuclear cells in both perivascular and perifollicular areas.

V. DISORDERS OF ELASTIN

A. GENERAL

Many of the inherited conditions and experimental agents which cause alterations in the formation of elastic fibers also affect collagen synthesis and maturation. Several of these disorders, including EDS (especially types VI and IX), Menke's syndrome, osteolathyrism, penicillamine, and copper deficiency, have been discussed in other parts of this chapter. Other disorders of elastin include solar elastosis and endocardial fibroelastosis.

B. SOLAR ELASTOSIS (ACTINIC ELASTOSIS, ACTINIC KERATOSIS)

1. People

Solar elastosis is a degenerative skin change produced by exposure to sunlight.[144,145]

The sun-damaged skin becomes wrinkled, atrophic, and hyperpigmented. Telangiectases and focal areas of hyperkeratosis may form on the skin. Pathologically, large masses of an amorphous basophilic material which stains positively for elastic tissue is present in the upper third of the dermis. Basophilic degeneration of collagen is also present in this region of the dermis. The origin of the elastoid material has been debated by some workers who believe that this material originates from elastic fibers. Others believe that the elastoid material is degenerating collagen.

2. Animals

A number of reports concerning ultraviolet exposure of the skin of mice and rats have indicated that changes compatible with solar elastosis can be experimentally produced.[145-151] These studies have usually employed albino rats and hairless mice. Ultraviolet-light-exposed skin has a marked increase in acid MPS content in the papillary dermis. During ultraviolet treatment, the elastic fibers of the upper dermis disappeared, but increased to abnormally high quantities after discontinuation of the ultraviolet treatment. Naturally occurring changes in the skin consistent with solar dermatitis have been reported in dogs and cats.[152-156] These changes usually occur in unpigmented areas of the skin and may be associated with skin neoplasms.

C. ENDOCARDIAL FIBROELASTOSIS

1. People

Primary endocardial fibroelastosis is a congenital cardiac defect in which the endocardium is diffusely thickened by fibrous and elastic tissue.[157-159] Both primary and secondary forms of endocardial fibroelastosis have been identified. Primary endocardial fibroelastosis occurs sporadically, and the occurrence of this condition among siblings has supported a genetic basis.[157] The mode of inheritance may be autosomal recessive. Secondary endocardial fibroelastosis may be associated with myocarditis, cardiac involvement in glycogen storage diseases, hypoxia, myocardial necrosis, and cardiovascular anomalies.[157-159]

The left side of the heart is the most commonly involved site, especially the left ventricle. The endocardium is thickened and has a white, porcelain appearance. The aortic and mitral valves may be thickened and misshapened. The left ventricle is usually dilated and hypertrophied. Less often, the left ventricle is contracted and unusually small in size. Histologically, there is a marked increase in thickness of the endocardium with an increase of both the outer collagenous layer and the inner elastic layers.[157,160] The fibroelastic tissue may invade the subendocardial myocardium. The morphology of the endocardial elastic fibers appears quite normal ultrastructurally, consisting of central amorphous elastin cores with microfibrils located on the periphery. It is believed that in this condition, a variety of factors may cause endocardial injury which result in the synthesis of abnormally large amounts of elastic and collagenous fibers.

2. Animals

A familial form of primary fibroelastosis has been identified in the Burmese cat.[161,162] The condition is inherited, but the mode of inheritance has not been firmly established. The condition is accompanied by diffuse fibrous and elastic endocardial thickening. Clinically, the condition is identified in young 1- to 4-month-old kittens by tachycardia, galloping cardiac rhythm, systolic murmur, cardiomegaly, and congestive heart failure with dyspnea and cyanosis. Severely affected kittens may die suddenly, but less severely affected kittens with few clinical signs may live to maturity. The endocardium is thickened and opaque, especially on the left side of the heart. The left ventricle and atrium are dilated, and the myocardium is hypertrophied.

The subendocardial changes are detected ultrastructurally several days after birth and initially involve a diffuse edema with dilated lymphatics. Later, the subendocardial changes are visible grossly and are morphologically characterized by proliferation of subendocardial fibroblasts with an increased synthesis of subendocardial collagen and elastic. The subendocardial fibroelastic proliferation may cause Purkinje cell entrapment and atrophy.

Cases of endocardial fibroelastosis have been identified and reported in other cat breeds[163] and in many other species including dogs,[163,164] pigs,[165] calves,[165] sea lions, hippopotamuses, elephants, and blue whales.[166] The cause of the endocardial fibrosis in these cases is not known. Levin has reported that puppies infected with canine enteric parvovirus had a cardiomyopathy, and endocardial fibroelastosis was found in some of the cases.[167]

VI. INFLAMMATORY AND IMMUNE-MEDIATED CONNECTIVE TISSUE DISEASES

A. GENERAL

A group of disorders with prominent inflammatory and immunologic changes of the connective tissues were formerly called "collagen diseases" and "collagen-vascular diseases".[168,169] However, more recent studies have shown that the connective tissue changes in these diseases are either secondary to or accompanied by inflammatory and/or immunologic changes and that the lesions are not confined to collagen or the vascular system.[168,169]

Many of these diseases are now believed to represent alterations in immunoregulatory mechanisms, and the connective tissue reaction, especially relating to changes in the collagen component, results largely from an altered immunologic response.[170] Antibodies are formed against normal body antigens, and these diseases have been termed autoimmune diseases.[171,172] A number of factors have been proposed to explain the cause of these autoimmune diseases, including altered lymphocyte clonal balance; genetic factors, especially those which control the major histocompatibility complex (MHC); viral agents; and drugs.[173]

Scleroderma, SLE, immune-mediated arthritis, and dermatomyositis are important prototypes of this group of disorders.[174,175] Lesions in these disorders may either be disseminated throughout the body or localized to a specific organ or region. This group of multisystem diseases has many clinical similarities, but at the same time the diseases are clinically very diverse.[175] Furthermore, signs of one disease may coexist with other similar connective tissue diseases. Cases in which several connective tissue diseases occur concomitantly have been termed "overlap syndromes".[174,175]

B. SCLERODERMA

1. People

Scleroderma (hard skin) involves either a localized hardening or thickening of the skin (morphea) or a more generalized and severe involvement of many organ systems (progressive systemic sclerosis).[175] This connective tissue disease is characterized by severe fibrosis and vascular abnormalities. The sites most frequently involved in the systemic form are the skin, esophagus, lungs, heart, and kidneys. The systemic condition most often occurs in females. The connective tissue changes, especially those involving collagen, result in edema and fibrosis. Inflammatory and degenerative changes occur in the vascular system throughout the body and lead to severe vasospasm with decreased vascular perfusion of tissues including the extremities, gastrointestinal system, heart, lungs, joints, kidney, and skeletal muscle.

The underlying cause of the fibrosis and vascular degeneration is not known. There are three popular theories regarding the primary cause of systemic scleroderma.[176-178] One theory proposes that the vascular change is the primary cause, while another theory is that the collagen deposition is primary and precedes the vascular damage. Still another theory proposes that the connective tissue changes are caused by an immunologic mechanism.

2. Animals

An important condition recently identified in white Leghorn chickens has many similarities to the progressive systemic sclerosis form of scleroderma.[179-181] The condition was identified in a selected line (line 200) of chickens and is inherited as an autosomal recessive trait with incomplete penetrance. Swelling and erythema of the comb and wattle are the first clinical signs.[179] Subsequently, affected chickens develop a polyarthritis of the peripheral joints. Later, the skin on the neck and back becomes indurated and tight, and feathers in these sites are shed. The comb, wattle, and feet become necrotic and slough as a result of severe vascular occlusion. Older birds develop pneumonitis, and vascular occlusive disease is present in the esophagus, heart, and kidney.

The pathological changes of the skin and comb are accompanied by severe and diffuse mononuclear cell infiltration in the skin and subcutaneous tissue.[179] Proliferation of small vessels in the involved tissues is accompanied by severe collagen deposition and mononuclear cell infiltration in perivascular regions. Affected chickens develop antinuclear antibodies, rheumatoid factor, antibodies to type II collagen, and have elevated levels of IgG.[181]

The clinical and pathological similarities of this chicken disorder to progressive systemic sclerosis of people indicate that it is a viable model for study of the human condition.[179-181] In spite of the similarities of the two conditions, several differences between the chicken model and the human disease have been identified.[180] One difference is that the mononuclear cell infiltration in the tissues in the chicken is primaily of B cell origin, whereas in people, the mononuclear cell infiltration is primarily of T cell origin. Another difference is that the vascular occlusion in the chicken model results from severe muscular hypertrophy of small vessels, while in the human condition vascular occlusion is associated with perivascular histocytic infiltration and intimal damage.[180]

A dominant mutant called tight skin (TDK) occurs in the mouse.[180,182,183] Changes in the connective tissue, particularly collagen, are similar to scleroderma.[182,183] In affected mice the skin is tight and inelastic due to an increased deposition of collagen in the dermis and subcutaneous tissue. The fibrous architecture of the papillary dermis is disrupted by hyalinization of collagen and disorganization of the normal weave pattern of collagen bundles. Fibroblasts in the papillary dermis are prominent and appear active. Vascular changes and involvement of internal organs found in the progressive systemic sclerosis of people and the chicken model of scleroderma are not seen in the TDK mouse.[180] The inflammatory and immunologic features of human scleroderma are also absent in this mouse disease.[183] The skin changes in the tight skin mouse mutant differ from human scleroderma in that they are more prominent in the papillary dermis, while changes in the reticular portion of the dermis predominate in the human conditions. Based on the distribution of the lesions, this condition in the mouse appears most similar to the localized form of human scleroderma.

C. SYSTEMIC LUPUS ERYTHEMATOSUS (SLE)

1. People

SLE is an inflammatory, multisystem disease in which an array of autoimmune antibodies are formed against host antigens and are deposited in tissues as fibrinoid deposits in structures containing collagen, including blood vessels and basement membranes.[172,174,175,184,185] Structures which are particularly susceptible to the deposition of these immune complexes are the skin, kidneys, spleen, serosal membranes, joints, and heart.[175] Bone marrow cells and blood leukocytes which contain phagocytized complexes of DNA and anti-DNA antibody are called LE cells, and their identification is useful in the diagnosis of SLE.[185,186] Antibodies to nuclear DNA, and in particular double-stranded DNA (Smith [Sm] antigen), are considered to be almost diagnostic for SLE.[184] Although the cause of the disease is not known, it is known to have a genetic predisposition.[184] Environmental influences, estrogens, and certain drugs, especially procainamide and hydralazine, may precipitate the disease.[174] This condition occurs four to ten times more often in adolescent and young adult women than in men.[184] The defect may reside in the immunologic system because both B cell hyperactivity and decreased T-suppressor cell activity have been found in SLE cases.[172,184]

2. Animals

A disease identified in the New Zealand black mice (NZB) (especially their Fl hybrid offspring) represented the first spontaneous autoimmune disease in animals to be identified.[187,188] Several recent reviews discuss the genetic, pathological, and immunologic aspects of the disease.[189,190] The autoimmune condition in mice is inherited as a dominant trait with varying phenotypic expression. Clinical and pathological changes of the skin, eye, nervous system, kidney, and joints which are similar to human SLE are seen in affected mice.[188,189] Circulating immune complexes, composed of autoantibodies and their respective tissue antigens, are deposited in the renal glomerulus and blood vessels of other target organs.[188-190] Autoantibodies develop to DNA, erythrocytes, lymphocytes, and other antigens. The underlying cause of this aberrant immunologic reaction is not known, but may involve a dysfunction of the B lymphocyte.[191]

Pathological changes in the kidney consist of a proliferative glomerulonephritis with the deposition of antigen-antibody complexes in the mesangium and capillary walls.[188,190,192] Deposition of osmiophilic material, first in the mesangium and later in the capillary wall of the kidney, are seen ultrastructurally. In addition to inflammatory cell infiltration of blood vessels, degenerative changes consisting of focal deposition of immune complexes in the intima and media of medium and small coronary arteries and arterioles (in some cases associated with myocardial infarction) have been reported.[193]

SLE has been described in dogs.[6,194,195] The condition is accompanied by hemolytic anemia, thrombocytopenia, membranoproliferative glomerulonephritis with proteinuria, polyarthritis, dermatitis, cardiac conduction abnormalities, and peribiliary cirrhosis. Clinical studies of SLE in select lines of dogs has revealed a high occurrence of other autoimmune diseases, including Sjogren's syndrome, Hashimoto's thyroiditis, and primary biliary cirrhosis.[195]

A genetic predisposition associated with this canine disease and a relatively high incidence has been observed both in a colony of inbred dogs and in certain families of Shetland sheepdogs.[195] The mode of inheritance appears to be complex and has not been defined.

The pathological changes observed in canine SLE include a membranous or membranoproliferative glomerulonephritis.[195-197] Complexes of antibody, antigen, and complement accumulate along subendothelial or subepithelial regions or within the mesangium, resulting in glomerular damage. Severe joint disease characterized by a nonerosive polyarthritis has been a constant feature associated with SLE.[195,198] Dermatologic signs consist of an alopecic maculopapular or ulcerative dermatitis which is more pronounced at the mucocutaneous junctions of the face and ears.[195] Microscopic changes in the skin include parakeratosis, telangiectasis of the vessels of the papillary dermis, and an accumulation of mononuclear inflammatory cells in the dermis.[195]

Antibodies directed against components of the nucleus, including DNA, RNA, nucleoproteins, and histone-related antigens; red blood cells; leukocytes; platelets; clotting factors; and IgG are produced in SLE apparently due to loss of immune regulation.[195] The antibodies form complexes with antigens and are deposited in tissues, forming a type III hypersensitivity reaction in various sites of the body, especially in association with small blood vessels and basement membranes.[198]

A disease resembling SLE has been described in domestic short-haired cats.[199] The condition is accompanied by fragility of red blood cells, hypergammaglobulinemia, cells positive for the LE test, and a membranous glomerulonephritis.

D. RHEUMATOID ARTHRITIS

1. People

Rheumatoid arthritis is a systemic disorder in which the autoimmune and inflammatory factors produce severe vascular damage which causes a chronic and recurrent synovitis and polyarthritis, especially of the small joints.[175,200] Synovial joints, bone, muscle, fascia, ligaments, and tendons may be involved. Cutaneous granulomas called rheumatoid nodules occur in the skin and are a characteristic change. A proliferative or necrotizing vasculitis involving small muscular arteries, arterioles, venules, and capillaries may be widespread throughout the body or may involve localized sites such as the extremities, skin, heart, intestines, eye, and central nervous system. Identification of the rheumatoid factor may be useful in the diagnosis of the condition. The disorder is approximately three to five times more prevalent in women than in men.

2. Animals

A mouse strain, MRL/l, developed by Murphy and Roths[201] has been shown to be a naturally occurring animal disease with an autoimmune disease composed of overlap features of both SLE and rheumatoid arthritis. This disease is quite unique in that it has the full components of human rheumatoid arthritis consisting of clinically manifested polyarthritis (involving primarily the hind limbs), rheumatoid factors involving both the IgM and IgG immunoglobulin fractions, polyarteritis, and an excessive proliferation of T lymphocytes with enhanced helper activity.[202] The immunologic defect in the MRL/l mice is genetically determined as an autosomal recessive trait, manifested by the lpr gene.[201,202] Pathologic changes in the synovium and periarticular connective tissue are not evident in 1-month-old mice, but are present in 75% of 5- to 6-month-old mice.[202] The proliferation of type B synovial cells in the affected mice is similar to the changes found in patients with early rheumatoid arthritis, and this proliferation of synovial cells is closely associated with the articular degeneration.[203]

Rheumatoid arthritis has been described in dogs of varying ages and breeds.[6,198,204-206] The arthritis is a severe clinically debilitating condition characterized by polyarthritis, joint pain, increased joint fluid, and swelling of the periarticular soft tissue. The joint changes are characterized by a proliferative synovitis with fibrin deposition, cell necrosis, and mononuclear cell infiltration in the synovial tissue. Rheumatoid factor may be present but at low levels.

A form of polyarthritis occurs in the rat and is caused by *Mycoplasma arthritidis*.[206-208] This condition has been reported as a naturally occurring and an experimentally induced disease. Polyarthritis and polyserositis cause joint swelling primarily of the carpus and tarsus. A mononuclear inflammatory cell reaction predominates in the periarticular tissues. In the chronic stages, ankylosis and exuberant periosteal new bone growth occurs.

Two causes of polyarthritis in pigs are *Erysipelothrix rhusiopathiae* and *Mycoplasma hyorhinis*. The most prevalent of the swine arthritides is caused by *E. rhusiopathiae*.[6,209-211] Pigs infected with this bacterial organism develop skin necrosis, endocarditis, and polyarthritis. The acute joint disease is accompanied by vascular disease involving the synovium. The media of small vessels undergo hyaline degeneration, and the perivascular tissues are infiltrated with mononuclear inflammatory cells, especially lymphocytes. In chronic stages, the joint tissue is thickened from granulation tissue formed on the synovium or the articular surface. The cartilage beneath the granulation tissue may undergo severe degeneration with loss of both cartilage and subchondral bone. Rheumatoid factors have been identified in arthritis produced by this agent.

An additional cause of arthritis in swine is *M. hyorhinis*. The organism produces both an arthritis and a polyserositis.[206] Clinically the condition resembles arthritis caused by *E. rhusiopathiae,* but there is less tissue reaction, including less synovial hypertrophy, less fibrosis, and less cartilage destruction. The primary lesion consists of villous synovitis which is characterized histologically by a nodular infiltration of lymphocytes and plasma cells, fibroblast proliferation, and hyperplasia of the synovial lining cells. Factors related to the immunologic mediation of this disease, including rheumatoid factors, have not been identified.

A retroviral disease of goats, termed caprine arthritis-encephalitis (CAE), is a naturally occurring persistent disease in which a progressive arthritis occurs in adult goats.[212-216] In adult goats, CAE is characterized by lameness, joint swelling, calcification, and inflammation of the periarticular soft tissues, degenerative joint disease with loss of articular cartilage, and, rarely, collapse of the joints. The carpus, hock, and stifle are most often affected. Pathological changes include infiltration of lymphocytes, macrophages, and plasma cells in the subsynovial tissues, hypertrophy of the synovial villi, and hyperplasia of the synovial lining cells. Perivascular infiltrates of lymphocytes and plasma cells were seen in the affected joints and in the subcutis, brain, kidney, and epicardium of approximately 50% of the infected goats studied. Autoantibodies, including rheumatoid factor, similar to those found in rheumatoid arthritis of people have not been reported in CAE.

E. DERMATOMYOSITIS

1. People

Dermatomyositis (DM) is an immune-mediated inflammatory disorder involving primarily the connective tissue of the skin and skeletal muscle.[175,217,218] Criteria for the diagnosis of DM include progressive symmetric muscle weakness, elevation of muscle enzymes in the serum, electromyographic changes indicating a myositis, characteristic muscle histological changes indicating degeneration and inflammation, and characteristic cutaneous manifestations.[217] The formation of Gottron's papules (flat-topped papules occurring over the dorsal surface of the joints of the fingers) and a macular type erythema, with or without edema, occurring on the dorsal skin surface over finger joints, elbows, knees, forehead, neck, and chest are the primary cutaneous changes.[175] These cutaneous changes may be accompanied by telangiectasis. Both juvenile and adult forms of this disease have been recognized.[217,218] In the adult form, most cases occur in the fifth to sixth decade of life, and females are affected twice as often as males. The juvenile form of DM is rare and occurs more frequently in females. Multiple autoantibodies and circulating immune complexes have been identified in DM patients.[217]

Histologically, the inflammatory response in the skin and skeletal muscle is often accompanied by cellular accumulations in perivascular regions. Perivascular inflammation, intimal hyperplasia, fibrin thrombi, and tissue infarction are noted in the blood vessels of the skin, muscles, gastrointestinal tract, and adipose tissue. Although the cause of either the adult or juvenile form of DM is not known, a number of immunologic abnormalities have been demonstrated, and the basis for the disorder may be a combination of a regulatory failure of the immune system with a genetic susceptibility.[175,217]

2. Animals

DM has been identified in the Collie and Shetland sheepdog breeds of dogs.[219-221] The condition is first noted in young dogs 7 to 11 weeks of age by erythema, alopecia, and ulcers of the ears, face, lips, tail, and bony prominences of the limbs. Muscle involvement, characterized by bilateral symmetrical muscle atrophy of the head, neck, trunk, and extremities, is noted later. Electromyographic changes consist of abnormalities associated with a myositis, including fibrillation potentials, positive sharp waves, and bizarre high frequency discharges. Vascular and inflammatory changes primarily involving the skin and skeletal muscle are seen histologically.[222] In severely affected dogs, the vascular and inflammatory changes were generalized throughout the body. Elevated serum levels of circulating immune complexes (CIC) have been found in the affected dogs, and the levels of the complexes appear to be correlated with the onset and severity of the dermatitis and myositis.[223,224]

F. POLYARTERITIS NODOSA

1. People

Polyarteritis nodosa is a condition in which small- to medium-sized arteries undergo a severe necrotizing periarterial inflammation which often leads to thrombosis and infarction.[174] The lesions have a segmental and nodular pattern, forming the basis for the term nodosa. Lesions may occur in vessels of the visceral organs, including the kidney, pancreas, liver, and gall bladder; heart; lungs; skin; and central nervous system. The disease is believed to result from an immune-mediated hypersensitivity reaction and is associated with the hepatitis B antigen in many cases.

2. Animals

Polyarteritis nodosa has been described in a number of animal species, especially older animals. The condition has been reported in rats,[225] mice,[226] dogs,[227] cats,[228] goats,[229] pigs,[229] and cattle.[213] Although the clinical and pathological changes depend on the location of the vasculitis, and any organ may be involved, the mesenteric arteries and their branches are the most frequently affected sites. The lesions involve fibrinoid necrosis of the muscular arteries with a subsequent severe inflammatory response and thrombosis.

G. AMYLOIDOSIS

1. People

Amyloidosis is a condition in which amyloid, an abnormal protein with a characteristic beta-pleated conformation, is deposited extracellularly in tissues of the body.[230-232] The major forms of amyloid are termed primary and secondary amyloid, depending on the source. Amyloid is composed of both major and minor proteins, and the major protein forms the fibril component of the amyloid.[184,230,231] In primary amyloidosis, the major protein is derived from the light chain of the immunoglobulin molecule and is called amyloid light chain (AL). Primary amyloidosis either occurs in the absence of a recognizable disease process or is associated with multiple myeloma. The major protein in secondary amyloidosis is not related to immunoglobulin but is derived from the protein component of high-density lipoprotein from serum. This major protein is termed amyloid A (AA), and because it is derived from serum, it is also termed serum amyloid A (SAA). SAA is an acute phase protein which is produced in inflammatory diseases. Secondary amyloidosis is found coexistent with, or as a result of, a variety of conditions including chronic infections, neoplasia, and heritable diseases in which the deposition of amyloid is a major disease component. A serum-derived minor protein component, termed serum amyloid P (SAP), has a structure similar to C-reactive protein and is always found in amyloid deposits.[184]

An additional form of amyloid, which differs in biochemical and histochemical characteristics from amyloid found in both primary and secondary amyloidosis, is associated with the focal deposition in the pancreatic islets and is termed insular amyloid.[233] The deposition of this type of amyloid is usually associated with diabetes, but it is also seen in nondiabetics.

In primary amyloidosis, the proteinaceous material is deposited in the cardiovascular system, gastrointestinal tract, lymphoid tissues, skin, and muscle.[184,232] Sites for deposition of amyloid in secondary forms include the liver, kidney, spleen, and adrenal glands.[184,232]

Amyloid is deposited extracellularly and has a hyaline, amorphous, homogeneous waxy appearance. The material is recognized histologically by its green birefringence when stained with Congo red and viewed with polarizing light.[184,232] Secondary amyloid loses its affinity for Congo red after potassium permanganate exposure, unlike primary amyloid and insular amyloid. The demonstration of amyloid is more specific when stained with fluorochromes, especially Thioflavin S and T, where it produces a secondary bright yellow fluorescence.[234] Amyloid also demonstrates metachromatic properties when stained with crystal violet and methyl violet. Amyloid stains pink with hematoxylin-eosin and violet with PAS stain. Ultrastructurally, amyloid is composed of characteristic 7.5- to 10-nm rigid, nonbranching, hollow-cored paired fibrils arranged in a loose meshwork.[184,231]

2. Animals

Primary amyloidosis in animals is rare, and those reported cases in dogs and mice have been associated with multiple myeloma.[235-237] An early onset form of primary amyloidosis of the Syrian hamster has been reported.[238-240] The literature is not entirely clear regarding the type of amyloid formed in this hamster condition, and both the AA and AL forms have been suggested.[239,240] In mice, amyloid deposits are associated with inherited amyloidosis[241] and with the senescence-accelerated mouse disorder in which a unique senile type of amyloid protein is deposited (AS,sam).[242-245] In these murine conditions, the amyloid deposits do not have the staining characteristics of the secondary amyloid AA, but may represent still another form of amyloid distinct from AL and AA.

A familial disorder in the Abyssinian breed of cats is accompanied by systemic amyloidosis in the absence of apparent inflammatory disease.[246-250] Based on staining characteristics, the amyloid appears to be the AA form, and the amyloid is deposited in the kidneys[247-249] and in other organ systems including endocrine, gastrointestinal, reticuloendothelial, and cardiovascular.[249]

Secondary amyloidosis is the most commonly encountered form of amyloidosis of animals[239,251] and is usually associated with chronic inflammation. Secondary amyloidosis has been observed in chronic inflammatory conditions in a number of species including the monkey,[252,253] mink,[254] mouse,[255] guinea pig,[256] horse,[251,257] and cow.[251] Of special interest is the formation and deposition of amyloid in the gray Collie syndrome[258] and in canine lupus erythematosus[259] in response to chronic inflammation. Pekins ducks are prone to develop secondary amyloidosis, especially under stressful conditions, without evidence of a chronic inflammatory response.[260,261]

Secondary amyloidosis, with the deposition of the AA form of amyloid, can be induced in laboratory animals, including mice and hamsters, by the administration of agents which induce an inflammatory response. The agents most commonly used to induce amyloidosis are casein and casein accompanied by lipopolysaccharide.[262,263] An induced form of generalized amyloidosis has been reported in cats, and hypervitaminosis A has been proposed as the cause.[264,265]

Insular amyloidosis, a condition in which a form of amyloid distinct from the primary and secondary forms is deposited locally in the pancreatic islets, has been most frequently found in people and cats,[266-268] but it also occurs in primates, raccoons, hyenas, and wild cats.[266] The condition is usually associated with mature onset diabetes mellitus.[266]

VII. OSTEOPOROSIS

A. PEOPLE

Osteoporosis, or bone atrophy, is a disorder of bone remodeling accompanied by a decrease in bone mass with normal bone mineral matrix.[269-275] Osteoporosis is a disturbance of homeostasis of bone in which either bone reabsorption is excessive and exceeds normal bone formation or bone formation is decreased and bone reabsorption is normal.[273] It is also possible that both bone formation could be decreased and bone reabsorption increased. Several types of osteoporosis occur, including postmenopausal osteoporosis, senile osteoporosis, osteoporosis associated with hyperparathyroidism, and osteoporosis associated with impaired bone formation.[270] Rarer types of osteoporosis include primary juvenile and young adult forms.[270]

The primary clinical manifestation of osteoporosis is fracture of bones, especially the bones of the wrist, vertebrae, and femur.[269,274] This is a serious medical problem, especially in older postmenopausal white women, and as many as 25% of women in this category have one or more fractures by the age of 65.[275] Pelvic and femoral fractures, a serious consequence of osteoporosis, occur in about 200,000 people in the U.S. per year. The social impact of this condition is staggering, and the medical cost is estimated to be in the billions of dollars.[274]

The cause of osteoporosis is unknown and may be multifactorial. Factors associated with the development of osteoporosis include sex; age; bone mass; exercise; nutrition, especially calcium and Vitamin D intake; and hormones, including estrogens, corticosteroids, insulin, parathyroid hormone, and calcitonin.[270,271,274,275]

B. ANIMALS

A number of experimentally induced conditions have been proposed as models to study human osteoporosis.[276] No single model appears to provide the conditions necessary for convincing comparisons, including similarities of the remodeling defect in mature animals involving the changes in bone formation and reabsorption. Frost[277] has presented criteria which are necessary to consider in selecting an animal model for comparative studies to human conditions including osteoporosis. The rat, mouse, hamster, chick, and the young growing animals of many species are not satisfactory models because their bone systems are undergoing both growth and remodeling, whereas human osteoporosis is a condition in adults involving bone remodeling but not growth.[277] Most experimental animal studies of osteoporosis have involved use of the rat and mouse, but dogs and primates, which are better model choices, have been used in a limited number of studies.

Bone changes were studied in older rats, including the Sprague-Dawley, Buffalo, and Fisher strains.[278] Osteoporotic changes were found in all three strains; however, the lesions in the Sprague-Dawley were primarily in the tibia, while all limb bones of the Buffalo strain showed osteoporotic changes. Diet change, limb immobilization, castration, and steroid administration have been used to induce osteoporosis-like bone changes in the rat.[279-282] A recent report indicated that osteoporotic bone changes were associated with spontaneously hypertensive rats (SHR).[283]

A recently developed murine model of accelerated senescence (senescence accelerated mouse, [SAM]) demonstrates age-related osseous changes similar to osteoporosis, especially in the SAM-R/3 and SAM-P/6 strains.[284] The studies indicated that older SAM mice of these two strains are prone to develop bone fractures and that histological changes of the tibia, femur, and lumbar spine showed an osteopenia. The osseous change was not predominantly a disorder of females as it is in people.

Changes have been induced in the tibias of rhesus monkeys by immobilizing the monkeys in a semirecumbent position for up to 7 months.[285] The authors indicated that bone loss from the tibia was a result of increased bone resorption and that is resembled senile osteoporosis.

Studies of beagle dogs showed bone loss of the distal part of the tibia after they reached 6 years of age.[286] This change involved a steady decline from 6 to 12 years. No differences in the severity of the bone loss were noted when comparing age-matched female dogs with male dogs. The advantage of using dogs for age-related bone-loss studies is that the skeletons of adult man and dog are identical in composition and in their bone remodeling processes.[286]

VIII. OSTEOPETROSIS

A. PEOPLE

Osteopetrosis (osteosclerosis, marble bones) is an inherited metabolic bone disorder which is accompanied by a generalized reduction in bone resorption resulting from a defect in the function or production of osteoclasts.[287-289] The condition in people is usually inherited as an autosomal recessive trait, but autosomal dominant forms have been reported.[289] A juvenile form and an adult form of the disease have been identified.[287] The juvenile form is more severe and debilitating than the adult form, but they appear similar radiographically. Radiographic changes include increased density of bone, lack of corticomedullary differentiation, and the formation of sclerotic foci termed endobones (bones within a bone). Clinically, the disorder is associated with frequent fractures, abnormal dentition, delayed growth, neurological abnormalities, leukoerythroblastic anemia, thrombocytopenia, and hepatosplenomegaly.[287] Serum calcium levels are normal in most human osteopetrosis cases.[287] Histological appearance of the bone indicates that there is an arrest of resorption of primary spongiosa and marked proliferation of fibrous tissue in the marrow space.[289] The underlying biochemical defect is not known.

B. ANIMALS

Four mouse mutants for osteopetrosis have been described, including the grey-lethal (gl), microphthalmic (mi), osteosclerotic (oc), and osteopetrotic (op) forms.[287,289,290] Although these mutants are discussed with considerable overlap of clinical and pathological descriptions, each most likely represents a different cellular and biochemical derangement. The longevity of the affected animals varies and the mi mutant may live as long as 1 year, while the gl mutants survive to only 6 weeks or less.[289] There is excessive accumulation of bone and calcified cartilage throughout the skeletal system in the murine osteopetroses. The most severe lesions are located in the proximal metaphyses of the humerus and tibia and in the distal metaphysis of the femur.[289] Histologically, the disorders are characterized by a persistence of endochondral primary and secondary trabeculae and continued formation of new trabecular bone. There is an increased activity of the growth plates in metaphyseal regions, and the growth plates consist of a core of calcified cartilage coated by bone matrix.[287,289] The osteoblast population is increased in number as compared to the osteoclast population, especially in the mi and gl mutants.[289] A severe deficiency of osteoclasts has been observed in the op, gl, and mi mutants, and morphological studies indicated that this deficiency is associated with a severe reduction in the numbers of osteoclast precursors in the op mutant.[290] The osteoclast precursors have large electron-dense intracytoplasmic granules. The rate of formation of bone collagen is increased in the mutants.[289] The blood calcium and phosphorous levels associated with these diseases is somewhat contradictory.[287,290] One report indicated that all of the mouse mutants are hypocalcemic while a later paper indicated that the op mutant was normocalcemic but hypophosphatemic. All of the mutants show a significantly reduced hypercalcemic response to administration of parathyroid extract, supporting the hypothesis of reduced bone resorption.[289] In addition to the skeletal changes, neurological changes and derangements of pigmentation may be noted.[287]

Three rat mutants with accompanying osteopetrosis have been described, including the obese osteopetrotic mutant (op), the Osborn-Mendel mutant (tl), and the incisors-absent mutant (ia).[287,289] The mutants are characterized by an increase in skeletal mass, bone growth retardation, and altered bone shape. Serum calcium levels are normal in the ia rat.[287] The op mutant survives approximately 1 year, while the ia and tl mutants have a normal life span.[289] All three mutants are sexually competent. The ia mutant may undergo spontaneous recovery.

The rabbit mutant (os) is an autosomal recessive trait identified in the Dutch breed of rabbits.[289] The longevity of this mutant is short, and the maximum life span is 4 to 5 weeks.[289] Serum calcium concentrations are abnormally low at birth, and this depression of serum calcium continues with age.[291] Serum phosphorous levels are depressed at birth but reach normal range at 4 weeks of age. The os mutant rabbits have delayed eruption of the incisors and an increased density of the bone.[289] The liver and spleen are markedly enlarged due to extramedullary hematopoiesis. A recent study indicates that osteoclasts from the os mutant rabbit have ultrastructural abnormalities, including the absence of a well-defined ruffled border,[291] indicating a morphological expression of an inability to actively resorb bone. Osteoblasts and osteocytes are present in early life and contain large, electron-dense inclusions.[291] Shortly after birth, both osteoclasts and osteoblasts are decreased in number, suggesting a stem cell defect.

Osteopetrosis has been identified in Aberdeen Angus, Hereford, Simmental, and Holstein breeds of cattle.[289,292-294] The pattern of inheritance is autosomal recessive in the Angus breed. Affected calves are usually premature and stillborn. The skeletal involvement is generalized, but the degree of severity differs in individual bones. The long bones appear normal but are fragile, and the marrow cavities are filled with unresorbed primary spongiosa. Microscopic changes especially involve the endochondral bone. The metaphyses have persistence both of cartilage (with focal distribution) and of the primary spongiosa. The diaphysis contains an abnormal deposition of coarse bone. The liver and spleen have extramedullary hematopoiesis.

Osteopetrosis has been reported in dachshund puppies[295] and a young adult Australian shepherd dog.[296] Both reports indicated that the disorders were accompanied by radiographic changes compatible with osteopetrosis and an anemia.

Osteopetrosis can be induced in chickens by avian leukosis viruses.[297,298] There are strain differences in the virus, and some strains induce osteopetrosis at a high frequency with a shortened latency period of 1 to 3 months. This viral-induced disorder appears to result in excessive osteoblast proliferation and may be accompanied by a fatal anemia.

IX. CARDIOVASCULAR DISORDERS

A. GENERAL

Cardiovascular diseases, especially atherosclerosis and hypertension, are the major causes of death in the U.S. and other developed countries.[299-301] Atherosclerosis of people is a vascular disease in which atheromatous plaques form in larger vessels, especially the lower abdominal aorta, coronary arteries, popliteal arteries, descending thoracic aorta, internal carotid arteries, and the circle of Willis. Several risk factors have been identified, including hypercholesterolemia, hypertension, cigarette smoking, and diabetes mellitus.[299,300] The plaques contain cells, including smooth muscle cells and blood-derived macrophages; connective tissue components; and lipids.[299-302] Connective tissue of the vessel plays an important role in the development of both atherosclerosis and hypertension.[303-305]

B. ATHEROSCLEROSIS

1. People

Collagen may contribute to several events involved in the formation of an atherosclerotic lesion.[303-309] Collagen is the major extracellular component of the fibrous plaque, a lesion found in advanced atherosclerosis. At least three major changes in the connective tissue of the diseased vessel have been proposed. The first involves the importance of the subendothelial connective tissue to maintain the integrity of the endothelial lining of the vessel. Maintenance of the proper integrity of this subendothelial connective tissue may play a critical role in the pathogenesis of atherosclerosis because a disruption of this subendothelial connective tissue may contribute to early damage of the endothelium.[305] A second change which occurs in the connective tissue during the formation of an atheroma is a change in the proportions of the various collagen types. Some reports indicate that there is an increase in the ratio of type I to type III collagen.[304,306-308] A more recent study indicates that type IV collagen is increased in association with the development of the vascular lesion and may be related to the initial change involving the intimal proliferation.[309] Third, following injury of the endothelium, the exposed collagens in the underlying intimal connective tissue are very reactive with platelets and cause platelet aggregation.[299,304] This early lesion involving alteration in the endothelial integrity and subsequent platelet aggregation and thrombus formation is believed to be a fundamental aspect of atherogenesis.

Glycosaminoglycans (GAGs) are a component of the vascular connective tissue and changes in GAGs have been noted in vessels in both naturally occurring and experimental atherosclerosis and in hypertension. Regarding GAGs, it has been observed that changes in the concentrations and the degree of sulfation of GAG has been correlated with the development of atherosclerosis.[310-312] GAGs have been associated with atherogenesis because they affect the transport of lipoprotein and the flow of fluid in the extravascular space and because they form insoluble complexes with lipoproteins, especially low-density lipoproteins, which are deposited in the developing atherosclerotic lesion.

2. Animals

The susceptibility of atherosclerosis varies among the animal species; some are more susceptible while others are more resistant. Pigs, rabbits, and chickens are quite sensitive to the development of atherosclerosis, while rats, cats, dogs, and cattle are relatively resistant to the development of this vascular change.[313] As in people, advanced lesions could be expected to be more commonly found in animals at the end of their life span.[314] The distribution of the lesions vary depending on the species: in people the lower abdominal aorta; the coronary arteries, especially the proximal left coronary artery; the descending thoracic aorta; and the cerebral vessels are the major sites of atheromatous involvement.[315] Studies of atherosclerosis of animals, including naturally occurring and experimental, have formed the basis for much of our understanding of the development of the lesion and for future direction for intervention and treatment.[316,317] By far, the majority of the studies have involved induced changes similar to the human atherosclerosis. This discussion will include both the naturally occurring and induced models.

The naturally occurring models with atherosclerotic changes have been identified, but this is not a common finding in animals.[313] Pigs are an especially suitable species for use in experimental cardiovascular research because they have many similarities to people, including the size and distribution of arteries, location of the normally thickened intima, and similarities of blood pressures, heart rates, and plasma lipoprotein patterns.[318-321] Pigs develop atherosclerosis with age, and these lesions are similar in distribution and appearance to those found in people.[318,320,321] Reports indicate that pigs develop early vascular lesions on nonatherogenic diets as early as 2 months of age and that the intimal thickening is pronounced as the pigs age.[320,322,323]

Naturally occurring atherosclerosis has been reported in several nonhuman primates, including cynomolgus monkeys and rhesus monkeys.[316,324] The vascular lesions are more commonly observed in cynomolgus monkeys than in rhesus monkeys. The distribution of the lesions in the cynomolgus monkey is most consistently found in the aortic arch and the thoracic aorta.[324,325] Some vascular lesions resemble the fibrous plaques seen in the human condition. The nature of the lesions in naturally occurring atherosclerosis of the rhesus monkey involves fatty streaks with an occasional fibrous plaque of the aortic and coronary arteries.[325]

Naturally occurring atherosclerosis has been recognized in several breeds of older dogs.[326] A familial hypothyroidism with hyperlipoproteinemia of the beagle is accompanied by atherosclerotic lesions, especially of the coronary vessels.[327] A form of hyperlipoproteinemia has been recognized in the miniature schnauzer breed.[328] Clinical changes include seizures and abdominal distress and, although vascular lesions are implied, the associated pathological changes have not been defined.

A type of arteriosclerosis with arterial degeneration with dissecting aneurysms of the coronary and renal vessels has been described in young racing greyhound dogs, especially in males.[329] The striking histological change was severe intimal damage with the formation of fibrous streaks. The cause of this vascular change is unknown, but a compensatory response to the severe hemodynamic changes resulting from the stress of racing has been suggested.

An autosomal, recessively inherited hypercholesterolemia has been identified in a strain of rabbits called the Watanabe heritable hyperlipidemic (WHHL).[299,330,331] The disorder is accompanied by severe atherosclerosis. The pattern of atherosclerosis in rabbits with this disorder is the same as that pattern found in the human familial hypercholesterolemia. The biochemical defect in this disorder has been identified as a defect in the low-density-lipoprotein (LDL) receptor.[332]

A genetically transmitted condition in the rat is accompanied by hypertension, severe vascular disease, hyperlipidemia, and obesity.[333] The condition is transmitted as an autosomal recessive trait. Grossly, the vascular lesions involved the pancreatic, superior mesenteric, and hepatic arteries. Microscopic changes in the vascular system were present in many vessels, including the coronary vessels.

Aortic atherosclerosis occurs as a natural disease in white Carneau pigeons.[316,334] Aortic lesions develop in the pigeon early in life and are present in all pigeons of this strain at 4 years of age. Lesions in the coronary vessels also develop, but later in life. The atherosclerotic plaques form mainly at the celiac bifurcation and contain lipid covered by a fibrous cap. Advanced plaques may calcify, and the media may undergo degeneration with ulceration and thrombosis.

Naturally occurring atherosclerosis occurs in both the male and female Japanese quail at approximately 2 years of age.[335,336] The disease is characterized by severe atherosclerosis involving the aortic arch, the thoracic and abdominal aortas, and the coronary vessels.

By far, most experimental studies of atherosclerosis have involved the induction of lesions using dietary methods. Those species which have been identified as naturally occurring models of human atherosclerosis have also been used most extensively in dietary studies to induce experimental atherosclerosis. The rabbit was the first species to be given atherosclerotic diets to study human atherosclerosis, and this species still remains an extensively studied model system.[316,337-340] The type and severity of the vascular lesion is greatly influenced by the atherogenic diet, especially the relationship of the cholesterol, fat, and carbohydrate content.[316,338] Although atherosclerotic lesions can be readily produced in rabbits on atherogenic diets, the lesions produced differ in distribution and in appearance compared to the human disease.[316] The rabbit develops lesions in the aortic arch and thoracic aorta rather than in the abdominal aorta as seen in people. The distribution of the coronary lesions differs between human beings and rabbits in that the large proximal coronary arteries are not as severely involved in the rabbit as in people. Also, the formation of fibromuscular caps with an accompanying increase in collagen synthesis is not a common finding in rabbits as it is in people, but this may be a reflection of the type of diet used to induce the lesions rather than a species resistance.[338]

Atherosclerosis has been induced in pigs by feeding a hyperlipemic diet.[316,318,320,341,342] Atherosclerotic lesions can be produced in the abdominal aorta and coronary arteries in young pigs after 90 d on a hyperlipemic diet.[342] Diet-induced atherosclerosis has been produced in both cynomolgus and rhesus monkeys, and the vascular changes involve an accelerated disease process similar to the human atherosclerosis.[324,343-348] Both naturally occurring and dietary-induced primate models of atherosclerosis research have been recently reviewed.[324] The white Carneau pigeon has been used as a model for cholesterol-aggravated atherosclerosis.[316,349,350] Intimal changes can be observed microscopically in newly hatched infant pigeons, which allows for studies of the development of the atherosclerotic lesion. Several strains of pigeons have been developed, and differences in the susceptibility of atherosclerosis among the strains have been noted.[316]

A form of atherosclerosis can be induced in chickens by a herpesvirus, Marek disease virus (MDV).[351] Chronic atherosclerosis can be produced in specific-pathogen-free, normocholesterolemic chickens. The disease appears to resemble atherosclerosis of people based on the distribution, morphological appearance, and biochemical nature of the lesions. Human beings are susceptible to several herpesvirus infections, and this avian model offers an opportunity to examine the role of infectious agents, especially herpesviruses, in the genesis of human atherosclerosis.

C. HYPERTENSION

1. People

Hypertension is an increase in blood pressure in excess of the levels normally expected in the general population. Hypertension appears to be caused by an interaction of genetic susceptibility with several environmental stimuli including stress, alcohol, smoking, salt, estrogens, and pregnancy.[352] Hypertension is an important antecedent in the development of atherosclerosis, and although there is convincing evidence that the connective tissue is involved in these disease processes, the role the connective tissue plays as a primary cause in the development of hypertension is not clear.[303,353] Several authors have reported that the hypertensive state is accompanied by an increase in the amount of collagen and elastin in blood vessels.[303,353-355] Results of some studies have indicated that the increase in collagen synthesis occurred after the blood pressure had increased, suggesting that the collagen synthesis was secondary to other factors such as mechanical stress on the vessel.[355] Nevertheless, the collagen accumulation in the blood vessel wall appears to perpetuate the condition. The synthesis of GAGs appears to be increased during hypertension.[353]

2. Animals

Naturally occurring animal models of hypertension have been identified in a number of species, including the rat, dog, mouse, chicken, and rabbit.[356,357] The most extensively studied species is the rat.[356-363] Several strains of hypertensive rats have been identified, including the spontaneously hypertensive rat (SHR),[356-359,363] the salt-sensitive and salt-resistant strains of rats developed by Dahl,[356,357,359] the New Zealand strain,[356,357,361] the Milan strain,[356,357,362,363] and others.[363] Hypertensive strains of mice,[364,365] rabbits,[366] chickens,[367] and turkeys[368] have been identified, but they have not been used as extensively as the rat models.

There are several reports of naturally occurring primary hypertension in dogs.[357,369-371] In a recent study, a colony of hypertensive dogs of Siberian husky descent was established, and although the condition was found to be inherited, the exact mode of inheritance was not determined. The identification of hypertension in the dog offers a unique opportunity to study hypertension in a species in which the physiology of the cardiovascular system has been so extensively studied.

ACKNOWLEDGMENTS

The authors gratefully acknowledge the intellectual support and encouragement of Drs. Robert Leader and George Padgett and their pioneering and enthusiastic philosophy of comparative medicine. The authors also are grateful to Ms. Patti Waldo, Sherie Wren, and Mary Lauver for editorial assistance; and to our animal care personnel, including Margaret Espe, Virgil Bullard, Sue Lyons, Richard Brown, and others who have assisted our program over the years.

REFERENCES

1. **Uitto, J. and Perejda, A. J., Eds.,** *Connective Tissue Disease. Molecular Pathology of the Extracellular Matrix,* Marcel Dekker, New York, 1987.
2. **Wagner, B. M., Fleischmajer, R., and Kaufman, N., Eds.,** *Connective Tissue Diseases,* Williams & Wilkins, Baltimore, 1983.
3. **Prockop, D. J. and Kivirikko, K. I.,** Heritable diseases of collagen, *N. Engl. J. Med.,* 311, 376, 1984.
4. **Pinnell, S. R. and Murad, S.,** Disorders of collagen, in *Metabolic Basis of Inherited Disease,* 5th ed., Stanbury, J. B., Wyngaarden, J. B., Fredrickson, D. S., Goldstein, J. L., and Brown, M. S., Eds., McGraw-Hill, New York, 1983, 1425.

5. **Uitto, J.,** UCLA conference. Biochemistry of collagen in diseases, *Ann. Intern. Med.*, 105, 740, 1986.
6. **Leader, R. W., Hegreberg, G. A., Padgett, G. A., and Wagner, B. M.,** Comparative pathology of connective tissue diseases, in *Connective Tissue Diseases*, Wagner, B. M., Fleischmajer, R., and Kaufman, N., Eds., Williams & Wilkins, Baltimore, 1983, 150.
7. **Minor, R. R., Wootton, J. A. M., Patterson, D. F., Uitto, J., and Bartel, D.,** Genetic diseases of collagen in animals, in *Connective Tissue Disease. Molecular Pathology of the Extracellular Matrix*, Uitto, J. and Perejda, A. J., Eds., Marcel Dekker, New York, 1987, 293.
8. **Hegreberg, G. A.,** Animal models of collagen disease, in *Animal Models of Inherited Metabolic Diseases*, Desnick, R. J., Patterson, D. F., and Scarpelli, D. G., Eds., Alan R. Liss, New York, 1982, 229.
9. **McKusick, V. A.,** *Heritable Disorders of Connective Tissue*, C. V. Mosby, St. Louis, 1972.
10. **Ehlers, E.,** Cutis laxa, Neigung zu Haemorrhagien in der haut, lockerung nehrerer Artikulationen, *Dermatol. Z.*, 8, 173, 1901.
11. **Danlos, M.,** Un cas de cutis laxa avec tumures par contusion chronique des genoux (xanthome juvenile pseudo-diabetique de M. M. Hallopeau et Mace de Lepinay), *Bull. Soc. Fr. Dermatol. Syphiligr.*, 19, 70, 1908.
12. **Summer, G. K.,** The Ehlers-Danlos syndrome, *Am. J. Dis. Child.*, 91, 419, 1956.
13. **Johnson, S. A. M. and Falls, H. F.,** Ehlers-Danlos syndrome: a clinical and genetic study, *Arch. Dermatol. Syphilol.*, 60, 82, 1949.
14. **Byers, P. H. and Holbrook, K. A.,** Molecular basis of clinical heterogeneity in the Ehlers-Danlos syndrome, *Ann. N.Y. Acad. Sci.*, 460, 298, 1985.
15. **Hollister, D. W., Byers, P. H., and Holbrook, K. A.,** Genetic disorders of collagen metabolism, *Adv. Hum. Genet.*, 12, 1, 1982.
16. **Kivirikko, K. I. and Kuivaniemi, H.,** Posttranslational modification of collagen and their alterations in heritable diseases, in *Connective Tissue Disease. Molecular Pathology of the Extracellular Matrix*, Uitto, J. and Perejda, A. J., Eds., Marcel Dekker, New York, 1987, 263.
17. **Hegreberg, G. A., Padgett, G. A., Gorham, J. R., and Henson, J. B.,** A connective tissue disease of dogs and mink resembling the Ehlers-Danlos syndrome of man. II. Mode of inheritance, *J. Hered.*, 60, 249, 1969.
18. **Hegreberg, G. A., Padgett, G. A., Ott, R. L., and Henson, J. B.,** A heritable connective tissue disease of dogs and mink resembling the Ehlers-Danlos syndrome of man. I. Skin tensile strength properties, *J. Invest. Dermatol.*, 54, 377, 1970.
19. **Hegreberg, G. A., Padgett, G. A., and Henson, J. B.,** A heritable connective tissue disease of dogs and mink resembling the Ehlers-Danlos syndrome of man. III. Histopathologic changes of the skin, *Arch. Pathol.*, 90, 159, 1970.
20. **Patterson, D. F. and Minor, R. R.,** Hereditary fragility and hyperextensibility of the skin of cats. A defect in collagen fibrillogenesis, *Lab. Invest.*, 37, 170, 1977.
21. **Freeman, L. J., Hegreberg, G. A., and Robinette, J. D.,** Ehlers-Danlos syndrome in dogs and cats, *Semin. Vet. Med. Surg.*, 2, 221, 1987.
22. **Freeman, L. J., Hegreberg, G. A., and Robinette, J. D.,** Cutaneous wound healing in Ehlers-Danlos syndrome, *Vet. Surg.*, 18, 88, 1989.
23. **Holbrook, D. A., Byers, P. H., Hegreberg, G. A., and Counts, D.,** Altered collagen fibrils in skin of animals with inherited connective tissue disorders, *Anat. Rec.*, 190, 424, 1978.
24. **Hegreberg, G. A., Padgett, G. A., and Page, R. C.,** The Ehlers-Danlos syndrome of dogs and mink, in *Symp. Proc. III. Animal Models for Biomedical Research*, National Academy of Sciences, Washington, D.C., 1970, 80.
25. **Counts, D. F., Knighten, P., and Hegreberg, G. A.,** Biochemical changes in the skin of mink with Ehlers-Danlos syndrome: increased collagen biosynthesis in the dermis of affected mink, *J. Invest. Dermatol.*, 69, 521, 1977.
26. **Siegel, R. C., Black, C. M., and Bailey, A. J.,** Cross-linking of collagen in the X-linked Ehlers-Danlos type V, *Biochem. Biophys. Res. Commun.*, 88, 281, 1979.
27. **Pinnell, S. R., Krane, S. M., Kenzora, J. E., and Glimcher, M. J.,** A heritable disorder of connective tissue: hydroxylysine-deficient collagen disease, *N. Engl. J. Med.*, 286, 1013, 1972.
28. **Krane, S. M., Pinnell, S. R., and Erbe, R. W.,** Lysyl-protocollagen hydroxylase deficiency in fibroblasts from siblings with hydroxylysine-deficient collagen, *Proc. Natl. Acad. Sci. U.S.A.*, 69, 2899, 1972.
29. **Ihme, A., Krieg, T., Nerlich, A., Feldmann, U., Rauterberg, J., Glanville, R. W., Edel, G., and Muller, P. K.,** Ehlers-Danlos syndrome type VI: collagen type specificity of defective lysyl hydroxylation in various tissues, *J. Invest. Dermatol.*, 83, 161, 1984.
30. **Englard, S. and Seifter, S.,** The biochemical functions of ascorbic acid, *Annu. Rev. Nutr.*, 6, 365, 1986.
31. **Levine, M.,** New concepts in the biology and biochemistry of ascorbic acid, *N. Engl. J. Med.*, 314, 892, 1986.
32. **Chaudhuri, C. R. and Chatterjee, I. B.,** L-ascorbic acid synthesis in birds: phylogenetic trend, *Science*, 164, 435, 1969.

33. **Tucker, B. W. and Halver, J. E.,** Vitamin C metabolism in rainbow trout, *Comp. Pathol. Bull.*, 18, 1, 1986.
34. **Lichtenstein, J. R., Martin, G. R., Kohn, L. D., Byers, P. H., and McKusick, V. A.,** Defect in conversion of procollagen to collagen in a form of Ehlers-Danlos syndrome, *Science*, 182, 298, 1973.
35. **Steinmann, B., Tuderman, L., Peltonen, L., Martin, G. R., McKusick, V. A., and Prockop, D. J.,** Evidence for a structural mutation of procollagen type I in a patient with the Ehlers-Danlos syndrome type VII, *J. Biol. Chem.*, 255, 8887, 1980.
36. **Hanset, R. and Ansay, M.,** Dermatosparaxie (peau dechiree) chez le veau: un default general du tissu conjonctif, de nature hereditaire, *Ann. Med. Vet.*, 111, 451, 1967.
37. **Lapiere, C. M., Lenaers, A., and Kohn, L. D.,** Procollagen peptidase: an enzyme excising the coordination peptides of procollagen, *Proc. Natl. Acad. Sci. U.S.A.*, 68, 3054, 1971.
38. **Helle, O. and Nes, N. N.,** A hereditary skin defect in sheep, *Acta Vet. Scand.*, 13, 443, 1972.
39. **Fjolstad, M. and Helle, O.,** A hereditary dysplasia of collagen tissues in sheep, *J. Pathol.*, 112, 183, 1974.
40. **Becker, U., Timpl, R., Helle, O., and Prockop, D. J.,** NH2-terminal extensions on skin collagen from sheep with a genetic defect in conversion of procollagen into collagen, *Biochemistry*, 15, 2853, 1976.
41. **Counts, D. F., Byers, P. H., Holbrook, K. A., and Hegreberg, G. A.,** Dermatosparaxis in a Himalayan cat. I. Biochemical studies of dermal collagen, *J. Invest. Dermatol.*, 74, 96, 1980.
42. **Holbrook, K. A., Byers, P. H., Counts, D. F., and Hegreberg, G. A.,** Dermatosparaxis in a Himalayan cat. II. Ultrastructural studies of dermal collagen, *J. Invest. Dermatol.*, 74, 100, 1980.
43. **Holbrook, K. A. and Byers, P. H.,** Structural abnormalities in the dermal collagen and elastic matrix from the skin of patients with inherited connective tissue disorders, *J. Invest. Dermatol.*, 79, 7s, 1982.
44. **Ramshaw, J. A. M.,** A mild form of ovine dermatosparaxis, *Collagen Rel. Res.*, 4, 441, 1984.
45. **Pierard, G. E., Le, T., Hermanns, J.-F., Nusgens, B. V., and Lapiere, C. M.,** Morphometric study of cauliflower collagen fibrils in dermatosparaxis of the calves, *Collagen Rel Res.*, 6, 481, 1986.
46. **Byers, P. H., Siegel, R. C., Holbrook, K. A., Narayanan, A. S., Bornstein, P., and Hall, J. G.,** X-linked cutis laxa: defective cross-link formation in collagen due to decreased lysyl oxidase activity, *N. Engl. J. Med.*, 303, 61, 1980.
47. **Kuivaniemi, H., Peltonen, L., Palotie, A., Kaitila, I., and Kivirikko, K. I.,** Abnormal copper metabolism and deficient lysyl oxidase activity in a heritable connective tissue disorder, *J. Clin. Invest.*, 69, 730, 1982.
48. **Peltonen, L., Kuivaniemi, H., Palotie, A., Horn, N., Kaitila, I., and Kivirikko, K. I.,** Alterations in copper and collagen metabolism in the Menkes syndrome and a new subtype of the Ehlers-Danlos syndrome, *Biochemistry*, 22, 6156, 1983.
49. **Menkes, J. H., Alter, M., Steigleder, G. K., Weakley, D. R., and Sung, J. H.,** A sex-linked recessive disorder with retardation of growth, peculiar hair, and focal cerebral and cerebellar degeneration, *Pediatrics*, 29, 764, 1962.
50. **Danks, D. M.,** Hereditary disorders of copper metabolism in Wilson's disease and Menkes' disease, in *Metabolic Basis of Inherited Diseases*, 5th ed., Stanbury, J. B., Wyngaarden, J. B., Fredrickson, D. S., Goldstein, J. L., and Brown, M. S., Eds., McGraw-Hill, New York, 1983, 1251.
51. **Hunt, D. M.,** Primary defect in copper transport underlies mottled mutants in the mouse, *Nature (London)*, 249, 852, 1974.
52. **Rowe, D. W., McGoodwin, E. B., Martin, G. R., Sussman, M. D., Grahn, D., Faris, B., and Franzblau, C.,** A sex-linked defect in the cross-linking of collagen and elastin associated with the mottled locus in mice, *J. Exp. Med.*, 139, 180, 1974.
53. **Van Den Hamer, C. J. A., Prins, H. W., and Nooijen, J. L.,** Menkes' disease in mottled mice and man, in *Models for the Study of Inborn Errors of Metabolism*, Hommes, F. A., Ed., Elsevier, New York, 1979, 95.
54. **Danks, D. M. and Camakaris, J.,** Mutations affecting trace elements in humans and animals. A genetic approach to an understanding of trace elements, *Adv. Hum. Genet.*, 13, 149, 1983.
55. **Fisk, D. E. and Kuhn, C.,** Emphysema-like changes in the lungs of the blotchy mouse, *Am. Rev. Respir. Dis.*, 113, 787, 1976.
56. **Starcher, B. C., Madaras, J. A., and Tepper, A. S.,** Lysyl oxidase deficiency in lung and fibroblasts from mice with hereditary emphysema, *Biochem. Biophys. Res. Commun.*, 78, 706, 1977.
57. **Shields, G. S., Coulson, W. F., Kimball, D. A., Carnes, W. H., Cartwright, G. E., and Wintrobe, M. M.,** Studies on copper metabilism. XXXII. Cardiovascular lesions in copper-deficient swine, *Am. J. Pathol.*, 41, 603, 1962.
58. **Soskel, N. T., Watanabe, S., Hammond, E., Sandberg, L. B., Renzetti, A. D., Jr., and Crapo, J. D.,** A copper-deficient, zinc supplemented diet produces emphysema in pigs, *Am. Rev. Respir. Dis.*, 126, 316, 1982.
59. **Carnes, W. H.,** Copper and connective tissue metabolism, *Int. Rev. Connect. Tissue Res.*, 4, 197, 1968.
60. **Coulson, W. F.,** Copper deficiency, with special reference to the cardiovascular system, *Methods Achiev. Exp. Pathol.*, 6, 111, 1972.

61. **Miller, E. J., Martin, G. R., Mecca, C. E., and Piez, K. A.,** The biosynthesis of elastin cross-links. The effect of copper deficiency and a lathyrogen, *J. Biol. Chem.*, 240, 3623, 1965.
62. **Chou, W. S., Savage, J. E., and O'Dell, B. L.,** Role of copper in biosynthesis of intramolecular cross-links in chick tendon collagen, *J. Biol. Chem.*, 244, 5785, 1969.
63. **Tanzer, M. L.,** Experimental lathyrism, *Int. Rev. Connect. Tissue Res.*, 3, 91, 1965.
64. **Narayanan, A. S., Siegel, R. C., and Martin, G. R.,** On the inhibition of lysyl oxidase by B-aminopropionitrile, *Biochem. Biophys. Res. Commun.*, 46, 745, 1972.
65. **Nimni, M. E.,** A defect in the intramolecular and intermolecular cross-linking of collagen, caused by penicillamine. I. Metabolic and functional abnormalities in soft tissues, *J. Biol. Chem.*, 243, 1457, 1968.
66. **Nimni, M. E. and Bavetta, L. A.,** Collagen defect induced by penicillamine, *Science*, 150, 905, 1965.
67. **Ronchetti, I. P., Fornieri, C., Baccarani Contri, M., Quaglino, D., Jr., and Caselgrandi, E.,** Effect of DL-penicillamine on the aorta of growing chickens. Ultrastructural and biochemical studies, *Am. J. Pathol.*, 124, 436, 1986.
68. **Deshmukh, K. and Nimni, M. E.,** A defect in the intramolecular and intermolecular cross-linking of collagen caused by penicillamine. II. Functional groups involved in the interaction process, *J. Biol. Chem.*, 244, 1787, 1969.
69. **Keiser, H. R., Henkin, R. I., and Kare, M.,** Reversal by copper of the lathyrogenic action of D-penicillamine, *Proc. Soc. Exp. Biol. Med.*, 129, 516, 1968.
70. **Walshe, J. M.,** The liver in Wilson's disease (hepatolenticular degeneration), in *Diseases of the Liver*, 6th ed., Schiff, L. and Schiff, E. R., Eds., Lippincott, Philadelphia, 1987, 1037.
71. **Danks, D. M.,** Hereditary disorders of copper metabolism in Wilson's disease and Menkes' disease, in *Metabolic Basis of Inherited Diseases*, 5th ed., Stanbury, J. B., Wyngaarden, J. B., Fredrickson, D. S., Goldstein, J. L., and Brown, M. S., Eds., McGraw-Hill, New York, 1983, 1251.
72. **Frydman, M., Bonne-Tamir, B., Farrer, L. A., Conneally, P. M., Magazanik, A., Ashbel, S., and Goldwitch, Z.,** Assignment of the gene for Wilson disease to chromosome 13: linkage to the esterase D locus, *Proc. Natl. Acad. Sci. U.S.A.*, 82, 1819, 1985.
73. **Adams, R. D. and Sidman, R. L.,** in *Introduction to Neuropathology*, McGraw-Hill, New York, 1968, 249.
74. **De Santi, M. M., Lungarella, G., Luzi, P., Miracco, C., and Tosi, P.,** Ultrastructural features in active chronic hepatitis with changes resembling Wilson's disease, *Am. J. Clin. Pathol.*, 85, 365, 1986.
75. **Hultgren, B. D., Stevens, J. B., and Hardy, R. M.,** Inherited, chronic, progressive hepatic degeneration in Bedlington Terriers with increased liver copper concentrations: clinical and pathologic observations and comparison with other copper-associated liver diseases, *Am. J. Vet. Res.*, 47, 365, 1986.
76. **Su, L.-C., Owen, C. A., Jr., Zollman, P. E., and Hardy, R. M.,** A defect of biliary excretion of copper in copper-laden Bedlington terriers, *Am. J. Physiol.*, 243, G231, 1982.
77. **Su, L.-C., Ravanshad, S., Owen, C. A., Jr., McCall, J. T., Zollman, P. E., and Hardy, R. M.,** A comparison of copper-loading disease in Bedlington terriers and Wilson's disease in humans, *Am. J. Physiol.*, 243, G226, 1982.
78. **Thornburg, L. P., Shaw, D., Dolan, M., Raisbeck, M., Crawford, S., Dennis, G. L., and Olwin, D. B.,** Hereditary copper toxicosis in west Highland white terriers, *Vet. Pathol.*, 23, 148, 1986.
79. **Thornburg, L. P., Dennis, G. L., Olwin, D. B., Mclaughlin, C. D., and Gulbas, N. K.,** Copper toxicosis in dogs. II. The pathogenesis of copper-associated liver disease in dogs, *Canine Pract.*, 12, 33, 1985.
80. **Johnson, G. F., Sternlieb, I., Twedt, D. C., Grushoff, P. S., and Scheinberg, I. H.,** Inheritance of copper toxicosis in Bedlington terriers, *Am. J. Vet. Res.*, 41, 1865, 1980.
81. **Herrtage, M. E., Seymour, C. A., Jefferies, A. R., Blakemore, W. F., and Palmer, A. C.,** Inherited copper toxicosis in the Bedlington terrier: a report of two clinical cases, *J. Small Anim. Pract.*, 28, 1127, 1987.
82. **Sillence, D. O., Senn, A., and Danks, D. M.,** Genetic heterogeneity in osteogenesis imperfecta, *J. Med. Genet.*, 16, 101, 1979.
83. **Criscitiello, M. G., Ronan, J. A., Besterman, E. M., and Schoenwetter, W.,** Cardiovascular abnormalities in osteogenesis imperfecta, *Circulation*, 31, 255, 1965.
84. **Riley, F. C., Jowsey, J., and Brown, D. M.,** Osteogenesis imperfecta: morphologic and biochemical studies of connective tissues, *Pediatr. Res.*, 7, 757, 1973.
85. **Follis, R. H., Jr.,** Maldevelopment of the corium in the osteogenesis imperfecta syndrome, *Bull. Johns Hopkins Hosp.*, 93, 225, 1953.
86. **Byers, P. H., Shapiro, J. R., Row, D. W., David, K. E., and Holbrook, K. A.,** Abnormal alpha 2-chain in type I collagen from a patient with a form of osteogenesis imperfecta, *J. Clin. Invest.*, 71, 689, 1983.
87. **Prockop, D. J.,** Osteogenesis imperfecta: phenotypic heterogeneity, protein suicide, short and long collagen, *Am. J. Hum. Genet.*, 36, 499, 1984.

88. **Spranger, J.,** The developmental pathology of collagen in humans, in *Genetic Aspects of Developmental Pathology,* Gilbert, E. F. and Optiz, J. M., Eds., Alan R. Liss, New York, 1987, 1.
89. **Chu, M.-L., Williams, C. J., Pepe, G., Hirsch, J. L., Prockop, D. J., and Ramirez, F.,** Internal deletion in a collagen gene in a perinatal lethal form of osteogenesis imperfecta, *Nature (London),* 304, 78, 1983.
90. **Deak, S. B., Nicholls, A., Pope, F. M., and Prockop, D. J.,** The molecular defect in a nonlethal variant of osteogenesis imperfecta: synthesis of proalpha 2 (I) chains which are not incorporated into trimers of type I procollagen, *J. Biol. Chem.,* 258, 15192, 1983.
91. **Peltonen, L., Palotie, A., and Prockop, D. J.,** A defect in the structure of type I procollagen in a patient who had osteogenesis imperfecta: excess mannose in the COOH-terminal propeptide, *Proc. Natl. Acad. Sci. U.S.A.,* 77, 6179, 1980.
92. **Nicholls, A. C., Pope, F. M., and Craig, D.,** An abnormal collagen alpha chain containing cysteine in autosomal dominant osteogenesis imperfecta, *Br. Med. J.,* 288, 112, 1984.
93. **Steinmann, B., Rao, V. H., Vogel, A., Bruckner, P., Gitzelmann, R., and Byers, P. H.,** Cysteine in the triple-helical domain of one allelic product of the alpha 1 (I) gene of type I collagen produces a lethal form of osteogenesis imperfecta, *J. Biol. Chem.,* 259, 11129, 1984.
94. **Denholm, L. J. and Cole, W. G.,** Heritable bone fragility, joint laxity and dysplastic dentin in Friesian calves; a bovine syndrome of osteogenesis imperfecta, *Aust. Vet. J.,* 60, 9, 1983.
95. **Denholm, L. J., Hall, C. E., Minor, R. R., and Krook, L. P.,** Morphometric analysis of defective collagen fibril growth in bovine osteogenesis imperfecta (Australia type)[a], *Ann. N.Y. Acad. Sci.,* 460, 412, 1985.
96. **Termine, J. D., Robey, P. G., Fisher, L. W., Shimokawa, H., Drumm, M. A., Conn, K. M., Hawkins, G. R., Cruz, J. B., and Thompson, K. G.,** Osteonectin, bone proteoglycan, and phosphophoryn defects in a form of osteogenesis imperfecta, *Proc. Natl. Acad. Sci. U.S.A.,* 81, 2213, 1984.
97. **Jensen, P. T., Rasmussen, P. G., and Basse, A.,** Congenital osteogenesis imperfecta in Charollais cattle, *Nord. Veterinaermed.,* 28, 304, 1976.
98. **Kater, J. C., Hartley, W. J., Dysart, T. H., and Campbell, A. R.,** Osteogenesis imperfecta and bone resorption: two unusual skeletal abnormalities in young lambs, *N. Z. Vet. J.,* 11, 41, 1963.
99. **Holmes, J. R., Baker, J. R., and Davies, E. T.,** Osteogenesis imperfecta in lambs, *Vet. Rec.,* 76, 980, 1964.
100. **Guenet, J. L., Stanescu, R., Maroteaux, P., and Stanescu, V.,** Fragilitas ossium (fro): an autosomal recessive mutant in the mouse, in *Animal Models of Inherited Metabolic Diseases,* Desnick, R. J., Patterson, D. F., and Scarpelli, D. G., Eds., Alan R. Liss, New York, 1982, 265.
101. **Calkins, E., Kahn, D., and Diner, W. C.,** Idiopathic familial osteoporosis in dogs: "osteogenesis imperfecta", *Ann. N.Y. Acad. Sci.,* 64, 410, 1956.
102. **Rimoin, D. L.,** The chondrodystrophies, *Adv. Hum. Genet.,* 5, 1, 1975.
103. **Kozlowski, K., Maroteaux, P., Silverman, F., Kaufmann, H., and Spranger, J.,** Classification des dysplasies osseuses: table ronde, *Ann. Radiol.,* 12, 965, 1969.
104. **Rimoin, D. L.,** International nomenclature of constitutional diseases of bone, *J. Pediatr.,* 93, 614, 1978.
105. **Spranger, J. W., Langer, L. O. J., and Wiedemann, H. R.,** *Bone Dysplasias. An Atlas of Constitutional Disorders of Skeletal Development,* W. B. Saunders, Philadelphia, 1974.
106. **Gruneberg, H.,** *The Pathology of Development. A Study of Inherited Skeletal Disorders in Animals,* John Wiley & Sons, New York, 1963, 52.
107. **Pennypacker, J. P., Kimata, K., and Brown, K. S.,** Brachymorphic mice (bm/bm): a generalized biochemical defect expressed primarily in cartilage, *Dev. Biol.,* 81, 280, 1981.
108. **Rittenhuse, E., Dunn, L. C., Cookingham, J., Calo, C., Spiegelman, M., Dooher, G. B., and Bennett, D.,** Cartilage matrix deficiency (cmd): a new autosomal recessive lethal mutation in the mouse, *J. Embryol. Exp. Morphol.,* 43, 71, 1978.
109. **Sawyer, L. M. and Goetinck, P. F.,** Chondrogenesis in the mutant nanomelia. Changes in the fine structure and proteoglycan synthesis in high density limb bud cell cultures, *J. Exp. Zool.,* 216, 121, 1981.
110. **Sande, R. D., Alexander, J. E. and Padgett, G. A.,** Dwarfism in the Alaskan malamute: its radiographic pathogenesis, *J. Am. Vet. Radiol. Soc.,* 15, 10, 1974.
111. **Bingel, S. A., Sande, R. D., and Wight, T. N.,** Chondrodysplasia in the Alaskan Malamute. Characterization of proteoglycans dissociatively extracted from dwarf growth plates, *Lab. Invest.,* 53, 479, 1985.
112. **Riser, W. H., Haskins, M. E., Jezyk, P. F., and Patterson, D. F.,** Pseudochondroplastic dysplasia in miniature poodles: clinical, radiologic, and pathologic features, *J. Am. Vet. Med. Assoc.,* 176, 335, 1980.
113. **Jubb, K. V. F., Kennedy, P. C., and Palmer, N.,** in *Pathology of Domestic Animals,* Vol. 1, 3rd ed., Academic Press, New York, 1985, 1.
114. **Whitbread, T. J., Gill, J. J. B., and Lewis, D. G.,** An inherited enchondrodystrophy in the English Pointer dog. A new disease, *J. Small Anim. Pract.,* 24, 399, 1983.
115. **Gregory, P. W., Tyler, W. S., and Julian, L. M.,** Bovine achondroplasia: the reconstitution of the Dexter components from non-Dexter stocks, *Growth,* 30, 393, 1966.

116. **Gregory, P. W., Julian, L. M., and Tyler, W. S.,** Bovine achondroplasia: possible reconstitution of the Telemark lethal, *J. Hered.*, 58, 220, 1967.
117. **Julian, L. M., Tyler, W. S., and Gregory, P. W.,** The current status of bovine dwarfism, *J. Am. Vet. Med. Assoc.*, 135, 104, 1959.
118. **Johnson, L. E., Harshfield, G. S., and McCone, W.,** Dwarfism, an hereditary defect in beef cattle, *J. Hered.*, 41, 177, 1950.
119. **Chang, T. K.,** Morphological study of the skeleton of Aneon sheep, *Growth*, 13, 269, 1949.
120. **Wray, C., Mathieson, A. O., and Copland, A. N.,** An achondroplastic syndrome in South Country Cheviot sheep, *Vet. Rec.*, 88, 521, 1971.
121. **McCaskey, P. C., Rowland, G. N., Page, R. K., and Minear, L. R.,** Focal failures of endochondral ossification in the broiler, *Avian Dis.*, 26, 701, 1982.
122. **Hargest, T. E., Leach, R. M., and Gay, C. V.,** Avian tibial dyschondroplasia. I. Ultrastructure, *Am. J. Pathol.*, 119, 175, 1985.
123. **Cordy, D. R. and Wind, A. P.,** Transverse fracture of the proximal humeral articular cartilage in dogs (so-called osteochondritis dessecans), *Pathol. Vet.*, 6, 424, 1969.
124. **Rooney, J. R.,** Osteochondrosis in the horse, *Mod. Vet. Pract.*, 56, 41, 1975.
125. **Nakano, T., Aheren, F. X., and Thompson, J. R.,** Mineralization of normal and osteochondrotic bone in swine, *Can. J. Anim. Sci.*, 61, 343, 1981.
126. **Farnum, C. E., Wilsman, N. J., and Hilley, H. D.,** An ultrastructural analysis of osteochondritic growth plate cartilage in growing swine, *Vet. Pathol.*, 21, 141, 1984.
127. **McKusick, V. A. and Neufeld, E. F.,** The mucopolysaccharide storage diseases, in *Metabolic Basis of Inherited Diseases*, 5th ed., Stanbury, J. B., Wyngaarden, J. B., Fredrickson, D. S., Goldstein, J. L., and Brown, M. S., Eds., McGraw-Hill, New York, 1983, 751.
128. **Fluharty, A. L.,** Diseases of glycosaminoglycan and proteoglycan metabolism, in *Connective Tissue Diseases. Molecular Pathology of the Extracellular Matrix*, Uitto, J. and Perejada, A. J., Marcel Dekker, New York, 1987, 491.
129. **Haskins, M. E., Jezyk, P. F., Desnick, R. J., McDonough, S. K., and Patterson, D. F.,** Mucopolysaccharidosis in a domestic short-haired cat — a disease distinct from that seen in the Siamese cat, *J. Am. Vet. Med. Assoc.*, 175, 384, 1979.
130. **Haskins, M. E., Jezyk, P. F., Desnick, R. J., McGovern, M. M., Vine, D. T., and Patterson, D. F.,** Animal models of mucopolysaccharidosis, in *Animals Models of Inherited Metabolic Diseases*, Desnick, R. J., Patterson, D. F., and Scarpelli, D. G., Eds., Alan R. Liss, New York, 1982, 177.
131. **Shull, R. M., Helman, R. G., Spellacy, E., Constantopoulous, G., Munger, R. J., and Neufeld, E. F.,** Morphologic and biochemical studies of canine mucopolysaccharidosis 1, *Am. J. Pathol.*, 114, 487, 1984.
132. **Spellacy, E., Shull, R. M., Constantopoulos, G., and Neufeld, E. F.,** A canine model of human alpha-L-iduronidase deficiency, *Proc. Natl. Acad. Sci. U.S.A.*, 80, 6091, 1983.
133. **Constantopoulos, G., Shull, R. M., Hastings, N., and Neufeld, E. F.,** Neurochemical characterizations of canine alpha-L-iduronidase deficiency disease (model of human mucopolysaccharidosis 1), *J. Neurochem.*, 45, 1213, 1985.
134. **Haskins, M. E., Jezyk, P. F., and Patterson, D. F.,** Mucopolysaccharide storage disease in three families of cats with arylsulfatase B deficiency: leukocyte studies and carrier identification, *Pediatr. Res.*, 13, 1203, 1979.
135. **Haskins, M. E., Jezyk, P. F., Desnick, R. J., and Patterson, D. F.,** Animal model of human disease. Mucopolysaccharidosis VI Maroteaux-Lamy syndrome. Arylsulfatase B-deficient mucopolysaccharidosis in the Siamese cat, *Am. J. Pathol.*, 105, 191, 1981.
136. **Haskins, M. E., Aguirre, G. D., Jezyk, P. F., and Patterson, D. F.,** The pathology of the feline model of mucopolysaccharidosis VI, *Am. J. Pathol.*, 101, 657, 1980.
137. **Jezyk, P. F., Haskins, M. E., Patterson, D. F., Mellman, W. J., and Greenstein, M.,** Mucopolysaccharidosis in a cat with arylsulfatase B deficiency: a model of the Maroteaux-Lamy syndrome, *Science*, 198, 834, 1977.
138. **Gasper, P. W., Thrall, M. A., Wenger, D. A., Macy, D. W., Ham, L., Dornsife, R. E., McBiles, K., Quackenbush, S. L., Kesel, M. L., Gillette, E. L., and Hoover, E. A.,** Correction of feline arylsulfatase B deficiency (mucopolysaccharidosis VI) by bone marrow transplantation, *Nature (London)*, 312, 467, 1984.
139. **Fine, J. D. and Moschella, S. L.,** Diseases of nutrition and metabolism, in *Dermatology*, Vol. 2, 2nd ed., Moschella, S. L. and Hurley, Y. J., Eds., W. B. Saunders, Philadelphia, 1985, 1470.
140. **Truhan, A. P. and Roenigk, H. H., Jr.,** The cutaneous mucinoses, *J. Am. Acad. Dermatol.*, 14, 1, 1986.
141. **Westheim, A. I. and Lookingbill, D. P.,** Plasmapheresis in a patient with scleromyxedema, *Arch. Dermatol.*, 123, 786, 1987.
142. **Johnson, W. C. and Helwig, E. B.,** Cutaneous focal mucinosis, *Arch. Dermatol.*, 93, 13, 1966.

143. **Dillberger, J. E. and Altman, N. H.,** Focal mucinosis in dogs: seven cases and review of cutaneous mucinoses of man and animals, *Vet. Pathol.*, 23, 132, 1986.
144. **Caro, W. A. and Bronstein, B. R.,** Tumors of the skin, in *Dermatology*, Vol. 2, Moschella, S. L. and Hurley, H. J., Eds., W. B. Saunders, Philadelphia, 1985, 1546.
145. **Johnston, J. J., Oikarinen, A. I., Lowe, N. J., and Uitto, J.,** Ultraviolet-induced connective tissue changes in the skin: models of actinic damage and cutaneous aging, in *Models in Dermatology*, Vol. 1, Maibach, H. I. and Lowe, N. J., Eds., S. Karger, Basel, 1985, 69.
146. **Sams, W. M., Smith, J. G., and Burk, P. G.,** The experimental production of elastosis with ultraviolet light, *J. Invest. Dermatol.*, 43, 467, 1964.
147. **Berger, H., Tsambaos, D., and Mahrle, G.,** Experimental production of elastosis with ultraviolet light, *Arch. Dermatol. Res.*, 269, 29, 1980.
148. **Nakamura, K. and Johnson, W. G.,** Ultraviolet light induced connective tissue changes in rat skin. A histopathologic and histochemical study, *J. Invest. Dermatol.*, 51, 253, 1968.
149. **Vogel, H. G., Alpermann, H. G., and Futterer, E.,** Prevention of changes after UV-irradiation by sun screen products in skin of hairless mice, *Arch. Dermatol. Res.*, 270, 421, 1981.
150. **Kligman, L. H., Akin, F. J., and Kligman, A. M.,** Sunscreens promote repair of ultraviolet radiation-induced dermal damage, *J. Invest. Dermatol.*, 81, 98, 1983.
151. **Johnston, K. J., Oikarinen, A. I., Lowe, N. J., Clark, J. G., and Uitto, J.,** Ultraviolet radiation-induced connective tissue changes in the skin of hairless mice, *J. Invest. Dermatol.*, 82, 587, 1984.
152. **Dorn, C. R., Taylor, D. O. N., and Schneider, R.,** Sunlight exposure and risk of developing cutaneous and oral squamous cell carcinomas in white cats, *J. Natl. Cancer Inst.*, 46, 1073, 1971.
153. **Stannard, A.,** Actinic dermatoses, in *Current Veterinary Therapy*, Kirk, R. W., Ed., W. B. Saunders, Philadelphia, 1974, 402.
154. **Candlin, F. T.,** Chronic solar dermatitis of the dog, *Vet. Med.*, 51, 523, 1956.
155. **Hargis, A. M.,** A review of solar-induced lesions in domestic animals, *Comp. Cont. Ed.*, 3, 287, 1981.
156. **Knowles, D. P. and Hargis, A. M.,** Solar elastosis associated with neoplasia in two dalmatians, *Vet. Pathol.*, 23, 512, 1986.
157. **Douglas Moore, A. A. and Lambert, E. C.,** Endocardial fibro-elastosis, in *Pediatric Cardiology*, Watson, H., Ed., C. V. Mosby, St. Louis, 1968, 731.
158. **Schryer, M. J. P. and Karnauchow, P. N.,** Endocardial fibroelastosis. Etiologic and pathogenetic considerations in children, *Am. Heart J.*, 88, 557, 1974.
159. **Ursell, P. C., Neill, C. A., Anderson, R. H., Ho, S. Y., Becker, A. E., and Gerlis, L. M.,** Endocardial fibroelastosis and hypoplasia of the left ventricle in neonates without significant aortic stenosis, *Br. Heart J.*, 51, 492, 1984.
160. **Fishbein, M. C., Ferrans, V. J., and Roberts, W. C.,** Histologic and ultrastructural features of primary and secondary endocardial fibroelastosis, *Arch. Pathol. Lab. Med.*, 101, 49, 1977.
161. **Paasch, L. H. and Zook, B. C.,** The pathogenesis of endocardial fibroelastosis in Burmese cats, *Lab. Invest.*, 42, 197, 1980.
162. **Zook, B. C., Paasch, L. H., Chandra, R. S., and Casey, H. W.,** The comparative pathology of primary endocardial fibroelastosis in Burmese cats, *Virchows. Arch. A:*, 390, 211, 1981.
163. **Eliot, T. S., Jr., Eliot, F. P., Lushbaugh, C. C., and Slager, U. T.,** First report of the occurrence of neonatal endocardial fibroelastosis in cats and dogs, *J. Am. Vet. Med. Assoc.*, 133, 271, 1958.
164. **Wegelius, O. and von Essen, R.,** Endocardial fibroelastosis in dogs, *Acta Pathol. Microbiol. Scand.*, 77, 66, 1969.
165. **Blood, D. C., Radostits, O. M., and Henderson, J. A.,** *Veterinary Medicine*, 6th ed., Balliere Tindall, London, 1983, 300.
166. **Black-Shaffer, B., Grinstead, C. E., II, and Braunstein, J. N.,** Endocardial fibroelastosis of large mammals, *Circ. Res.*, 16, 383, 1965.
167. **Levin, S.,** Parvovirus: a possible etiologic agent in cardiomyopathy and endocardial fibroelastosis, *Hum. Pathol.*, 11, 404, 1980.
168. **Wagner, B. M.,** Connective tissue disease-historical perspective, in *Connective Tissue Diseases*, Wagner, B. M., Fleischmajer, R., and Kaufman, N., Eds., Williams & Wilkins, Baltimore, 1983, 1.
169. **Cruse, J. M., Whitcomb, D., and Lewis, R. E., Jr.,** Autoimmunity-historical perspective, in *Concepts in Immunopathology*, Vol. 1, Cruse, J. M. and Lewis, R. E., Eds., S. Karger, New York, 1985, 32.
170. **Gay, S. and Kresina, T. F.,** Immunological disorders of collagen, in *Collagen in Health and Disease*, Weiss, J. B. and Jayson, M. I., Eds., Churchill Livingstone, New York, 1982, 269.
171. **Cruse, J. M. and Lewis, R. E., Jr., Eds.,** *Concepts in Immunopathology*, Vol. 1, S. Karger, New York, 1985.
172. **Alarcon-Segovia, D., Alcocer-Varela, J., and Diaz-Jouanen, E.,** The connective tissue diseases as disorders of immune regulation, *Clin. Rheumat. Dis.*, 11, 451, 1985.
173. **Cruse, J. M. and Lewis, R. E., Jr., Eds.,** Contemporary concepts of autoimmunity, in *Concepts in Immunopathology*, Vol. 1, S. Karger, New York, 1985, 1.

174. **Hughes, G. R. V.,** *Connective Tissue Diseases*, Blackwell Scientific, London, 1977.
175. **Gilliam, J. N., Cohen, S. B., Sontheimer, R. D., and Moschella, S. L.,** Connective tissue diseases, in *Dermatology*, Vol. 2, Moschella, S. L. and Hurley, H. J., Eds., W. B. Saunders, Philadelphia, 1985, 1087.
176. **Krieg, T., Perlish, J. S., Mauch, C., and Fleischmajer, R.,** Collagen synthesis by scleroderma fibroblasts, *Ann. N.Y. Acad. Sci.*, 460, 375, 1985.
177. **Fleischmajer, R., Perlish, J. S., and Duncan, M.,** Scleroderma. A model for fibrosis, *Arch. Dermatol.*, 119, 957, 1983.
178. **Kahari, V. M., Heino, J., Larjava, H., and Vuorio, E.,** Alterations in scleroderma fibroblast surface glycoproteins associated with increased collagen synthesis, *Acta Derm. Venereol.*, 67, 199, 1987.
179. **Gershwin, M. E., Abplanalp, H., Castles, J. J., Ikeda, R. M., Van der Water, J., Edlunk, J., and Haynes, D.,** Characterization of a spontaneous disease of white Leghorn chickens resembling progressive systemic sclerosis (scleroderma), *J. Exp. Med.*, 153, 1640, 1981.
180. **Van de Water, J. and Gershwin, M. E.,** Animal models of scleroderma, in *Models in Dermatology*, Vol. 1, Maibach, H. I. and Lowe, N. J., Eds., S. Karger, Basel, 1985, 210.
181. **Haynes, D. C. and Gershwin, M. E.,** Diversity of autoantibodies in avian scleroderma: an inherited fibrotic disease of white Leghorn chickens, *J. Clin. Invest.*, 73, 1557, 1984.
182. **Green, M. C., Sweet, H. W., and Bunder, L. E.,** Tight-skin, a new mutation of the mouse causing excessive growth of connective tissue and skeleton, *Am. J. Pathol.*, 82, 493, 1976.
183. **Jimenez, S. A., Millan, A., and Bashey, R. I.,** Scleroderma-like alterations in collagen metabolism occurring in the TSK (tight skin) mouse, *Arthritis Rheum.*, 27, 180, 1984.
184. **Robbins, S. L. and Kumar, V.,** *Basic Pathology*, 4th ed., W. B. Saunders, Philadelphia, 1987, 152.
185. **Hahn, B. H.,** Systemic lupus erythematosus, in *Clinical Immunology*, Vol. 1, Parker, C. W., Ed., W. B. Saunders, Philadelphia, 1980, 583.
186. **Hargraves, M. M., Richmond, H., and Morton, R.,** Presentation of two bone marrow elements: The "tart" cell and the "L. E." cell, *Proc. Staff Meet. Mayo Clin.*, 23, 25, 1948.
187. **Howie, J. B. and Helyer, B. J.,** The immunology and pathology of NZB mice, *Adv. Immunol.*, 9, 215, 1968.
188. **Milich, D. R. and Gershwin, M. E.,** The pathogenesis of autoimmunity in New Zealand mice, in *Immunologic Defects in Laboratory Animals 2*, Gershwin, M. E. and Merchant, B., Plenum Press, New York, 1981, 77.
189. **Theofilopoulos, A. N. and Dixon, F. J.,** Murine models of systemic lupus erythematosus, *Adv. Immunol.*, 37, 269, 1985.
190. **Kyogoku, M., Nose, M., Sawai, T., Miyazawa, M., Tachiwaki, and Kawashima, M.,** Immunopathology of murine lupus-overview, SL/Ni and MRL/Mp-lpr/lpr-, *Prog. Clin. Biol. Res.*, 229, 95, 1987.
191. **Waegell, W. O., Gershwin, M. E., and Castles, J. J.,** The use of congenital immunologic mutants to probe autoimmune disease in New Zealand mice, *Prog. Clin. Biol. Res.*, 229, 175, 1987.
192. **Comerford, F. R., Cohen, A. S., and Desai, R. G.,** The evolution of the glomerular lesion in NZB mice, *Lab. Invest.*, 19, 643, 1968.
193. **Accinni, L. and Dixon, F. J.,** Degenerative vascular disease and myocardial infarction in mice with lupus-like syndrome, *Am. J. Pathol.*, 96, 477, 1979.
194. **Lewis, R. M., Schwartz, R. S., and Henry, W. B., Jr.,** Canine systemic lupus erythematosus, *Blood*, 25, 143, 1965.
195. **Quimby, F. W.,** Canine systemic lupus erythematosus, in *Immunologic Defects in Laboratory Animals 2*, Gershwin, M. E. and Merchant, B., Eds., Plenum Press, New York, 1981, 175.
196. **Slauson, D. O. and Lewis, R. M.,** Comparative pathology of glomerulonephritis in animals, *Vet. Pathol.*, 16, 134, 1979.
197. **Muller, G. H., Kirk, R. W., and Scott, D. W.,** *Small Animal Dermatology*, 3rd ed., W. B. Saunders, Philadelphia, 1983, 181.
198. **Bennett, D.,** Immune-based non-erosive inflammatory joint disease of the dog. I. Canine systemic lupus erythematosus, *J. Small Anim. Pract.*, 28, 871, 1987.
199. **Quimby, F. W.,** Systemic lupus erythematosus, in *Spontaneous Animal Models of Human Disease*, Vol. 1, Andrews, E. J., Ward, B. C., and Altman, N. H., Eds., Academic Press, New York, 1979, 300.
200. **Pope, R. M. and Talal, N.,** Autoimmunity in rheumatoid arthritis, in *Concepts in Immunopathology*, Vol. 1, Cruse, J. M. and Lewis, R. E., Jr., S. Karger, Basel, 1985, 219.
201. **Murphy, E. D. and Roths, J. B.,** Autoimmunity and lymphoproliferation: induction by mutant gene *lpr*, and acceleration by a male-associated factor in strain BXSB mice, in *Genetic Control of Autoimmune Disease*, Rose, N. R., Bigazzi, P. E., and Warner, N. L., Eds., Elsevier, New York, 1978, 207.
202. **Hang, L., Theofilopoulos, N., and Dixon, F. J.,** A spontaneous rheumatoid arthritis-like disease in MRL/l mice, *J. Exp. Med.*, 155, 1690, 1982.
203. **Tanaka, A., O'Sullivan, F. X., Koopman, W. J., and Gay, S.,** Etiopathogenesis of rheumatoid arthritis-like disease in MRL/l mice. II. Ultrastructural basis of joint destruction, *J. Rheumatol.*, 15, 1, 1988.

204. **Newton, C. D., Lipowitz, A. J., Halliwell, R. E., Allen, H. L., Biery, D. N., and Schumacher, H. R.,** Rheumatoid arthritis in dogs, *J. Am. Vet. Med. Assoc.*, 168, 113, 1976.
205. **Lipowitz, A. J. and Newton, C. D.,** Laboratory parameters of rheumatoid arthritis of the dog: a review, *J. Am. Anim. Hosp. Assoc.* 11, 600, 1975.
206. **Newton, C. D., Schumacher, H. R., and Halliwell, R. E.,** Arthritis-rheumatoid-like, in *Spontaneous Animal Models of Human Disease,* Vol. 2, Andrews, E. J., Ward, B. C., and Altman, N. H., Eds., Academic Press, New York, 1979, 253.
207. **Sokoloff, L.,** Animal model of human disease: arthritis due to mycoplasma in rats and swine, *Am. J. Pathol.*, 73, 261, 1973.
208. **Kirchhoff, H., Heitmann, J., Dubenkropp, H., and Schmidt, R.,** Antigenic cross-reactions between Mycoplasma arthritidis and rat tissues, *Vet. Microbiol.*, 9, 237, 1984.
209. **Kurosaki, Y.,** Distribution of Erysipelothrix rhusiopathiae in arthritic swine and the serovars of the isolates, *J. Jpn. Vet. Med. Assoc.*, 37, 302, 1984.
210. **Turner, G. V. S., I.,** The pathology of infectious polyarthritis in slaughter pigs. II. A microbiological study of polyarthritis in slaughter pigs, *S. Afr. Vet. Assoc. J.*, 53, 95, 1982.
211. **Sikes, D., Crimmins, L. T., and Fletcher, O. J.,** Rheumatoid arthritis of swine: a comparative pathologic study of clinical spontaneous remissions and exacerbations, *Am. J. Vet. Res.*, 30, 753, 1969.
212. **Crawford, T. B., Adams, D. S., Cheevers, W. P., and Cork, L. C.,** Chronic arthritis in goats caused by a retrovirus, *Science,* 207, 997, 1980.
213. **Crawford, T. B., Adams, D. S., Sande, R. D., Gorham, J. R., and Henson, J. B.,** The connective tissue component of the caprine arthritis-encephalitis syndrome, *Am. J. Pathol.*, 100, 443, 1980.
214. **Adams, D. S., Crawford, T. B., and Klevjer-Anderson, P.,** A pathogenetic study of the early connective tissue lesions of viral caprine arthritis-encephalitis, *Am. J. Pathol.*, 99, 257, 1980.
215. **Al-Ani, F. K. and Vestweber, J. G. E.,** Caprine arthritis-encephalitis syndrome (CAE): a review, *Vet. Res. Commun.*, 8, 243, 1984.
216. **Banks, K. L., Jacobs, C. A., Michaels, F. H., and Cheevers, W. P.,** Lentivirus infection augments concurrent antigen-induced arthritis, *Arthritis Rheum.*, 30, 1046, 1987.
217. **Callen, J. P.,** Dermatomyositis, *Dis. Mon.*, 33, 242, 1987.
218. **Mastaglia, F. L. and Ojeda, V. J.,** Inflammatory myopathies. I and II, *Ann. Neurol.*, 17, 215, 1985.
219. **Hargis, A. M., Haupt, K. H., Hegreberg, G. A., Prieur, D. J., and Moore, M. P.,** Familial canine dermatomyositis: initial characterization of the cutaneous and muscular lesions, *Am. J. Pathol.*, 116, 234, 1984.
220. **Haupt, K. H., Prieur, D. J., Moore, M. P., Hargis, A. M., Hegreberg, G. A., Gavin, P. R., and Johnson, R. S.,** Familial canine dermatomyositis: clinical, electrodiagnostic, and genetic studies, *Am. J. Vet. Res.*, 46, 1861, 1985.
221. **Hargis, A. M., Prieur, D. J., Haupt, K. H., and Collier, L. L.,** Post-mortem findings in a Shetland sheepdog with dermatomyositis, *Vet. Pathol.*, 23, 509, 1986.
222. **Hargis, A. M., Prieur, D. J., Haupt, K. H., Collier, L. L., Evermann, J. F., and Ladiges, W. C.,** Postmortem findings in four litters of dogs with familial canine dermatomyositis, *Am. J. Pathol.*, 123, 480, 1986.
223. **Haupt, K. H., Prieur, D. J., Hargis, A. M., Cowell, R. L., McDonald, T. L., Werner, L. L., and Evermann, J. F.,** Familial canine dermatomyositis: clinicopathologic, immunologic, and serologic studies, *Am. J. Vet. Res.*, 46, 1870, 1985.
224. **Hargis, A. M., Prieur, D. J., Haupt, K. H., McDonald, T. L., and Moore, M. P.,** Prospective study of familial canine dermatomyositis. Correlation of the severity of dermatomyositis and circulating immune complex levels, *Am. J. Pathol.*, 123, 465, 1986.
225. **Skold, B. H.,** Chronic arteritis in the laboratory rat, *J. Am. Vet. Med. Assoc.*, 138, 204, 1961.
226. **Upton, A. C., Conklin, J. W., Cosgrove, G. E., Gude, W. D., and Darden, E. B.,** Necrotizing polyarteritis in aging RF mice, *Lab. Invest.*, 16, 483, 1967.
227. **Kelly, D. F., Grunsell, C. S. G., and Kenyon, C. J.,** Polyarteritis in the dog: a case report, *Vet. Rec.*, 92, 363, 1973.
228. **Altera, K. P. and Bonasch, H.,** Perarteritis nodosa in a cat, *J. Am. Vet. Med. Assoc.*, 149, 1307, 1966.
229. **Walvoort, H. C., Gruys, E., and van Dijk, J. E.,** Polyarteritis nodosa, *Tijdschr. Diergeneeskd.*, 112, 204, 1987.
230. **Glenner, G. G.,** Amyloid deposits and amyloidosis. The B-fibrilloses. I, *N. Engl. J. Med.*, 302, 1283, 1980.
231. **Kisilevsky, R.,** Amyloidosis: a fimiliar problem in the light of current pathogenetic developments, *Lab. Invest.*, 49, 381, 1983.
232. **Breathnach, S. M.,** Amyloid and amyloidosis, *J. Am. Acad. Dermatol.*, 18, 1, 1988.
233. **Ehrlich, J. C. and Ratner, I. M.,** Amyloidosis of the islets of Langerhans; a restudy of islet hyalin in diabetic and nondiabetic individuals, *Am. J. Pathol.*, 38, 49, 1961.
234. **Saeed, S. M. and Fine, G.,** Thioflavin-T for amyloid detection, *Am. J. Clin. Pathol.*, 47, 588, 1967.

235. **Jennings, A. R.,** Plasma-cell myelomatosis in the dog, *J. Comp. Pathol. Ther.*, 59, 113, 1949.
236. **Ebbesen, P. and Rask-Nielsen, R.,** On amyloidosis and paraproteinemia in seven transplantation sublines of a murine plasma cell leukemia, *U.S. Natl. Cancer Inst. J.*, 38, 723, 1967.
237. **Osborne, C. A., Johnson, K. H., Perman, V., and Schall, W. D.,** Renal amyloidosis in the dog, *J. Am. Vet. Med. Assoc.*, 153, 669, 1968.
238. **Gleiser, C. A., Van Hoosier, G. L., Sheldon, W. G., and Read, W. K.,** Amyloidosis and renal paramyloid in a closed hamster colony, *Lab. Anim. Sci.*, 21, 197, 1971.
239. **Mezza, L. E., Quimby, W., Durham, S. K., and Lewis, R. M.,** Characterization of spontaneous amyloidosis of Syrian hamsters using the potassium permanganate method, *Lab. Anim. Sci.*, 34, 376, 1984.
240. **Murphy, J. C., Fox, J. G., and Niemi, S. M.,** Nephrotic syndrome associated with renal amyloidosis in a colony of Syrian hamsters, *J. Am. Vet. Med. Assoc.*, 185, 1359, 1984.
241. **Heston, W. E. and Deringer, M. K.,** Hereditary renal disease and amyloidosis in mice, *Arch. Pathol.*, 46, 49, 1948.
242. **Takeshita, S., Hosokawa, M., Irino, M., Higuchi, K., Shimizu, K., Yasuhira, K., and Takeda, T.,** Spontaneous age-associated amyloidosis in senescence-accelerated mouse (SAM), *Mech. Ageing Dev.*, 20, 13, 1982.
243. **Matsumura, A., Higuchi, K., Shimizu, K., Hosokawa, M., Hashimoto, K., Yasuhira, K., and Takeda, T.,** A novel amyloid fibril protein isolated from senescence-accelerated mice, *Lab. Invest.*, 47, 270, 1982.
244. **Higuchi, K., Matsumura, A., Honma, A., Takeshita, S., Hashimoto, K., Hosokawa, M., Yasuhira, K., and Takeda, T.,** Systemic senile amyloid in senescence-accelerated mice: a unique fibril protein demonstrated in tissues from various organs by the unlabeled immunoperoxidase method, *Lab. Invest.*, 48, 231, 1983.
245. **Yonezu, T., Tsunasawa, S., Higuchi, K., Kogishi, K., Naiki, H., Hanada, K., Sakiyama, F., and Takeda, T.,** A molecular-pathologic approach to murine senile amyloidosis, *Lab. Invest.*, 57, 65, 1987.
246. **Chew, D. J., DiBartola, S. P., Boyce, J. T., and Gasper, P. W.,** Renal amyloidosis in related Abyssinian cats, *J. Am. Vet. Med. Assoc.*, 181, 139, 1982.
247. **Boyce, J. T., DiBartola, S. P., Chew, D. J., and Gasper, P. W.,** Familial renal amyloidosis in Abyssinian cats, *Vet. Pathol.*, 21, 33, 1984.
248. **DiBartola, S. P., Benson, M. D., Dwulet, F. E., and Cornacoff, J. B.,** Isolation and characterization of amyloid protein AA in the Abyssinian cat, *Lab. Invest.*, 52, 485, 1985.
249. **DiBartola, S. P., Tarr, M. J., and Benson, M. D.,** Tissue distribution of amyloid deposits in Abyssinian cats with familial amyloidosis, *J. Comp. Pathol.*, 96, 387, 1986.
250. **DiBartola, S. P., Hill, R. L., Fechheimer, N. S., and Powers, J. D.,** Pedigree analysis of Abyssinian cats with familial amyloidosis, *Am. J. Vet. Res.*, 47, 2666, 1986.
251. **Jakob, W.,** Spontaneous amyloidosis of mammals, *Vet. Pathol.*, 8, 292, 1971.
252. **Banks, K. L. and Bullock, B. C.,** Naturally occurring secondary amyloidosis of a squirrel monkey, *Saimiri sciureus*, *J. Am. Vet. Med. Assoc.*, 151, 839, 1967.
253. **Blanchard, J. L., Baskin, G. B., and Watson, E. A.,** Generalized amyloidosis in Rhesus monkeys, *Vet. Pathol.*, 23, 425, 1986.
254. **Cheema, A., Henson, J. B., and Gorham, J. R.,** Aleutian disease of mink. Prevention of lesions by immunosuppression, *Am. J. Pathol.*, 66, 543, 1972.
255. **Eriksen, N., Ericsson, L. H., Pearsall, N., Lagunoff, D., and Benditt, E. P.,** Mouse amyloid protein AA: homology with nonimmunoglobulin protein of human and monkey amyloid substance, *Proc. Natl. Acad. Sci. U.S.A.*, 73, 964, 1976.
256. **Skinner, M., Cathcart, E. S., Cohen, A. S., and Benson, M. D.,** Isolation and identification by sequence analysis of experimentally induced guinea pig amyloid fibrils, *J. Exp. Med.*, 140, 871, 1974.
257. **Shaw, D. P., Gunson, D. E., and Evans, L. H.,** Nasal amyloidosis in four horses, *Vet. Pathol.*, 24, 183, 1987.
258. **Cheville, N. F., Cutlip, R. C., and Moon, H. W.,** Microscopic pathology of the gray collie syndrome, *Pathol. Vet.*, 7, 225, 1970.
259. **Grindem, C. B. and Johnson, K. H.,** Amyloidosis in a case of canine systemic lupus erythematosus, *J. Comp. Pathol.*, 94, 569, 1984.
260. **Cowan, D. F. and Johnson, W. C.,** Amyloidosis in the white Peking duck. I. Relation to social environmental stress, *Lab. Invest.*, 23, 551, 1970.
261. **Gorevic, P. D., Greenwald, M., Frangione, B., Pras, M., and Franklin, E. C.,** The amino acid sequence of duck amyloid A (AA) protein, *J. Immunol.*, 118, 1113, 1977.
262. **Cohen, A. S. and Shirahama, T.,** Animal model: spontaneous and induced amyloidosis, *Am. J. Pathol.*, 68, 441, 1972.
263. **Hol, P. R., Snel, F. W. J. J., Niewold, Th. A., and Gruys, E.,** Amyloid-enhancing factor (AEF) in the pathogenesis of AA-amyloidosis in the hamster, *Virchows Arch. B:*, 52, 273, 1986.
264. **Clark, L. and Seawright, A. A.,** Generalized amyloidosis in seven cats, *Pathol. Vet.*, 6, 117, 1969.
265. **Clark, L. and Seawright, A. A.,** Amyloidosis associated with chronic hypervitaminosis A in cats, *Aust. Vet. J.*, 44, 584, 1968.

266. **Yano, B. L., Hayden, D. W., and Johnson, K. H.,** Feline insular amyloid. Ultrastructural evidence for intracellular formation by nonendocrine cells, *Lab. Invest.*, 45, 149, 1981.
267. **Johnson, K. H., Westermark, P., Nilsson, G., Sletten, K., O'Brien, T. D., and Hayden, D. W.,** Feline insular amyloid: immunohistochemical and immunochemical evidence that the amyloid is insulin-related, *Vet. Pathol.*, 22, 463, 1985.
268. **Johnson, K. H., O'Brien, T. D., Hayden, D. W., Jordan, K., Ghobrial, H. K., Mahoney, W. C., and Westermark, P.,** Immunolocalization of islet amyloid polypeptide (IAPP) in pancreatic beta cells by means of peroxidase-antiperoxidase (PAP) and protein A-gold techniques, *Am. J. Pathol.*, 130, 1, 1988.
269. **Nordin, B. E. C.,** Osteoporosis with particular reference to the menopause, in *The Osteoporotic Syndrome. Detection, Prevention, and Treatment*, Avioli, L. V., Ed., Grune & Stratton, New York, 1983, 13.
270. **Riggs, B. L.,** Evidence for etiologic heterogeneity of involutional osteoporosis, in *Osteoporosis*, Menczel, J., Robin, G. C., Makin, M., and Steinberg, R., Eds., John Wiley & Sons, New York, 1982, 3.
271. **Meuleman, J.,** Beliefs about osteoporosis, a critical appraisal, *Arch. Intern. Med.*, 147, 762, 1987.
272. **Parfitt, A. M.,** Definition of osteoporosis: age-related loss of bone and its relationship to increased fracture risk *NIH Consensus Development Conference: Osteoporosis*, 1984, 15.
273. **Krane, S. M.,** Connective tissue, in *Pathophysiology. The Biological Principles of Disease*, Smith, L. H. and Thier, S. O., Eds., W. B. Saunders, Philadelphia, 1985, 787.
274. **Lane, J. M. and Vigorita, V. J.,** Osteoporosis, *Orthop. Clin. North Am.*, 15, 711, 1984.
275. **Marx, J. L.,** Osteoporosis: new help for thinning bones, *Science*, 207, 628, 1980.
276. **Krook, L., Whalen, J. P., Lesser, G. V., and Berens, D. L.,** Experimental studies on osteoporosis, *Methods Achiev. Exp. Pathol.*, 7, 72, 1975.
277. **Frost, H. M.,** *Intermediary Organization of the Skeleton*, Vol. 2, CRC Press, Boca Raton, FL, 1986, 178.
278. **Simon, M. R.,** The rat as an animal model for the study of senile idopathic osteoporosis, *Acta Anat.*, 119, 248, 1984.
279. **Sones, A. D., Wolinsky, L. E., and Kratochvil, F. J.,** Osteoporosis and mandibular bone reabsorption in the Sprague Dawley rat, *Calcif. Tissue Int.*, 39, 267, 1986.
280. **Izawa, Y., Makita, T., Hino, S., Hashimoto, Y., Kushida, K., Inoue, T., and Orimo, H.,** Immobilization osteoporosis and active vitamin D: effect of active vitamin D analogs on the development of immobilization osteoporosis in rats, *Calcif. Tissue Int.*, 33, 623, 1981.
281. **Wink, C. S.,** Scanning electron microscopy of castrate rat bone, *Calcif. Tissue Int.*, 34, 547, 1982.
282. **Lo Cascio, V., Bonucci, E., Imbimbo, B., Ballanti, P., Tartarotti, D., Galvanini, G., Fuccella, L., and Adami, S.,** Bone loss after glucocorticoid therapy, *Calcif. Tissue Int.*, 36, 435, 1984.
283. **Izawa, Y., Sagara, K., Kadota, T., and Makita, T.,** Bone disorders in spontaneously hypertensive rat, *Calcif. Tissue Int.*, 37, 605, 1985.
284. **Matsushita, M., Tsuboyama, T., Kasai, R., Okumura, H., Yamamuro, T., Higuchi, K., Higuchi, K., Kohno, A., Yonezu, T., Utani, A., Umezawa, M., and Takeda, T.,** Age-related changes in bone mass in the senescence-accelerated mouse (SAM), *Am. J. Pathol.*, 125, 276, 1986.
285. **Young, D. R., Niklowitz, W. J., and Steele, C. R.,** Tibial changes in experimental disuse osteoporosis in the monkey, *Calcif. Tissue Int.*, 35, 304, 1983.
286. **Martin, R. K., Albright, J. P., Jee, W. S. S., Taylor, G. N., and Clarke, W. R.,** Bone loss in the Beagle tibia: influence of age, weight, and sex, *Calcif. Tissue Int.*, 33, 233, 1981.
287. **Marks, S. C. and Walker, D. G.,** Mammalian osteopetrosis — a model for studying cellular and humoral factors in bone reabsorption, in *The Biochemistry and Physiology of Bone*, Vol. 4, 2nd ed., Bourne, G. H., Ed., Academic Press, New York, 1976, 227.
288. **Beighton, P., Horan, F., and Hamersma, H.,** A review of the osteopetroses, *Postgrad. Med. J.*, 53, 507, 1977.
289. **Walker, D. G., Reeves, J. D., and Fox, R. R.,** Osteopetrosis, in *Spontaneous Animal Models of Human Disease*, Vol. 2, Andrews, E. J., Ward, B. C., and Altman, N. H., Eds., Academic Press, New York, 1979, 228.
290. **Marks, S. C., Jr.,** Morphological evidence of reduced bone resorption in osteopetrotic (op) mice, *Am. J. Anat.*, 163, 157, 1982.
291. **Marks, S. C., Jr., MacKay, C. A., and Seifert, M. F.,** The osteopetrotic rabbit: skeletal cytology and ultrastructure, *Am. J. Anat.*, 178, 300, 1986.
292. **Greene, H. J., Leipold, H. W., Hibbs, C. M., and Kirkbride, C. A.,** Congenital osteopetrosis in Angus calves, *J. Am. Vet. Med. Assoc.*, 164, 389, 1974.
293. **Ojo, S. A., Leipold, H. W., Cho, D. Y., and Guffy, M. M.,** Osteopetrosis in two Hereford calves, *J. Am. Vet. Med. Assoc.*, 166, 781, 1975.
294. **Jubb, K. V. F., Kennedy, P. C., and Palmer, N.,** *Pathology of Domestic Animals*, Vol. 1, 3rd ed., Academic Press, New York, 1985, 24.
295. **Riser, W. H. and Frankhauser, R.,** Osteopetrosis in the dog: a report of three cases, *J. Am. Vet. Radiol. Soc.*, 11, 29, 1970.

296. **Lees, G. E. and Sautter, J. H.,** Anemia and osteopetrosis in a dog, *J. Am. Vet. Med. Assoc.*, 175, 820, 1979.
297. **Schmidt, E. V., Crapo, J. D., Harrelson, J. M., and Smith, R. E.,** A quantitative histologic study of avian osteoporotic bone demonstrating normal osteoclast numbers and increased osteoblastic activity, *Lab. Invest.*, 44, 164, 1981.
298. **Schmidt, E. V. and Smith, R. E.,** Animal model of human disease. Skeletal hyperostoses. Viral induction of avian osteopetrosis, *Am. J. Pathol.*, 106, 297, 1982.
299. **Ross, R.,** The pathogenesis of atherosclerosis — an update, *N. Engl. J. Med.*, 314, 488, 1986.
300. **Ross, R. and Glomset, J. A.,** The pathogenesis of atherosclerosis, *N. Engl. J. Med.*, 295, 369, 1976.
301. **Manger, W. M. and Page, I. H.,** An overview of current concepts regarding the pathogenesis and pathophysiology of hypertension, in *Arterial Hypertension: Pathogenesis, Diagnosis and Therapy*, Rosenthal, J., Ed., Springer-Verlag, New York, 1982, 1.
302. **Moore, S.,** Pathogenesis of atherosclerosis, *Metabolism*, 34 (Suppl 1), 13, 1985.
303. **Rhodes, R. K.,** The blood vessel, in *Collagen in Health and Disease*, Weiss, J. B. and Jayson, M. I. V., Churchill Livingston, New York, 1982, 376.
304. **Barnes, M. J.,** Collagens in atherosclerosis, *Collagen Rel. Res.*, 5, 65, 1985.
305. **Mayne, R.,** Vascular connective tissue. Normal biology and derangement in human diseases, in *Connective Tissue Disease. Molecular Pathology of the Extracellular Matrix*, Uitto, J. and Perejda, A. J., Eds., Marcel Dekker, New York, 1987, 163.
306. **Bihari-Varga, M.,** Collagen, aging and atherosclerosis, *Atherosclerosis Rev.*, 14, 171, 1986.
307. **Madri, J. A. and Stenn, K. S.,** Aortic endothelial cell migration. I. Matrix requirements and composition, *Am. J. Pathol.*, 106, 180, 1982.
308. **McCullagh, K. G.,** Increased type I collagen in human atherosclerotic plaque, *Atherosclerosis*, 46, 247, 1983.
309. **Murata, K., Motayama, T., and Kotake, C.,** Collagen types in various layers of the human aorta and their changes with the atherosclerotic process, *Atherosclerosis*, 60, 251, 1986.
310. **Sanwald, R., Ritz, E., and Wiese, G.,** Acid mucopolysaccharide metabolism in early atherosclerotic lesions, *Atherosclerosis*, 13, 247, 1971.
311. **Iverius, P.,** The interaction between human plasma lipoproteins and connective tissue glycosaminoglycans, *J. Biol. Chem.*, 247, 2607, 1972.
312. **Sparks, J. D., Sparks, C. E., and Kritchevsky, D.,** Hypercholesterolemia and aortic glycosaminoglycans of rabbits fed semi-purified diets containing sucrose and lactose, *Atherosclerosis*, 60, 183, 1986.
313. **Robinson, W. F. and Maxie, M. G.,** The cardiovascular system, in *Pathology of Domestic Animals*, Vol. 3, 3rd ed., Jubb, K. V. F., Kennedy, P. C., and Palmer, N., Eds., Academic Press, New York, 1985, 1.
314. **Jones, T. C. and Hunt, R. D.,** *Veterinary Pathology*, 5th ed., Lea & Febiger, Philadelphia, 1983, 1250.
315. **Buja, L. M.,** The vascular system, in *Basic Pathology*, 4th ed., Robbins, S. L. and Kumar, V., Eds., W. B. Saunders, Philadelphia, 1987, 285.
316. **Stills, H. F., Jr. and Clarkson, T. B.,** Atherosclerosis, in *Spontaneous Animal Models of Human Disease*, Vol. 1, Andrews, E. J., Ward, B. C., and Altman, N. H., Eds., Academic Press, New York, 1979, 70.
317. **Bird, R. P., Mercer, N. J. H., and Draper, H. H.,** Animal models for the study of nutrition and human disease: colon cancer, atherosclerosis, and osteoporosis, *Adv. Nutr. Res.*, 7, 155, 1985.
318. **Ratcliffe, H. L. and Luginbuhl, H.,** The domestic pig: a model for experimental atherosclerosis, *Atherosclerosis*, 13, 133, 1971.
319. **Reitman, J. S., Mahley, R. W., and Fry, D. L.,** Yucatan miniature swine as a model for diet-induced atherosclerosis, *Atherosclerosis*, 43, 119, 1982.
320. **Lee, K. T.,** Swine as animal models in cardiovascular research, in *Swine in Biomedical Research*, Vol. 3, Tumbleson, M. E., Ed., Plenum Press, New York, 1986, 1481.
321. **Hughes, H. C.,** Swine in cardiovascular research, *Lab. Anim. Sci.*, 36, 348, 1986.
322. **Stout, L. C.,** Pathogenesis of diffuse intimal thickening (DIT) in aortas and coronary arteries of 2 $^1/_2$-year old miniature pigs, *Exp. Mol. Pathol.*, 37, 427, 1982.
323. **Jennings, M. A., Florey, H. W., Stehbens, W. E., and French J. E.,** Intimal changes in the arteries of a pig, *J. Pathol. Bacteriol.*, 81, 49, 1961.
324. **Clarkson, T. B., Weingand, K. W., Kaplan, J. R., and Adams M. R.,** Mechanisms of atherogenesis, *Circulation*, 76 (Suppl I), I-20, 1987.
325. **Prathap, K.,** Spontaneous aortic lesions in wild adult Malaysian long-tailed monkeys *(Macaca irus)*, *J. Pathol.*, 110, 135, 1973.
326. **Liu, S., Tilley, L. P., Tappe, J. P., and Fox, P. R.,** Clinical and pathologic findings in dogs with atherosclerosis: 21 cases (1970—1983), *J. Am. Vet. Med. Assoc.*, 189, 227, 1986.
327. **Manning, P. J., Corwin, L. A., Jr., and Middleton, C. C.,** Familial hyperlipoproteinemia and thyroid dysfunction of Beagles, *Exp. Mol. Pathol.*, 19, 378, 1973.
328. **Rogers, W. A., Donovan, E. F., and Kociba, G. J.,** Idiopathic hyperlipoproteinemia in dogs, *J. Am. Vet. Med. Assoc.*, 166, 1087, 1975.

329. **Bjotvedt, G.,** Spontaneous renal arteriosclerosis in greyhounds, *Canine Pract.*, 13, 26, 1986.
330. **Goldstein, J. L., Kita, T., and Brown, M. S.,** Mechanisms of disease. Defective lipoprotein receptors and atherosclerosis. Lesions from an animal counterpart of familial hypercholesteroemia, *N. Engl. J. Med.*, 309, 288, 1983.
331. **Watanabe, Y.,** Serial inbreeding of rabbits with hereditary hyperlipidemia (WHHL-rabbit): incidence and development of atherosclerosis and xanthoma, *Atherosclerosis*, 36, 261, 1980.
332. **Kita, T., Brown, M. S., Watanabe, Y., and Goldstein, J. L.,** Deficiency of low density lipoprotein receptors in liver and adrenal gland of the WHHL rabbit, an animal model of familial hypercholesterolemia, *Proc. Natl. Acad. Sci. U.S.A.*, 78, 2268, 1981.
333. **Koletsky, S.,** Pathologic findings and laboratory data in a new strain of obese hypertensive rats, *Am. J. Pathol.*, 80, 129, 1975.
334. **St. Clair, R. W.,** Metabolic changes in the arterial wall associated with atherosclerosis in the pigeon, *Fed. Proc.*, 42, 2480, 1983.
335. **Wexler, B. C.,** Spontaneous atherosclerosis in the Japanese quail, *Artery*, 3, 507, 1977.
336. **Shih, J. C. H.,** Atherosclerosis in Japanese quail and the effect of lipoic acid, *Fed. Proc.*, 42, 2494, 1983.
337. **Strickberger, S. A., Russek, L. N., and Phair, R. D.,** Evidence for increased aortic plasma membrane calcium transport caused by experimental atherosclerosis in rabbits, *Circ. Res.*, 62, 75, 1988.
338. **Ehrhart, L. A. and Holdenbaum, D.,** Aortic collagen, elastin and non-fibrous protein synthesis in rabbits fed cholesterol and peanut oil, *Atherosclerosis*, 37, 423, 1980.
339. **Wright, P. L., Smith, K. F., Day, W. A., and Fraser, R.,** Small liver fenestrae may explain the susceptibility of rabbits to atherosclerosis, *Arteriosclerosis*, 3, 344, 1983.
340. **Pietila, K. and Nikkari, T.,** Enhanced synthesis of collagen and total protein by smooth muscle cells from atherosclerotic rabbit aortas in culture, *Atherosclerosis*, 37, 11, 1980.
341. **Kim, D. N., Imai, H., Schmee, J., Lee, K. T., and Thomas, W. A.,** Intimal cell mass derived atherosclerotic lesions in the abdominal aorta of hyperlipemic swine. I. Cell of origin, cell divisions and cell losses in first 90 days on diet, *Atherosclerosis*, 56, 169, 1985.
342. **Thomas, W. A., Lee, K. T., and Kim, D. N.,** Pathogenesis of atherosclerosis in the abdominal aorta and coronary arteries of swine in the first 90 days on a hyperlipidemic diet, in *Swine in Biomedical Research*, Vol. 3, Tumbleson, M. E., Ed., Plenum Press, New York, 1986, 1511.
343. **Armstrong, M. L.,** Atherosclerosis in rhesus and cynomolgus monkeys, *Primates Med.*, 9, 16, 1976.
344. **Weingand, K. W., Clarkson, T. B., Adams, M. R., and Bostrom, A. D.,** Effects of age and/or puberty on coronary artery atherosclerosis of cynomolgus monkeys, *Atherosclerosis*, 62, 137, 1986.
345. **Kramsch, D. M. and Hollander, W.,** Occlusive atherosclerotic disease of the coronary arteries of monkeys (Macaca irus) induced by diet, *Exp. Mol. Pathol.*, 9, 1, 1968.
346. **Manning, P. J. and Clarkson, T. B.,** Development, distribution and lipid content of diet-induced atherosclerotic lesions of rhesus monkeys, *Exp. Mol. Pathol.*, 17, 38, 1972.
347. **Taylor, C. B., Manalo-Estrella, P., and Cox, G. E.,** Atherosclerosis in rhesus monkeys. V. Marked diet-induced hypercholesterolemia with xanthomatosis and severe atherosclerosis, *Arch. Pathol.*, 76, 239, 1963.
348. **Scott, R. F., Morrison, E. S., Jarmolych, J., Nam, S. C., Kroms, M., and Coulston, F.,** Experimental atherosclerosis in rhesus monekys. I. Gross and light microscopic features and lipid values in serum and aortas, *Exp. Mol. Pathol.*, 7, 11, 1967.
349. **St. Clair, R. W., Toma, J. J., Jr., and Lofland, H. B.,** Chemical composition of atherosclerotic lesions of aortas from pigeons with naturally occurring or cholesterol-aggravated atherosclerosis, *Proc. Soc. Exp. Biol. Med.*, 146, 1, 1974.
350. **Jerome, W. G. and Lewis, J. C.,** Early atherogenesis in white Carneau pigeons. II. Ultrastructural and cytochemical observations, *Am. J. Pathol.*, 119, 210, 1985.
351. **Fabricant, C. G., Fabricant, J., Minick, C. R., and Litrenta, M. M.,** Herpesvirus-induced atherosclerosis in chickens, *Fed. Proc.*, 42, 2476, 1983.
352. **Peart, W. S.,** General review of hypertension, in *Hypertension: Physiopathology and Treatment*, Genest, J., Kuchel, D., Hamet, P., and Cantin, M., Eds., McGraw-Hill, New York, 1983, 3.
353. **Falcy, C., Farjanel, J., Rattner, A., Grochulski, A., Gilson, N., Borsos, A. M., Cristeff, N., Dupeyron, J. P., Peyroux, J., and Sternberg, M.,** Aorta collagen metabolism in spontaneously hypertensive and aortic-constricted rats: variations in enzyme activities with disaccharide unit synthesis and degradation according to blood pressure and age, *Collagen Rel. Res.*, 5, 519, 1985.
354. **Wolinsky, H.,** Long term effects of hypertension in the rat aortic wall and their relation to concurrent aging changes. Morphologic and chemical studies, *Circ. Res.*, 30, 301, 1972.
355. **Newman, R. A. and Langner, R. O.,** Age-related changes in the vascular collagen metabolism of the spontaneously hypertensive rat, *Exp. Gerontol.*, 13, 83, 1978.
356. **De Jong, W.,** Experimental and genetic models of hypertension, in *Handbook of Hypertension*, Vol. 4, Elsevier, New York, 1984.

357. **Bishop, S. P., Kawamura, K., and Detweiler, D. K.,** Systemic hypertension, in *Spontaneous Animal Models of Human Disease,* Vol. 1, Andrews, E. J., Ward, B. C., and Altman, N. H., Eds., Academic Press, New York, 1979, 50.
358. **Zicha, J., Kunes, J., and Jelinek, J.,** Experimental hypertension in young and adult animals, *Hypertension,* 8, 1096, 1986.
359. **Okamoto, K.,** Spontaneous hypertension in rats, *Int. Rev. Exp. Pathol.,* 7, 227, 1969.
360. **Dahl, L. K., Heine, M., and Tassinari, L.,** Effects of chronic excess salt ingestion. Evidence that genetic factors play an important role in susceptibility to experimental hypertension, *J. Exp. Med.,* 115, 1173, 1962.
361. **Phelan, E. L.,** The New Zealand strain of rats with genetic hypertension, *N. Z. Med. J.,* 67, 334, 1968.
362. **Bianchi, G., Fox, U., DiFrancesco, G. F., Giovannetti, A. M., and Pagetti, D.,** Blood pressure changes produced by kidney cross-transplantation between spontaneously hypertensive rats (SHR) and normotensive rats (NH), *Clin. Sci. Mol. Med.,* 47, 435, 1974.
363. **Yamori, Y.,** Physiopathology of the various strains of spontaneously hypertensive rats, in *Hypertension: Physiopathology and Treatment,* 2nd ed., Genest, J., Kuckel, O., Hamet, P., and Cantin, M., Eds., McGraw-Hill, New York, 1983, 556.
364. **Schlager, G. and Weibust, R. S.,** Genetic control of blood pressure in mice, *Genetics,* 55, 497, 1967.
365. **Elias, M. F., Sorrentino, R. N., Pentz, C. A., Florini, J. R., and Schlager, G.,** "Spontaneously" hypertensive mice: a potential genetic model for the study of the relationship between heart size and blood pressure, *Explor. Aging Res.,* 1, 251, 1975.
366. **Alexander, N., Hinshaw, L. B., and Drury, D. R.,** Development of a strain of spontaneously hypertensive rabbits, *Proc. Soc. Exp. Biol. Med.,* 86, 855, 1954.
367. **Sturkie, P. D., Weiss, H. S., Ringer, R. K., and Sheahan, M. M.,** Heritability of blood pressure in chickens, *Poult. Sci.,* 38, 333, 1959.
368. **Krista, L. M., Waibel, P. E., Shoffner, R. N., and Sautter, J. H.,** Naturally dissecting aneurysm (aortic rupture) and blood pressure in the turkey, *Nature (London),* 214, 1162, 1967.
369. **Spangler, W. L., Gribble, D. H., and Weiser, M. G.,** Canine hypertension: a review, *J. Am. Vet. Med. Assoc.,* 170, 995, 1977.
370. **Tippet, F. E., Padgett, G. A., Eyster, G., Blanchard, G., and Bell, T.,** Primary hypertension in a colony of dogs, *Hypertension,* 9, 49, 1987.
371. **Slaughter, J. B., II, Padgett, G. A., Blanchard, G., Eyster, G., Bell, T. G., and Grace, J. A.,** Canine essential hypertension: probable mode of inheritance, *J. Hypertension,* 4 (Suppl. 5), S170, 1986.

Chapter 3

NONINVASIVE METHODS FOR DETECTION OF ORGAN FIBROSIS

Leila Risteli and Juha Risteli

TABLE OF CONTENTS

I. INTRODUCTION

The development of a fibrosis is usually a slow process and cannot be directly predicted on the basis of the conventional laboratory tests that reflect cell injury of the parenchymal organ. The invasive assessment of the extent of fibrosis — the histological evaluation of a biopsy sample — has obvious drawbacks which limit its repeated use for follow-up purposes. The activity of the fibrotic process can only be estimated indirectly, e.g., by counting the number of inflammatory cells. Such invasive methods as the assay of the tissue activity of collagen biosynthetic enzymes or the measurement of the amount of mRNA for a certain collagen or other connective tissue protein have not yet found indications for routine use.

During the last 10 years the interest in measuring different connective tissue-related components in biological fluids has increased. Most of the new methods suggested are based on immunological detection of different domains of well-defined chemical constituents of the extracellular matrix.

Despite intensive studies into details of the biochemical reactions involved, little is still known about the normal phases and regulation of the synthesis and degradation of the extracellular matrix in various parts of the human body. When radioimmunological methods for procollagen propeptides (see below) were introduced, a more or less direct relationship between the rate of collagen deposition in a tissue and the concentration of the propeptide in serum seems to have been implicitly assumed. Recently the different contributions of various tissues to the blood pools of these antigens and the further metabolism and the elimination of the metabolites have attracted attention.

This review deals with the methods for determining the activity of connective tissue metabolism, mainly in humans. In the first part, general aspects concerning the interpretation of the results are presented, together with comments on individual assays. In the second part, the clinical usefulness of these methods is summarized on the basis of current information.

We have not used the term fibrosis strictly, when choosing material to be dealt with, but have included a number of other conditions in which connective tissue is either laid down or degraded. We found this necessary, since for the correct interpretation of the results in a fibrosing condition it is essential to be aware of the other factors that can contribute to the results. Assays for fibronectin will not be dealt with, as those presently in use do not distinguish between the form which is a genuine plasma protein and the other, normally obviously small, fraction which is derived from the cellular fibronectin of tissues. It is possible, however, that in the near future specific serum assays will be introduced for the latter.[1]

The possibility of monitoring development of a hepatic fibrosis by noninvasive means has been reviewed by a number of authors.[2-5] There are also reviews on radioimmunoassays for connective tissue components,[6] for type IV collagen and laminin,[7] and another on the extracellular matrix proteins found in physiological fluids.[8]

II. GENERAL ASPECTS OF ASSESSING ORGAN FIBROSIS

A. FIBROSIS, FIBROGENESIS, OR FIBROLYSIS?

Fibrosis is basically a histological term, usually indicating an increased amount of connective tissue visible in a biopsy specimen (Table 1). This amount can be biochemically detected by analyzing the collagen content of a tissue sample. Fibrosis is most often used as a static description, i.e., it indicates a certain state of the tissue at the time of the biopsy.

The corresponding dynamic terms, indicating changes in the connective tissue content of a tissue, are fibrogenesis and fibrolysis, the increase and decrease of connective tissue, respectively (Table 1).

It cannot be taken for granted that, e.g., collagen synthesis as such is equivalent to

TABLE 1
Definitions of Fibrosis, Fibrogenesis, and Fibrolysis

Term	Histological description	Biochemical counterpart
Fibrosis	Increased amount of connective tissue with certain structural alterations	Increased amount of a certain component of connective tissue, e.g., type I collagen
Fibrogenesis	Net accumulation of connective tissue between two biopsies	Synthesis of a certain component of connective tissue larger than its degradation
Fibrolysis	Net loss of connective tissue between two biopsies	Degradation of a certain component of connective tissue larger than its synthesis

fibrogenesis. In fact, increased synthesis and deposition of connective tissue are in most situations accompanied by accelerated degradation.[9] Thus, fibrogenesis, a genuine increase in the extracellular matrix, will only take place when the degrading mechanisms cannot keep pace with the enhanced synthesis. For estimating fibrogenesis, one would, in fact, need two different methods: one that gives information on the synthesis and another which indicates degradation of a certain component of connective tissue. The estimation of fibrolysis could follow the same principle.

Since there are a number of components in connective tissue and a specific assay could, in principle, be developed for each of them, it is not surprising that the number of published methods and their applications to different clinical situations is increasing rapidly. The general aim of those active in the field should be to select those methods which give the best possible answers to well-defined clinical questions.

B. BIOCHEMICAL BASIS OF COLLAGEN TYPE-SPECIFIC DETERMINATIONS

The interstitial collagens (types I to III and V), of which the metabolism is best known, are synthesized and secreted as larger precursor molecules called procollagens. These contain additional peptide sequences, usually called propeptides, at both the amino-terminal and the carboxy-terminal ends. The function of the carboxy-terminal propeptide domain is to direct the assembly of the three polypeptide chains and to initiate their winding into a triple-helical conformation. The carboxy-terminal propeptides have a noncollagenous amino acid composition and a globular conformation. The amino-terminal propeptides contain a short collagenous sequence in a triple-helical conformation, flanked by noncollagenous sequences.

The propeptides are cleaved off from the collagen molecules before these associate to form fibers. Thus, the amount of the free propeptides reflects stoichiometrically the amount of collagen molecules deposited, a relationship analogous to that between the C-peptide of proinsulin and the endogenously produced insulin. However, in reality the situation is more complex. Since the propeptides of different collagen types are immunologically different, their assays are collagen type-specific.

Cleavage of the carboxy-terminal propeptide seems to be a prerequisite for the assembly of the collagen fibers.[10] However, at least in the case of type III collagen, the removal of the amino-terminal propeptide is incomplete, a certain proportion of the type III collagen molecules retaining these propeptides also in the fibers. Accordingly, the amino-terminal propeptides can be detected on the surface of type III collagen fibers. There seems to be a profound difference between the collagen types I and III in this respect.[11] It has recently been suggested that the propeptides, once set free from the procollagen molecules, could have additional functions of their own, as genuine matrix proteins. The suggestions were based on studies of the amino-terminal propeptide of type I procollagen in developing bone[12] and of the carboxy-terminal propeptide of type II procollagen in calcifying cartilage.[13,14]

Not all collagen types lose propeptide domains before associating into supramolecular structures. Type IV collagen is an example of these. In this collagen type, both ends of the molecule are needed to establish covalent links between neighboring molecules in the type IV collagen network. When parts of the type IV collagen molecule are detected in body fluids, they can either represent molecules that have leaked from tissues during the biosynthesis of type IV collagen or be derived from the degradation of the network. A number of other collagen types have been described recently, which can also contain nonhelical domains between the ends of the molecules. Their biosynthetic processing is not known in detail.

A number of other tests, not involving the collagenous proteins themselves, have also been suggested for evaluating the activity of fibrogenesis. They can be based on other connective tissue components, small breakdown products of collagen, or enzymes involved in connective tissue metabolism.

C. CHARACTERIZATION OF BIOCHEMICAL TESTS

We will describe the characteristics and the development of the assay for the amino-terminal propeptide of type III procollagen in great detail. This is due to the fact that this method is the one most studied so far: it was the first assay to be seriously considered as a possible serum indicator of fibrosis or fibrogenesis. It has gradually become evident that, in addition to the activity of type III collagen synthesis in a tissue, a number of additional factors affect the serum concentration of the propeptide and must thus be taken into account in the interpretation of the results. This experience will no doubt prove valuable also when new assays are introduced and evaluated.

The serum concentration of a substance depends on several factors. These include the tissue origin and rate of release of the compound, the route along which it reaches the blood (e.g., via lymph or directly through endothelial fenestrations), and its volume of distribution. At least some soluble connective tissue components are known to be distributed between several pools in the body, e.g., synovial and ascitic fluids having much higher concentrations than the blood.

Another important aspect is the elimination of the measured compound from the circulation: what organs are involved; what is the rate of uptake; is the substance further metabolized and perhaps set free into the blood in modified form or excreted in antigenic form, e.g., in urine?

When a new test is taken into clinical use in hospital laboratories, its analytical performance is systematically evaluated. A prerequisite is that the technology needed is practical for such laboratories. Radioimmunoassays, enzyme immunoassays, and simple enzyme determinations with chromogenic substrates fulfill this criterion. Good reproducibility of the test is essential. In this respect, the interassay variation is most important, since it is the most realistic assessment of the performance that the clinicians observe when they use the test for follow-up purposes. Use of blood samples, instead of urine, is a general trend in modern laboratory medicine.

We will try to evaluate the methods suggested in the literature with respect to how reliably they can be considered to reflect either synthesis or degradation of extracellular matrix. If the method is available commercially, the producer is given for obtaining further information.

III. COMMENTS ON INDIVIDUAL ASSAYS

A. AMINO-TERMINAL PROPEPTIDE OF TYPE III PROCOLLAGEN

Type III collagen is a major component of interstitial connective tissue. It is preferentially synthesized in the early phases of a fibroproliferative response, which makes the determination of its metabolism biologically relevant.

The method for assessing the amino-terminal propeptide of type III procollagen[15] was the first connective tissue-related radioimmunoassay to become commercially available.

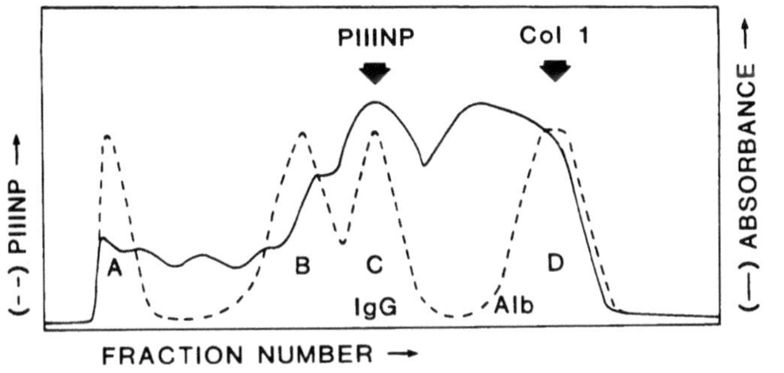

FIGURE 1. Serum antigens related to the amino-terminal propeptide of type III procollagen (PIIINP) as separated by gel filtration (schematic presentation). The dashed line indicates the propeptide immunoreactivity, the continuous line the protein profile of a normal human serum sample. Different immunoassays developed for PIIINP recognize the antigens A to D to variable extents; see Table 3. (From Risteli, J., Risteli, L., and Niemi, S., *Farmos Diagnostica Newsl.*, 1988. With permission.)

Accordingly, the assay has been used in a number of clinical studies. In different publications, the antigen measured has been called and abbreviated in several different ways (e.g., procollagen III peptide, type III procollagen peptide, type III collagen amino-propeptide, Col 1-3, PIIIP, PIIINP, NP3, PCIII). As the complete name, amino-terminal propeptide of type III procollagen, is not significantly longer than many of the others suggested, its use is preferable to avoid confusion. When an abbreviation is needed, PIIINP is better than PIIIP, since it is possible that assays for the carboxy-terminal propeptide of this procollagen will be developed in the near future.

The problems associated with the analytical behavior of the PIIINP method and with the interpretation of the results have been presented on several occasions.[6,11,16]

1. Heterogeneity of the PIIINP Antigen in Serum

In human serum, antigenicity related to the amino-terminal propeptide of type III procollagen is found in several different forms, with different molecular sizes. To facilitate the comparison of different assays we have named them here peaks A to D; their elution positions in gel filtration are schematically presented in Figure 1. The apparent proportions of the different forms in any serum sample vary according to the antisera and the assay conditions used. In addition, their proportions can vary in a patient's serum during different phases of a disease (Figure 2).[17,18]

The largest form (A in Figure 1) may represent intact type III pN-collagen or procollagen molecules. It is eluted in the void volume of, e.g., a Sephacryl S-300 column. Form B is somewhat larger than the propeptide; its exact chemical composition and origin are not known. Form C has the same size as the authentic propeptide, when cleaved off *en bloc* during the conversion of procollagen to collagen. The concentrations of forms A to C are normally quite low in serum samples of adults. The smallest, form D, which is obviously derived from further degradation of the others, predominates in normal adult serum.

The degradation of the propeptide antigen, which gives rise to form D, could, in principle, be artefactual, i.e., take place during processing and storage of the blood sample.[4] This is unlikely, however. The concentration of the larger forms in plasma corresponds to that in serum.[19] The proportions of the different forms in individual serum samples remain constant despite repeated freezing and thawing and are different in samples from healthy people (or patients in an inactive phase of their disease) and in those from patients with active illness (Figure 2).

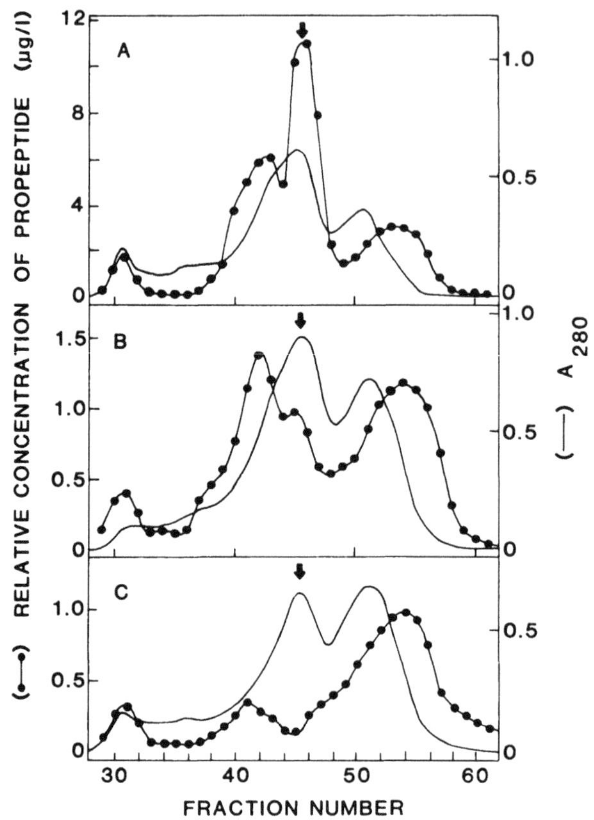

FIGURE 2. Gel filtration analysis of human serum on a Sephacryl® S-300 column. The solid circles indicate the concentration of PIIINP (assay No. 1 in Table 3) and the thin line the protein content of the fractions. The arrows indicate the elution position of PIIINP (peak C in Figure 1). The serum samples were taken from a patient with acute alcoholic hepatitis (A) and during recovery from the disease (B and C). (Reprinted with permission from Niemelä, O., Risteli, L., Sotaniemi, E. A., and Risteli, J., *Gastroenterology*, 85, 254, 1983. Copyright 1983 by The American Gastroenterological Association.)

The distribution of the total PIIINP antigenicity between the various antigen forms in different biological fluids is summarized in Table 2.

2. Different Methods for PIIINP

The structure of the amino-terminal propeptide of type III procollagen is schematically presented in Figure 3. It contains three different domains, termed Col 1, Col 3, and Col 2, read from the amino terminus toward the carboxy terminus.

The earliest assays reported for PIIINP utilize the immunological cross-reaction between the human and the bovine antigens (Table 3). The immunogen was isolated from fetal calf skin in the form of pN-collagen, from which the amino-terminal segment (containing both the propeptide and the amino-terminal, nonhelical telopeptide of the collagen proper) was liberated by digestion with bacterial collagenase.[15]

In assays set up in this way, serial dilutions of human serum samples typically give inhibition curves that are not parallel with those of the standard antigens. The difference is due to the smallest antigen, form D, which binds the antibodies with a reduced affinity.[29]

TABLE 2
Heterogeneity of the PIIINP Antigens in Biological Fluids

Fluid	Size distribution[a]	Concentration (assay[b])	Ref.
Amniotic fluid	C > BD > A	1—2 mg/l (No.1)	20
Ascitic fluid	C > BD	1—10 mg/l (No.1)	15, 21
Bronchoalveolar lavage fluid	Not known	Not detectable (No. 1) in healthy persons	22
Bile	C > D	Not given	23
Cerebrospinal fluid	ACD	4 ± 2.5 (No. 1)	24
Seminal fluid	Not known	4—58 μg/l (No. 1)	25
Serum	D > BC > A	[b]	[b]
Synovial fluid	Not known	1.1 ± 1.2, mg/l (No. 1)	26
Urine	D	20—187 μg/l (No. 4)	15, 27
Wound fluid	C > D	0.1—4 mg/l (No. 7)	28

[a] Relative abundance of the antigen forms described in Figure 1.
[b] See Table 3 for different assays, reference values, and references.

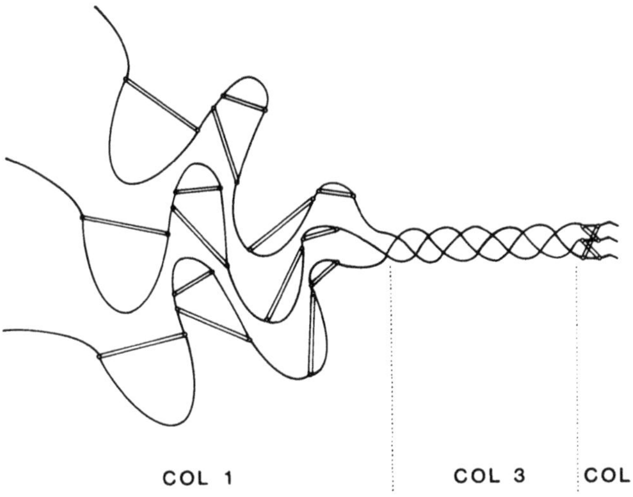

FIGURE 3. Schematic drawing of the amino-terminal propeptide of type III procollagen (PIIINP) and its structural domains Col 1, Col 2, and Col 3. The open circles indicate the location of cysteine residues and the bars the disulfide bonds (their exact arrangement is not known). The molecular weight of PIIINP is about 42,000 and that of its monomeric Col 1 domain about 10,000. (From Risteli, J., Risteli, L., and Niemi, S., *Farmos Diagnostica Newsl.*, 1988. With permission.)

In such an assay, different dilutions of the same sample give different apparent concentrations. This is why the use of the 50% inhibition method in determining the concentration has been suggested. There is little theoretical support for such a procedure, however.[6,8] The same increased inhibitions can result from different, unpredictable combinations of the antigenic forms A to D and, thus, may have different pathophysiological backgrounds.

In addition, both the serial diluting of the sample and the nonparallelism as such introduce inaccuracy to the measurements. Typical intra- and interassay variations with this type of method are up to 10 and up to 25%, respectively. Although the coefficients of variation can perhaps be decreased by careful laboratory work, the tediousness of the procedure makes these methods unsuitable for the clinical laboratory.

TABLE 3
Assays for the Amino-Terminal Propeptide of Type III Procollagen

Antigen and species of origin	Antigen form specificity (see Figure 1 for peaks)	Reference interval for healthy adults (μg/l serum)[a]	Ref.
1. Amino-terminal segment (bovine)	(A)BC > D	2—12	15[b]
2. Procollagen (goat)	Not known	45—81	30
3. Amino-terminal propeptide (human)	(A)BC > D	13—33	31
4. As 1, assay with Fab fragments of antibodies (bovine)	D > (A)BC	15—63	27
5. Collagenase digest of aminoterminal segment (bovine)	D > (A)BC	13—125	32
6. Col 1 domain of propeptide (human)	D > (A)BC	19—71	31
7. Amino-terminal propeptide (human)	BC	1.7—4.2	19
8. Type III collagen (human)	A		33

[a] When not given by authors, we have calculated this using the data provided in the publication.
[b] Assays having similar characteristics have also been described.[11,34,35]

Another assay, based on the use of goat type III pN-collagen, seems to resemble the bovine antigen assay, although it has not been described in detail.[30] Neither does the use of the human propeptide solve the problem of nonparallelism between serum samples and standards.[31]

There are two approaches to decreasing the nonparallelism in the assays: either to make all the serum antigen forms, including D, react with equal affinity with the antibodies or to reduce further the reactivity of form D.

The former can be achieved by using monovalent Fab fragments of the antibodies instead of the antiserum[27] or by treating the propeptide standard with bacterial collagenase and substituting such degradation products for the standard.[32] Similar results are obtained when a radioimmunoassay is based on the purified, monomeric Col 1 domain of the propeptide (see Figure 3).[31] These three assays are all titratable, i.e., serum samples and standard inhibitors give parallel inhibition curves. The intra- and interassay variations of these methods are smaller than those of the 50% intercept methods.

An example of the second approach to the problem of nonparallelism is a new assay, developed recently in our laboratory. The reactivity of the small degradation products (peak D of Figure 1) was reduced by selection of the antiserum and the reaction conditions.[19] In this assay, which is based on highly purified human PIIINP, the slopes of the inhibition curves obtained with human serum samples do not significantly differ from those given by the standards. Consequently, it is enough to use only one fixed volume of serum sample to obtain the result. The intra- and interassay variations are significantly smaller than with the 50% intercept methods, both being around 5%.

A radioimmunoassay developed for the helical domain of type III collagen detects exclusively peak A.[33]

Three of the above methods are currently available commercially as radioimmunoassay kits. They are Nos. 1 (available from Behringwerke, Marburg, West Germany, as RIA-Gnost P-III-P), 4 (Behringwerke, as Fab-P-III-P), and 7 (Farmos-Diagnostica, Oulunsalo, Finland, as PIIINP).

The reference intervals (defined as mean ±2 SD) for the different assays are given in Table 3. The lowest apparent concentration is obtained with method No. 7, which measures the peaks B and C, and the highest with methods Nos. 4, 5, and 6, measuring preferentially peak D. In Nos. 2 and 8, the concentrations are based on type III procollagen and collagen standards, respectively.

Radioimmunoassays analogous to method No. 1 (Table 3) have been developed for the amino-terminal segment of rat type III procollagen.[36,37] Growth and chemical liver injury

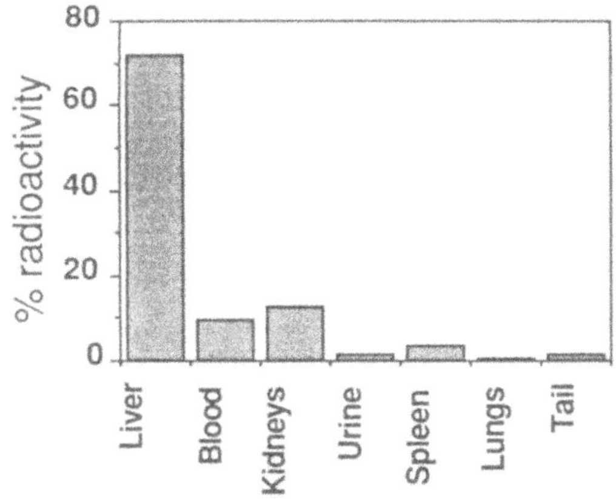

FIGURE 4. Anatomic distribution of intravenously injected PIIINP in rats. The amino-terminal segment (containing both the amino-terminal propeptide and the amino-terminal, nonhelical telopeptide of the collagen molecule) was prepared by collagenase digestion of bovine type III pN-collagen and coupled to ^{125}I-labeled tyramine-cellobiose. An aliquot of 10^6 cpm, corresponding to 3.3 μg of PIIINP, was injected into the tail vein, and the radioactivity of the organs was counted 30 min later. The total recovery of the radioactivity was 80%. (From Smedsrød, B., *Collagen Relat. Res.*, 8, 375, 1988. With permission.)

increase the serum concentration in rats; the serum antigenicity resolves into four different peaks upon gel filtration.[36] Also, an ELISA assay has been developed for rat PIIINP.[38]

3. Metabolism and Excretion of PIIINP

It can be shown by immunohistochemical and immunoelectron-microscopic methods that type III collagen fibers are covered with a layer of molecules that have retained their amino-terminal propeptide. The presence of such fibers has been reported for skin, liver, and lymph node tissue.[11,39-41] Significant amounts of type III pN-collagen can also be extracted from both fetal and adult tissues.[40] The proportion of such molecules has not been estimated in any tissue, however. The amino-terminal propeptide has been recently found to contain a substrate site for transglutaminase,[42] but so far it is not known whether this cross-linking mechanism is active *in vivo*. The propeptide could be involved in cross-linking between two neighboring collagen molecules or between the pN collagen and some other connective tissue component or blood coagulation factor.

It is probable that when such type III collagen molecules are degraded *in vivo* that have retained their amino-terminal propeptides, the degradation products still bear the antigenic determinants of PIIINP. This could give rise to peak B and perhaps peak A of the serum antigens. However, this hypothesis has not been verified.

During stimulated type III collagen synthesis, e.g., in a healing surgical wound, the major antigen species to be found in soluble form *in situ* is the intact, trimeric propeptide.[28] No larger forms can be detected. Small degradation products, corresponding to peak D in serum, are present, but their proportion decreases with increasing type III collagen synthesis. Thus, it seems that the further degradation of the propeptide does not take place *in situ* immediately after its cleavage from the procollagen molecule.

Several lines of evidence indicate that the liver is involved in the metabolism of circulating PIIINP antigens. When iodine-labeled PIIINP was injected into the tail vein of rats, over 70% of the radioactivity could be recovered from the liver (Figure 4).[43] Isolated liver endothelial cells were subsequently shown to be able to bind specifically PIIINP on their surface and to internalize it.[43] The role of the liver in metabolizing the larger propeptide antigen forms (A, B, and C in Figure 1) is also emphasized by experiments in which removal of the liver from healthy pigs resulted in a rapid increase in the serum concentration of PIIINP, when measured with methods Nos. 1 or 7 in Table 3.[44] A propeptide assay that mainly detects the smallest form D (e.g., No. 4 in Table 3) does not detect this increase, suggesting that the liver is not responsible for the clearance of the smallest degradation products of PIIINP.

It has been reported that a PIIINP antigen form equal in size to the authentic propeptide is excreted into bile.[23] This excretion may be enhanced in alcoholic cirrhosis.

Net production and release into the blood of PIIINP by fibrotic liver tissue was proposed when the serum concentrations of PIIINP in cubital and hepatic veins were compared.[45] However, when the concentrations of PIIINP in the portal and hepatic veins of pigs were directly compared, uptake of PIIINP by the liver was found.[46] In the latter study, no concentration difference was noted between portal vein and hepatic artery.[46] Similar results have also been obtained in humans with alcoholic liver disease, the hepatic vein having a lower concentration than the portal vein.[47]

In urine, small degradation products of PIIINP are found corresponding to peak D in serum (Figure 1).[15,27] Recent catheterization studies suggest that the kidneys extract ten times more of these Col 1 domain-related antigens than they excrete into the urine.[46] The daily urinary excretion in adults is about 30 to 110 μg,[27] and it seems to be decreased in alcoholic cirrhosis.[23,31] The exact chemical composition of the urinary propeptide antigen has not been studied. In gel filtration the material is eluted clearly ahead of the monomeric Col 1 domain of PIIINP, indicating a larger molecular size.[31]

B. PROPEPTIDES OF TYPE I PROCOLLAGEN

1. Carboxy-Terminal Propeptide

According to present knowledge, the carboxy-terminal propeptides of the interstitial, fibrillar collagen types are without exception cleaved off when the corresponding collagen molecules are to be associated into fibrils. Thus, their assessment should provide a means of estimating the amount of collagen that is being synthesized.

The radioimmunoassay for the carboxy-terminal propeptide of type I procollagen (suggested abbreviation PICP), was already developed in 1974.[48] The antigen was isolated from the culture medium of human fibroblasts. The inhibition curves given by serial dilutions of serum samples are parallel to those given by the standard.[49,50] The intra- and interassay variations of the method are within 10%.[51] The reference range of the serum concentration of PICP is about 50 to 150 μg/l in adults[50-52] and about 200 to 600 μg/l in children.[50]

The PICP assay has been used in studies of bone diseases, alcoholic liver disease, and growth disorders. Since type I collagen is the only collagen type present in mature bone, most of the serum PICP antigenicity has been presumed to originate in this tissue. However, in alcoholic liver disease the changes in serum PICP are similar to those in PIIINP.[52] It is not known whether the increase in the serum concentration of the PICP in this case is due more to release from the diseased tissue or to an impaired clearance. A recent study suggests that hepatic dysfunction significantly increases the serum PICP.[53] No studies are currently available on the further metabolism and elimination of this serum antigen. It is not found in urine, however.[49]

We have recently prepared the carboxy-terminal segment of human type I procollagen and set up the radioimmunoassay.[54] This method detects only one antigen species in human serum, the size of which is identical to that of the reference antigen. It can also be detected in a healing wound, where its concentration increases about 35-fold during the first week of healing. Simultaneously, there is a twofold increase in the serum concentration of PICP.[55]

The original procedure used in preparing the carboxy-terminal propeptide does not contain a step for separating types I and III procollagen. Thus, the assay[48] may also detect the carboxy-terminal propeptide of type III procollagen.[51] This need not affect the interpretation of the results, however, as the concentration of type I procollagen-derived antigen is probably much higher. The use of a specific monoclonal antibody[56] to PICP may solve this problem in the future.

2. Amino-Terminal Propeptide

The amino-terminal propeptide of type I procollagen (suggested abbreviation PINP) has been isolated from bovine[11,57] and rat[38] tissues, but not from human sources. One group has found the cross-reaction between the corresponding human and bovine antigens sufficient for immunoassay,[11] whereas others have not been able to measure the antigen concentration in fetal or adult human serum with this kind of assay.[57] The size distribution of the PINP antigen in human serum has not been reported.

Most human sera have been reported to produce inhibition curves that are parallel to that of the standard in the ELISA assay for bovine PINP.[11] In those cases which presented nonparallelism the ELISA result was not considered interpretable by the authors. The mean serum concentration of healthy adults ($n = 13$) was about 29 ± 3.3 (SEM) μg/l. In contrast to PIIINP, no increases in serum PINP were found in a group of patients with primary biliary cirrhosis.[11] The further metabolism and elimination of the PINP antigen have not been studied.

In rats, the serum PINP concentration is affected by their age and growth rate.[38] Acute liver injury, induced by CCl_4, is associated with a transient increase in serum PINP, as well as in PIIINP.[38]

C. BASEMENT MEMBRANE-DERIVED ANTIGENS

Basement membranes are thin sheet-like structures which separate parenchymal cells from the connective tissue proper. Although these membranes are usually not considered in association with parenchymal fibrosis, the metabolism of basement membrane components and that of the interstitial connective tissue are interrelated. Development of a liver fibrosis also involves an increase in the amount of type IV (basement membrane) collagen in the tissue.[58] This can be histologically detected both as the formation of real basement membranes in the sinusoids (capillarization) and as accumulation of type IV collagen and laminin, along with other matrix molecules, in the fibrous septae.[59] The formation of new basememt membranes and/or thickening of existing ones can affect the nutrition of the surrounding tissue and thus probably aggravate the interstitial fibrotic process as well.

1. Type IV Collagen-Derived Antigens

No propeptides are released prior to the deposition of type IV collagen. The amino- and carboxy-terminal parts of a type IV collagen molecule are known as the 7S collagen domain and the NC1 domain, respectively.[7,60]

The 7S domain was originally isolated as a short rod-like particle, which turned out to contain the amino-terminal regions of four type IV collagen molecules. Depending on the extent of the proteolytic treatment used to isolate this structure from the type IV collagen network, either the short or the long form of the 7S domain (the latter containing additional sequences at the carboxy-terminal ends of the component chains) is obtained.[7] Antibodies to the 7S region can also be raised by immunizing with more complete type IV collagen molecules.[61] By using different forms of the 7S domain as standard, as labeled antigen, and as immunizing agent, it is possible to develop several radioimmunoassays which all recognize determinants in the 7S region but show different characteristics.[62,63]

There is little if any 7S domain in normal human serum, when measured with an assay

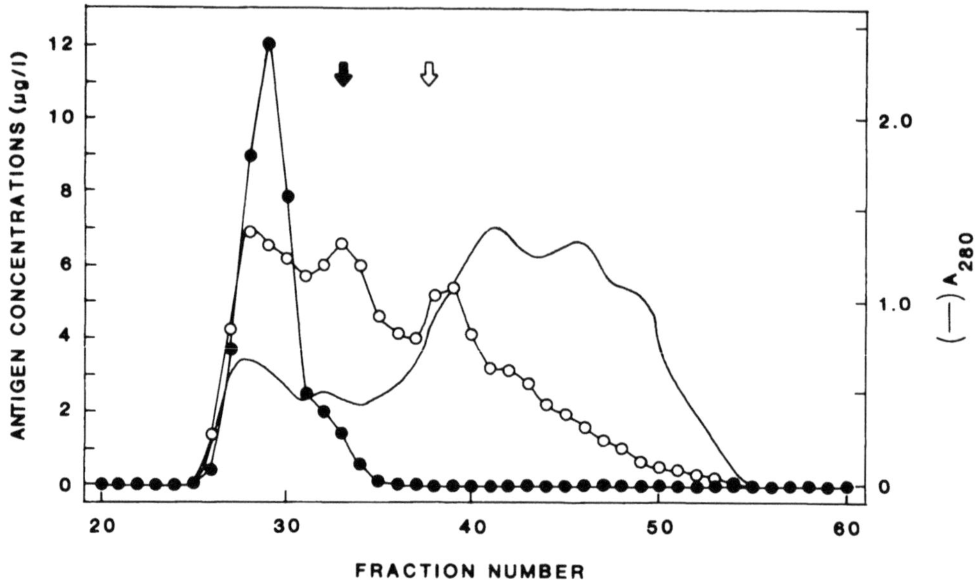

FIGURE 5. Gel filtration analysis of human serum on a column of Sephacryl® S-300. The solid circles indicate the concentration of 7S domain of type IV collagen, and the open circles that of laminin P1, as assessed by specific radioimmunoassays. The arrows denote the elution positions of the corresponding standard antigens. The 7S domain antigen is homogeneous and much larger than the standard in size, possibly representing almost intact type IV collagen molecules. The laminin antigen is heterogeneous, most likely representing a series of degradation products of laminin. (From Niemelä, O., Risteli, L., Sotaniemi, E. A., and Risteli, J., *Eur. J. Clin. Invest.*, 15, 132, 1985. With permission.)

based on its long form (reference values less than 10 μg/l).[63] When measurable, the antigenic material in serum is homogenous, has a relatively large molecular size (Figure 5), and gives an inhibition curve parallel to that of the standard.[63]

Another immunoassay which is based on the short form of bovine 7S domain and an antiserum against the same species has also been used for analyzing human serum samples.[62] The size distribution or other characteristics of this serum antigen have not been reported.

In mice and rats the serum concentration of 7S domain is readily measurable, being about 20 μg/l in adult rats.[64] Also here, serum samples give inhibition curves parallel to that of the standard. In rat serum, the total 7S antigenicity is a sum of two forms with different molecular sizes.[65,66]

There are specific radioimmunoassays for the NC1 domains of both human[67] and murine[66] type IV collagen. The assay developed for the human protein detects one major form of antigen in serum that has a molecular size of about 60,000.[67] It has been suggested that such fragments are derived from the degradation rather than from the synthesis of type IV collagen. There is no direct evidence for this, however. The inhibition curves obtained with serum samples are parallel to that of the standard. Some NC1 antigenicity is also found in urine.

The serum concentration of the NC1 antigen was reported to increase significantly in fibrotic liver disease and slightly in metastatic cancers and lupus erythematosus.[67] After the original report on the assay, no further studies have been performed for evaluating its clinical usefulness.

An assay for the NC1 domain of murine type IV collagen detects two forms of antigens in rodent sera. The smaller resembles the small NC1 antigen, as detected by the human NC1 assay (see above), whereas the larger also carries the 7S determinant.[66] The latter antigen form may represent intact type IV collagen molecules. The large serum antigen can be degraded with bacterial collagenase, giving rise to smaller 7S and NC1 antigens.[66] The mouse NC1 assay has been used in studies of experimental diabetes[66] and toxic liver injury[37] in rats. Both conditions cause an increase in the total concentration of the NC1 antigen and in the proportion of its larger form in serum.

The elimination routes and further metabolism of type IV collagen-derived serum antigens are currently unknown.

2. Laminin Fragments

Laminin is a major noncollagenous glycoprotein of basement membranes. Its most antigenic parts are located in two relatively protease-resistant areas of the molecule, termed fragments 1 and 2 (the latter also being called fragment 4, depending on the extent of the digestion). Fragment 1 is derived from the central, heavily disulfide-linked part of the cross-shaped laminin molecule. This fragment has been prepared from both murine and human laminins, and the radioimmunoassays for these show basically similar characteristics. Fragment 2 (or 4), derived from the end of one of the short arms of the laminin molecule, has been isolated from protease-digested mouse laminin. The radioimmunoassay is suitable for studies on experimental animals, but no corresponding human antigen assay is available at present.

The human serum antigens related to the laminin fragment 1 (usually termed P1, indicating that the protease used for generating the fragment was pepsin) are heterogenous in size (Figure 5), thus resembling the situation seen in the PIIINP assay.[63,68,69] The assays also present similar nonparallelism between standards and serum samples as does the original PIIINP assay. Two different modifications of the laminin P1 assay have been used: one ordinary radioimmunoassay, including use of the 50% intercept method for calculating the concentration in the unknown sample,[63] and another, in which Fab fragments substitute for the intact antilaminin antibodies.[68] The latter assay is available commercially (Behringwerke, Marburg, West Germany).

The reference interval for healthy adults has been reported to be 15 to 37 μg/l of serum with the ordinary assay[63] and 0.81 to 1.41 U/ml with the Fab modification, one arbitrary unit being defined as the mean antigen content of a certain panel of 55 normal human sera.[68] Neither of the ways of expressing the concentration seems optimal, the former because different samples can give rise to inhibition curves that differ in their slopes, making the quantification arbitrary, and the latter because these units are completely artificial and do not indicate any defined biological activity (as do, e.g., units of enzyme activity). There is an obvious need for a new assay that would not present the same technical problems.

The changes in the serum concentration of the human laminin P1 antigens in any condition seem to be relatively small, up to about twofold the upper reference limit. Although such changes can be statistically significant, when groups of patients are compared, one has to question the usefulness of such an assay for monitoring a single patient. The reproducibility of the present assays is not good enough for follow-up purposes.

When studying murine samples, the unsatisfactory behavior of the P1 assay can be easily overcome by using the fragment P2 as basis for another assay.[65] The corresponding serum antigens, although multiple, produce inhibition curves that are parallel to that of the standard. In mice and rats, the P2 assay shows greater dynamics than does the P1 assay in the same experimental diseases.[64,70] The P2 antigen is also found in urine, where no P1 antigenicity can be demonstrated.[70]

D. HYALURONAN

Hyaluronate (present name hyaluronan) is a polymer of alternating *N*-acetylglucosamine and glucuronic acid units.[71] It is mainly found in loose connective tissue.

The serum assay for hyaluronan is not an immunoassay, but based on the use of radiolabeled specific binding proteins that have been extracted from bovine cartilage[72] or from human brain tissue (hyaluronectin).[73] The former assay can be obtained commercially (Pharmacia, Uppsala, Sweden; HA Test 50).

In serum both high and low molecular weight forms of hyaluronan are found.[74] The serum concentration of the glycan in adults ranges from about 10 to 100 μg/l, with no difference between men and women.[72,73] No studies are available on the concentration in children. In people older than 60 years of age the level seems to increase, with large interindividual variation.[72]

Hyaluronan reaches the blood via the lymph.[74,75] Radioactively labeled hyaluronan, when injected intravenously in rats, is rapidly cleared from the circulation, most of the uptake of the high molecular weight glycan taking place in the endothelial cells of the liver.[76] These bind hyaluronan by specific, multiple-site attachment, whereas the subsequent internalization of the glycan also has features of nonspecific endocytosis.[76] The endothelial cells preferentially bind high molecular weight hyaluronan.[76]

In humans, both the liver and the kidneys are responsible for the clearance of circulating endogenous hyaluronan. The net hepatosplanchnic extraction ratio is 33% and the net renal extraction ratio 20%.[77] The kidneys presumably remove the low molecular weight form of circulating hyaluronan, which they partially excrete into the urine and partially metabolize. Unless diseases are present that affect hyaluronan metabolism, its serum concentration correlates with the glomerular filtration rate.[77] In subjects with renal failure there is a relationship between the serum hyaluronan and creatinine concentrations.[78] Thus, increased serum concentrations of hyaluronan can result from enhanced production in inflamed connective tissue[79,80] or from impaired removal from the circulation in diseases of liver[73,81] or the kidneys.[77,78]

E. OTHER ASSAYS

A specific radioimmunoassay has been developed for the triple-helical parts of human *type VI collagen*, isolated from pepsin-digested placental tissue.[82] The antigen can be detected in serum, bile, ascitic fluid, and concentrated urine. The serum antigen is heterogenous, resolving into five distinct peaks upon gel filtration, most of the antigenicity being due to material having the largest molecular weight. The reference interval for adult human serum is about 16 to 28 μg/l. The form of the inhibition curves that serum samples produce has not been reported, but those for bile, ascitic fluid, and concentrated urine are less steep than the inhibition curve of the standard.[82] No clinical studies have been carried out with this assay, neither are the origin and further metabolism of the antigens measured known. Of the seven patients with liver cirrhosis, who were included in the first report on the assay, most had serum type VI collagen concentrations that were below the reference range.[82]

A solid-phase radioimmunoassay has been developed for *breakdown products of elastin*.[83] Similar antigens are present in serum, though they have not been further characterized. The concentration range of 16 healthy nonsmokers was reported to be about 5 to 30 mg/l. A significant elevation was seen in a group of patients with chronic obstructive pulmonary disease.[83] There have been no further studies applying this assay, and nothing is known about the further metabolism and elimination of the circulating antigens.

Of the enzymes involved in the posttranslational processing of procollagen, *immunoreactive prolyl 4-hydroxylase* and the *activity of galactosylhydroxylysyl glucosyltransferase* can be measured in serum. Both of these have been shown to correlate with the corresponding enzyme activities in the affected tissue in several inflammatory and fibrotic conditions. Such results should be interpreted with caution, however. It has recently been found that the so-called β-subunit of prolyl 4-hydroxylase functions as another enzyme, protein disulfide isomerase.[84,85] The same polypeptide also seems to serve as a cellular thyroid hormone binding protein.[86] In fact, most of this subunit in tissues is never incorporated into active prolyl 4-hydroxylase. Unfortunately, all the radioimmunoassays reported for prolyl 4-hydroxylase recognize determinants in the β-subunit. Thus, taking the present knowledge of this immunoreactive protein into account, one cannot presume that the material the assay detects in serum has any direct relationship with collagen synthesis. This is true despite the good correlations between the serum concentration of the immunoreactive protein and the activity of prolyl 4-hydroxylase in tissues.

Galactosylhydroxylysyl glucosyltransferase is responsible for the formation of hydroxylysine-linked disaccharide units that are found in collagens and a few other proteins, e.g., the C1q component of complement and the acetylcholinesterase enzyme. The synthesis of complement components increases in many inflammatory conditions. In such situations the increases in the activity of the glucosyltransferase in serum are most likely not derived from collagen synthesis only. Furthermore, the assessment of the activity is methodologically not suitable for a routine clinical chemistry laboratory.

Of the enzymes involved in the metabolism of proteoglycans, the *activity of N-acetyl-β-D-glucosaminidase* has been found to increase in serum in chronic liver diseases, but the predictive values for fibrosis and cirrhosis are low.[87]

PZ-peptidase is an enzyme supposed to be involved in the further catabolism of collagenous amino acid sequences. The activity is measured with a synthetic peptide substrate and is readily measurable in serum samples.[88]

Several small degradation products of collagen can be detected both in serum and urine. *Urinary hydroxyproline* has been used as a general indicator of collagen turnover. The clinical applications of this determination have been extensively reviewed.[89] In adults, the rapid turnover of C1q seems to contribute significantly to the basal level of urinary hydroxyproline.[90] This assay still has considerable use in the follow-up of malignant tumors, especially when metastatic into the bone.[91,92] Hydroxyproline excretion is increased by the collagen and gelatin present in the diet.

The contribution of C1q is probably still more pronounced in the case of the *urinary hydroxylysine-linked saccharides,*[90] as C1q is rich in glucosylgalactosylhydroxylysine. The assessment of the hydroxylysine glycosides has not found indications in clinical use.

The urinary excretion of a collagen cross-link, *pyridinoline,* can be measured both with conventional chromatographic procedures[93,94] and by immunochemical means.[95] This cross-link is most prevalent in cartilage collagen; its urinary excretion has been suggested to reflect specifically degradation of the cartilage matrix.[96] Although the concentration of this compound in bone collagen is much lower than in cartilage collagen, its total amount in bone can be significant, however. Without further studies, e.g., in patients with active bone diseases, it is thus impossible to judge whether the assay is specific for breakdown of cartilage.

IV. USE OF THE ASSAYS IN CLINICAL SITUATIONS

None of the serum assays for specific connective tissue components that are currently available can be considered specific for any single organ. Thus, additional information on the clinical manifestations of the disease, on the results of laboratory tests and radiological and other investigations, as well as on the age and sex of the patient must be taken into account when interpreting the results of the connective tissue assays. In this respect these tests resemble any other laboratory parameter.

A. PHYSIOLOGICAL CONDITIONS WITH INCREASED COLLAGEN METABOLISM

Physiological states which increase the serum concentrations of extracellular matrix components are quite common: e.g., growth, pregnancy, and recent surgical operations with

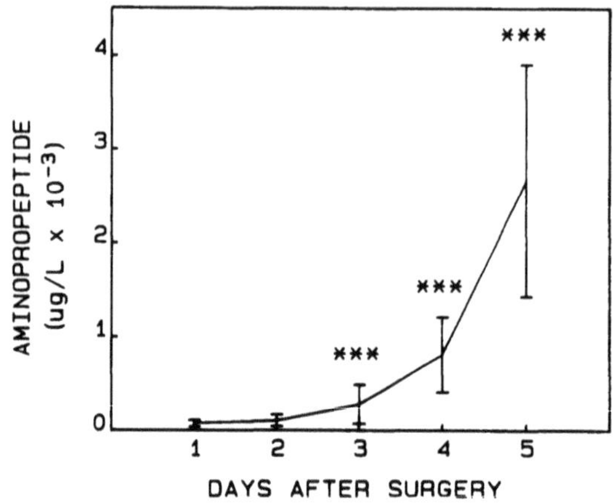

FIGURE 6. Concentration of the amino-terminal propeptide of type III procollagen in surgical abdominal wounds in humans (assay No. 7 in Table 3). The wound fluid was collected via thin silicone rubber tubes left in the wound during operation. In gel filtration analysis only forms C and D (see Figure 1) were found. It was the form C representing intact PIIINP which increased most. (From Haukipuro, K., Risteli, L., Kairaluoma, M. I., and Risteli, J., *Ann. Surg.*, 206, 752, 1987. With permission.)

subsequent wound healing are associated with enhanced connective tissue metabolism. On the other hand, when increases in the serum concentrations of collagen metabolites are observed in these conditions, known to affect collagen metabolism, this further confirms the validity of these methods.

1. Wound Healing

Relatively small, but statistically significant increases can be observed in the serum concentrations of the amino-terminal propeptide of type III procollagen, the 7S domain of type IV collagen, and the P1 fragment of laminin during the early phases of healing of a surgical wound in patients.[97] At least for PIIINP, the magnitude and the time-course of the change depend on the size and location of the wound.[98] Thus, after an uncomplicated cholecystectomy the concentration remains within the reference interval. After operations involving the hip joint, the increase takes place more slowly than after abdominal operations.

The connective tissue antigens can be assayed *in situ* in a healing wound. Aliquots of interstitial fluid are collected via a thin silicone rubber tube that has been left in the wound, and the antigens are analyzed in this fluid with the relevant radioimmunoassays. The concentration of the amino-terminal propeptide of type III procollagen increases dramatically at day 3, with a further increase taking place till at least day 5 (Figure 6).[28] At day 5 the PIIINP concentration in the wound is about 1000-fold, compared with that in serum. Most of this increase is due to the antigen similar in size to the intact, cleaved propeptide (peak C in Figure 1), although a smaller increase also takes place in the concentration of the small degradation products (peak D) in the wound.[28] At least during the first 5 d of healing, no PIIINP antigens are found in the wound that would be larger than the liberated propeptide. This is in contrast to the situation in serum, where the increases are due to forms B and C.[98] The origin of form B is not clear at present. However, as this antigen form is not found free in the wound, it is more probably derived from later turnover of the granulation tissue matrix or from other tissues, rather than directly from the stimulated collagen synthesis in the wound.

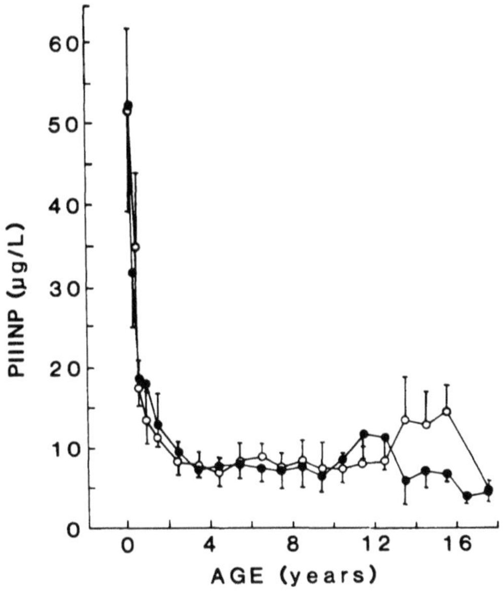

FIGURE 7. Serum concentration of the amino-terminal propeptide of type III procollagen (assay No. 7 in Table 3) in infants and children. (●) Girls (n = 125); (○) boys (n = 146). (From Risteli, J., Niemi, S., Trivedi, P., Mäentausta, O., Mowat, A. P., and Risteli, L., *Clin. Chem.*, 34, 715, 1988. With permission.)

The concentration of the carboxy-terminal propeptide of type I procollagen also increases in the wound during the first week of healing. This is reflected in a significant, although smaller, increase in serum. In both locations, the increase is due to one antigen species that has the same size as the intact propeptide.[55]

2. Growth

Somatic growth results from the proliferation of cells and from the deposition of extracellular matrix between the cells. This takes place both in the skeleton and in soft tissues. In addition to *de novo* synthesis, existing matrix material must be modified and removed. Thus, during phases of active growth, both the synthesis and the turnover of collagen and other extracellular matrix constituents are active.

The concentrations of the carboxy-terminal propeptide of type I procollagen[50] and of the amino-terminal propeptide of type III procollagen[19,99] are much higher in growing infants and children than in adults. The age-dependence curve of the latter, in fact, closely resembles the charts showing the growth velocities at different ages (Figure 7). After very high levels in infancy, a second peak is reached at puberty, at 12 years of age in girls, and at 14 to 15 years of age in boys.[19,99]

In contrast to the metabolites of interstitial collagens, the serum concentrations of two basement membrane-derived antigens, the 7S domain of type IV collagen and the fragment P1 of laminin, show little or no elevations during phases of active growth.[100]

3. Pregnancy

In the adult, pregnancy is a unique situation in terms of the rate of extracellular matrix synthesis and turnover in a growing organ, the uterus. In addition, the endocrinologic changes taking place during pregnancy may affect the properties of other connective tissues as well.

The serum concentration of the amino-terminal propeptide of type III procollagen increases in most women during the second half of the third trimester of pregnancy. The highest values, reached at term, are about threefold as high as the upper reference limit for nonpregnant subjects.[101] However, in some women the concentration never increases beyond the reference interval, obviously indicating interindividual differences in type III collagen metabolism or in the efficiency of metabolizing or excreting the propeptide.

The concentration of the fragment P1 of laminin in serum increases during pregnancy, reaching a maximum of nearly twice the level of nonpregnant women close to term.[102] This is in contrast to the serum concentration of the 7S domain of type IV collagen, which, when analyzed with our assay (large form), remains constant throughout pregnancy.

In preeclampsia, and also in primarily hypertensive, pregnant patients, the serum concentration of the amino-terminal propeptide of type III procollagen tends to increase more than in healthy pregnant women.[101] This may also be true for laminin P1 in preeclampsia, although the difference from normals is small.[102]

B. LIVER FIBROSIS AND CIRRHOSIS

Liver fibrosis, both human and experimental, has been the model condition when new methods for detecting changes in connective tissue metabolism have been applied. Consequently, there is an extensive body of literature concerning the serum analysis of extracellular matrix components in liver fibrosis and cirrhosis. In many cases, the results have been interpreted in terms of increased collagen synthesis and deposition in a straightforward manner. However, the possible involvement of the liver itself in the further metabolism and clearance of connective tissue-derived substances, which has become evident only lately, should now be seriously taken into consideration when interpreting serum concentrations in liver disease. Another relevant fact to be kept in mind is the heterogeneity of fibrotic liver diseases, in terms of etiology, pathogenesis, and histological distribution of the fibrosis;[58] thus, the conclusions that seem relevant for one type of these diseases are not necessarily directly applicable to another disease.

Table 4 lists the liver diseases for which the effects on the serum concentration of the amino-terminal propeptide of type III procollagen have been documented. Most of the information has been obtained with the commercial RIA-Gnost assay (No. 1 in Table 3). Most studies have been cross-sectional.

In different investigations, the serum concentration of PIIINP has been found to correlate either with the extent of fibrosis[34] or with the degree of inflammation in liver tissue, or with none of these (see Table 4). Typically, very high serum PIIINP concentrations are seen when there is marked inflammation in the liver (e.g., acute hepatitis, alcoholic hepatitis), whereas silent accumulation of connective tissue without inflammation (e.g., idiopathic hemochromatosis) has relatively little effect on the serum propeptide (Figure 8).

The serum PIIINP concentration has been suggested to be a useful prognostic indicator in some liver diseases. In primary biliary cirrhosis, the propeptide concentration measured before treatment was reported to distinguish effectively the group with a rapidly developing, fatal disease from another with a slowly progressing disease.[125] This finding could be reproduced in another group of patients, where also a similar, but not statistically significant trend was noted for the laminin P1 antigen in serum.[126] The serum concentration of the 7S domain of type IV collagen was elevated in only a few patients. Part of the patients included in the latter study had already been treated. It should be noted, however, that all the increases reported for primary biliary cirrhosis are relatively small. A real value in predicting the outcome of an individual patient with this disease has not yet been proved for any connective tissue assay.

In resolving acute hepatitis the initially high serum PIIINP concentrations decrease, remaining at a higher level in chronic active hepatitis than if there is a complete recovery or a chronic persistent hepatitis develops (Figure 9).[117,118] Thus, the serum PIIINP concen-

TABLE 4
Serum Amino-Terminal Propeptide of Type III Procollagen in Liver Diseases

Disease	Findings and conclusions	Ref.
Alcoholic liver disease	Some or little increase in fatty liver or inactive cirrhosis, large increase in alcoholic hepatitis (with cirrhosis)	15, 17, 23, 27, 52, 88, 103—109
	Correlates with inflammation	15, 52, 104, 108—110
	Correlates with fibrosis	30, 52, 103, 107, 108, 111—113
	Does not correlate with extent of fibrosis	104, 105, 109
	Elevation mostly due to larger antigen forms	17, 114
	Assay for small antigen forms (No. 4 in Table 3) may differentiate between fatty livers with and without perivenular fibrosis	108, 112
Idiopathic hemochromatosis	Within the reference interval or slightly increased	104, 115
	Does not correlate with hepatic fibrosis or laboratory indices of the disease	104, 115
	Larger increases when additional hepatotoxic factors present	104
Acute viral hepatitis	Large increases in acute inflammation	15, 27, 116—118
	Resolution associated with a normalizing PIIINP concentration	117, 118
	In acute phase no difference between those who later resolve and those who develop chronic active hepatitis	117, 118
	Elevation mostly due to larger antigen forms	27, 118
Chronic active hepatitis	May reflect activity of the disease during treatment	15, 88, 99, 103, 105, 109, 111, 117—122
	Not useful in children	99
	Elevation mostly due to larger antigen forms	27, 118, 121
	Does not correlate with extent of fibrosis	105, 121
	Does not correlate with laboratory indices of inflammatory activity	121, 122
Chronic persistent hepatitis	Usually normal or close to upper reference limit	88, 103, 105, 109, 111, 117, 120
Primary biliary cirrhosis	Often elevated	11, 111, 120, 123—125
	Correlates with stage of disease	11, 124, 126
	May correlate with prognosis	125, 126
	Treatment has little effect on serum PIIINP	120, 123
Hepatocellular carcinoma	Large increases	127, 128
	Increase may predict development of a hepatoma in cirrhosis	128
Indian childhood cirrhosis	Large increases	100
Amoebic liver abscess	Large increases	116
Schistosomal liver fibrosis	Elevated in early stages of *S. mansoni* infection	116, 129
	Close to normal in advanced fibrosis	129
Methotrexate-induced fibrosis	With normal PIIINP little risk for fibrosis and cirrhosis	130
Biliary atresia	Elevated in about half of affected infants	131
	No prognostic value	131
Idiopathic hepatitis of infancy	As in biliary atresia	131
Alpha-1-antitrypsin deficiency	As in biliary atresia	131

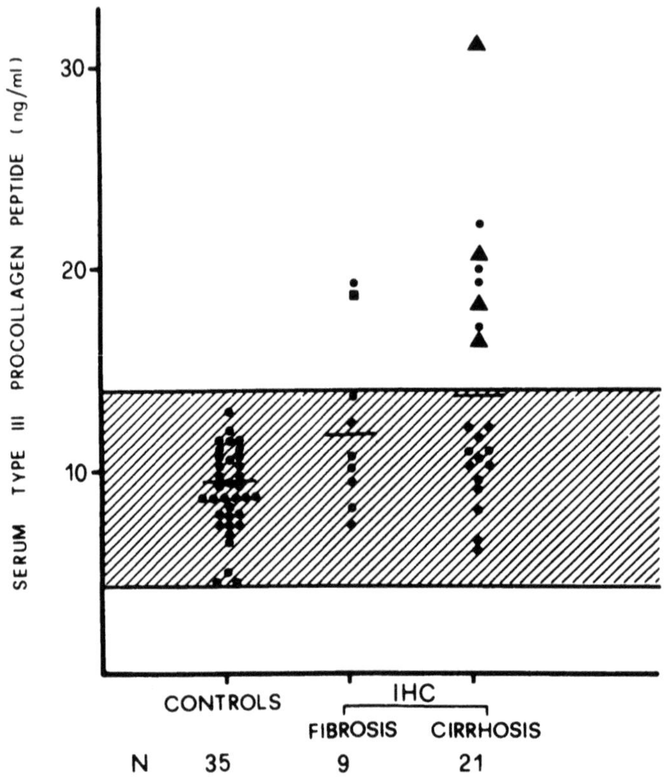

FIGURE 8. Serum concentration of amino-terminal propeptide of type III procollagen (assay No. 1 in Table 3) in patients with idiopathic hemochromatosis (IHC). The shaded area represents the reference interval (mean ± 2 SD). The triangles indicate patients with excessive alcohol intake (ETOH) and the square denotes a patient with HBsAg-positive chronic active hepatitis. (From Colombo, M., Annoni, G., Donato, M. F., Conte, D., Martines, D., Zaramella, M. G., Bianchi, P. A., Piperno, A., and Tiribelli, C., *Hepatology*, 5, 475, 1985. © by the American Association for the Study of Liver Diseases. With permission.)

tration could serve as a marker for the activity of chronic active hepatitis.[121] As they could find no correlation between the extent of liver fibrosis and the serum PIIINP concentration, one group even concluded that PIIINP has one important advantage over tests like bilirubin or the transaminases in this disease: its serum levels are not affected by the presence of cirrhosis.[121]

The serum concentration of the antigen related to the large 7S domain of type IV collagen is increased in alcoholic liver disease, in particular in alcoholic hepatitis, and changes, together with the activity of the hepatitis.[63,114] Also the concentration of the carboxy-terminal propeptide of type I procollagen in serum increases, its changes in the various forms of alcoholic liver disease resembling those of PIIINP in the same patients.[52]

The peripheral venous concentration of the laminin P1 antigen follows approximately the same pattern as does that of PIIINP in alcoholic liver disease.[63] The concentration of laminin P1 has been suggested to correlate strongly, and that of PIIINP weakly, with the elevated portal pressure in liver cirrhotic and fibrotic patients.[132] This conclusion was based on a study of 33 patients; the diagnoses included viral, alcoholic, and primary biliary cirrhosis. A highly significant correlation has also been found between the laminin P1 antigen concentration in the hepatic vein of cirrhotic subjects and portal venous pressure.[133] In

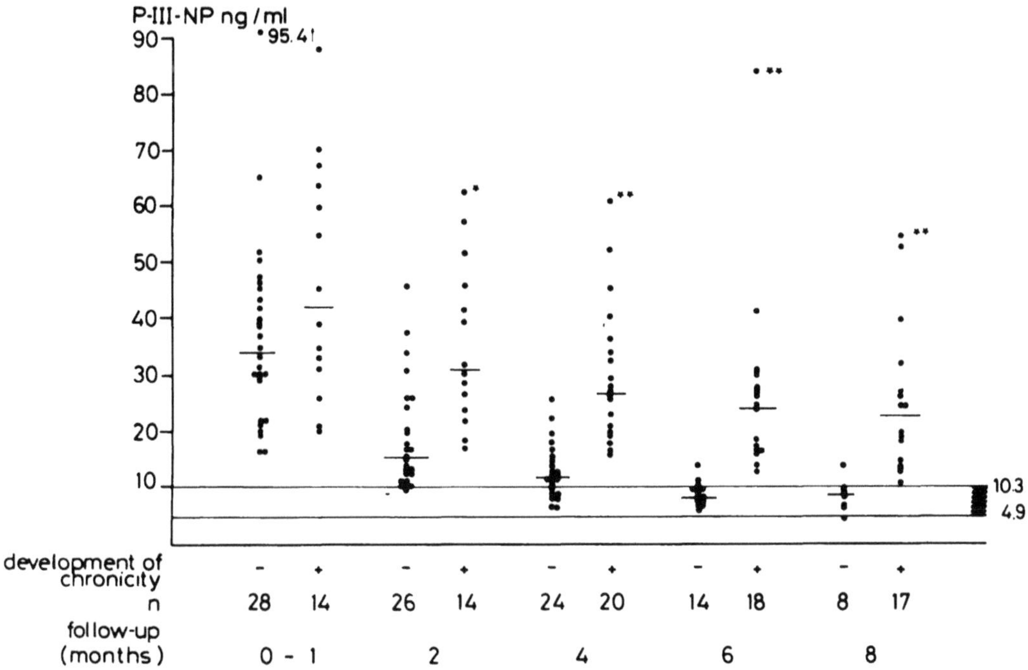

FIGURE 9. Serum concentration of amino-terminal propeptide of type III procollagen (assay No. 1 in Table 3) during follow-up from onset of symptoms in 33 patients with uncomplicated viral hepatitis (−) and in 30 patients developing chronic liver disease (+). The shaded area indicates the reference interval. Statistical significance: $*p < 0.05$, $**p < 0.01$ compared to uncomplicated hepatitis. (From Bentsen, K. D., Hørslev-Petersen, K., Junker, P., Juhl, E., Lorenzen, I., and the Copenhagen Hepatitis Acuta Programme, *Liver*, 7, 96, 1987. © The Munksgaard International Publishers Ltd., Copenhagen, Denmark. With permission.)

alcoholic liver disease, the concentration of the laminin P1 antigen in peripheral blood correlates with the impairment of antipyrine metabolism better than do the concentrations of the 7S domain of type IV collagen or PIIINP.[134]

The serum concentration of hyaluronan is increased in liver cirrhoses of varying etiologies.[73,81] Contrary to PIIINP in liver disorders, pure inflammation of the liver, as seen in acute hepatitis, has less effect on the serum concentration of the glycan. There is considerable overlap between the two types of liver disease, however, and, thus, the hyaluronan concentration probably cannot serve in the differential diagnosis of cirrhosis. However, this possibility deserves to be further studied, particularly with a follow-up approach.

In children with chronic fibrosing liver disease, the high serum levels of the PIIINP antigenicity due to growth tend to mask any changes that would reflect the state of the liver.[99,131] In a study of 62 infants with liver disease (extrahepatic biliary atresia, idiopathic hepatitis of infancy, or $\alpha 1$ antitrypsin deficiency) about half of them had serum PIIINP concentrations above the upper reference limit for age; there was no relationship between the elevated levels and prognosis.[131] Older children with chronic active hepatitis were constantly found to have serum PIIINP concentrations within the reference range for age.[99]

Indian childhood cirrhosis presents as a particularly aggressive fibrogenesis in the liver. In a cross-sectional study of 34 infants and children with this disease, the serum concentrations of PIIINP, laminin P1, and the 7S domain of type IV collagen were all significantly elevated.[100] The most sensitive indicator of the disease was the 7S domain, 95% of the patients having pathological values. The corresponding figures for laminin P1 and PIIINP were 91 and 76%, respectively. These assays have not been studied with respect to their possible value in predicting the occurrence of the disease in children who live in the area where the disease is endemic or in the follow-up of treatment.

The changes in extracellular matrix metabolites in serum have been studied in two well-known rat models for toxic liver injury. Administration of carbon tetrachloride results in an increase in the serum PIIINP concentration by day 45,[36,38] the change leveling off by 90 d.[38] A more transient increase takes place in the serum PINP antigen.[38] An increase in the serum concentration of the type IV collagen-related 7S and NC1 antigens is evident by 4 weeks.[135]

Liver injury induced by dimethylnitrosamine differs from the carbon tetrachloride injury in that there is a rapid, about tenfold increase in the serum concentrations of the 7S and NC1 antigens by 2 to 3 weeks. This is accompanied by a smaller increase in the serum PIIINP concentration. Both of these changes correlate closely with the amounts of mRNA for the respective collagen types in liver tissue.[37]

C. LUNG FIBROSIS

Lung fibrosis in humans is a heterogenous group of diseases, with respect to both etiology and prognosis.[136] A special characteristic of fibrotic lung tissue is the fact that the concentration of collagen or any other connective tissue component, when expressed, e.g., per unit of dry weight, DNA, or total protein, is not different from normal, but there is a definite increase in their total amounts per lung.[136] A number of experimental models have been described that mimic some of the human pathologies.

The tissue changes that lead to lung fibrosis seem to be hardly reflected in the serum concentrations of the connective tissue components PIIINP and hyaluronan. This is evident from cross-sectional studies of patients with sarcoidosis,[22,137] silicosis,[138] tuberculous fibrosis,[138] idiopathic pulmonary fibrosis,[137-139] and farmer's lung.[140,141] Accordingly, although no detailed follow-up studies are available, these serum assays obviously have no prognostic value in patients with a disease that may lead to pulmonary fibrosis.

In contrast to this, the fluid that has been collected by bronchoalveolar lavage during an acute phase of sarcoidosis,[22,137] silicosis,[138] tuberculous fibrosis,[138] idiopathic pulmonary fibrosis,[137,138] asbestosis,[142] or farmer's lung[141] has a high concentration of PIIINP. This is true both when the value is expressed per unit volume and when it is related to the albumin concentration of the fluid. The antigen forms present in lavage fluid have not been characterized. No measurable PIIINP antigenicity is released from the lungs of healthy individuals by the lavage procedure.[22,141]

During recovery from an attack of farmer's lung disease, the PIIINP level in bronchoalveolar lavage fluid decreases, but may remain above normal for at least 6 to 14 months (Figure 10).[141] No studies are available on the potential of the PIIINP measurement for indicating the development of a fibrosis during follow-up in individual patients.

Hyaluronan can be detected in the bronchoalveolar lavage fluid of some healthy persons.[141] In sarcoidosis[143] and in acute farmer's lung[141] the concentration is highly increased, both in absolute terms and in relation to albumin. During recovery from acute farmer's lung, the concentration decreases but is not necessarily normalized during a follow-up of 6 to 14 months (Figure 10).[141] In this condition both PIIINP and hyaluronan in the lavage fluid were found to have a negative correlation with diffusion capacity.[141]

There is one report on experimental lung fibrosis, induced with bleomycin in rabbits, where an early increase in the concentration of PIIINP is shown to take place both in serum and in bronchoalveolar lavage fluid.[144] Unfortunately, the primary antibody of the commercial assay (No. 1 in Table 3) which was used in this study was also raised in rabbits; thus, the authors have probably been titrating the anti-rabbit second antibody rather than really measuring the rabbit PIIINP. To avoid ambiguities like this, it is essential that the antigen to be measured is carefully characterized in each new system where one intends to use an inhibition assay.[8]

The serum concentration of elastin degradation products is increased in chronic obstruc-

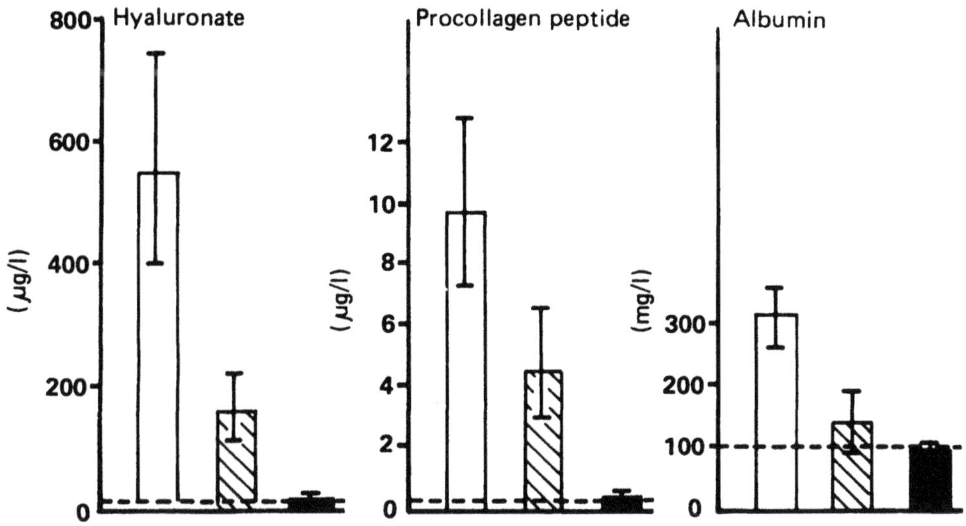

FIGURE 10. Bronchoalveolar lavage concentrations of hyaluronan, PIIINP (assay No. 1 in Table 3) and albumin (mean ± SEM) in patients with various stages of farmer's lung. The open columns: on admission with acute symptoms (n = 10); shaded columns: during recovery phase 4 to 10 weeks after admission (n = 7); and solid columns: at clinical remission 6 to 14 months after admission (n = 7). (From Bjermer, L., Engström-Laurent, A., Lundgren, R., Rosenhall, L., and Hällgren, R., *Br. Med. J.*, 295, 803, 1987. With permission.)

tive pulmonary disease.[83] This is in accordance with increased activity of elastase as a suggested pathogenetic basis for emphysema. However, the validity of the elastin fragment assay for detecting the development of pulmonary emphysema has not been tested in a follow-up study; neither has this assay been used in any fibrotic lung disease. Elastase is able to degrade other connective tissue components as well, e.g., type III and IV collagens. No method that could reveal fragmentation of these proteins in the lung tissue has been applied to pulmonary diseases.

D. FIBROTIC CONDITIONS OF THE BONE MARROW

Myelofibrosis, the replacement of normal bone marrow by fibrous connective tissue, can be primary (agnogenic myeloid metaplasia) or result from another hematologic or malignant disease. Histologically it is sometimes divided into reticulin fibrosis (early) and collagen fibrosis (late). The development of myelofibrosis involves both an increased deposition of interstitial collagens and a process resembling the capillarization of the sinusoids, as seen in liver cirrhosis.[145]

The serum concentration of the amino-terminal propeptide of type III procollagen is increased in myelofibrosis (Figure 11).[146-150] In other myeloproliferative diseases, such as polycythemia vera and chronic myelogenous leukemia, it seems to correlate roughly with the extent of bone marrow fibrosis. In a study of patients with polycythemia vera, higher PIIINP concentrations were initially seen in those who developed myeloid metaplasia during a follow-up period of 12 or 24 months, suggesting that the assay could have value in predicting the transformation in this disease.[150] Patients with a myelofibrosis of recent onset typically have higher serum concentrations of PIIINP than those with chronic fibrosis or osteomyelosclerosis.[148-150] It has been suggested that serial determinations of PIIINP could be used in follow-up of treatment in myelofibrosis, although the assay cannot be considered to replace the bone marrow biopsy.[151] In the same study, the authors expressed the need both for technically better PIIINP assays and for reliable markers of type I collagen synthesis in this disease.

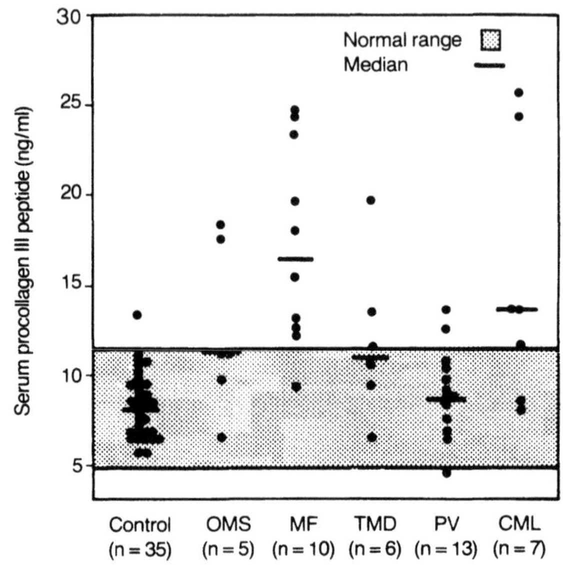

FIGURE 11. Serum concentration of the amino-terminal propeptide of type III procollagen in 41 patients with chronic myeloproliferative disorders. The PIIINP assay used was No. 1 in Table 3. OMS, osteomyelosclerosis; MF, myelofibrosis; TMD, transitional myeloproliferative disorder; PV, polycythemia vera; CML, chronic myelogenous leukemia. (From Hasselbalch, H., Junker, P., Lisse, I., Bentsen, K. D., Risteli, L., and Risteli, J., *Am. J. Hematol.*, 23, 101, 1986. With permission.)

The serum concentration of 7S domain of type IV collagen, as detected with the assay for the large antigen, is also increased in myelofibrosis.[152] These findings are remarkably similar to those on liver fibrosis.

E. CONNECTIVE TISSUE DISEASES

In patients with active rheumatoid arthritis, the serum concentration of PIIINP is significantly higher than in healthy individuals or in subjects with an inactive disease. This can be demonstrated both with an assay which preferentially measures the larger antigen forms (e.g., No. 1 in Table 3) and with an assay (e.g., No. 4 in Table 3) that mainly detects the small degradation products of PIIINP. However, the increase in the active stage of the disease is more due to the larger antigen forms than to the degradation products.[153] Taking the extent of the disease process into account, the increases seen in serum PIIINP in active rheumatoid arthritis can be considered relatively small.

The PIIINP concentration in synovial fluid from inflamed joints is 50 to 100 times as high as in serum. However, there seems to be little difference between the synovial fluids taken from inflamed and noninflamed joints in this respect.[26] Also, the concentration of the 7S domain of type IV collagen is higher in rheumatic synovial fluid than in serum, whereas that of the laminin P1 antigen is similar to its concentration in serum.[154]

In a cross-sectional study of patients with progressive systemic sclerosis where two different assays for PIIINP were applied (Nos. 1 and 4 in Table 3), the latter was more often increased (Figure 12).[155] The increases were not large, however. The mean serum level of the small PIIINP antigens correlated with the degree of fibrosis in internal organs.

The serum concentration of hyaluronan is increased in active rheumatoid arthritis[79] and in scleroderma.[80] In rheumatic patients physical activity causes a significant increase in the

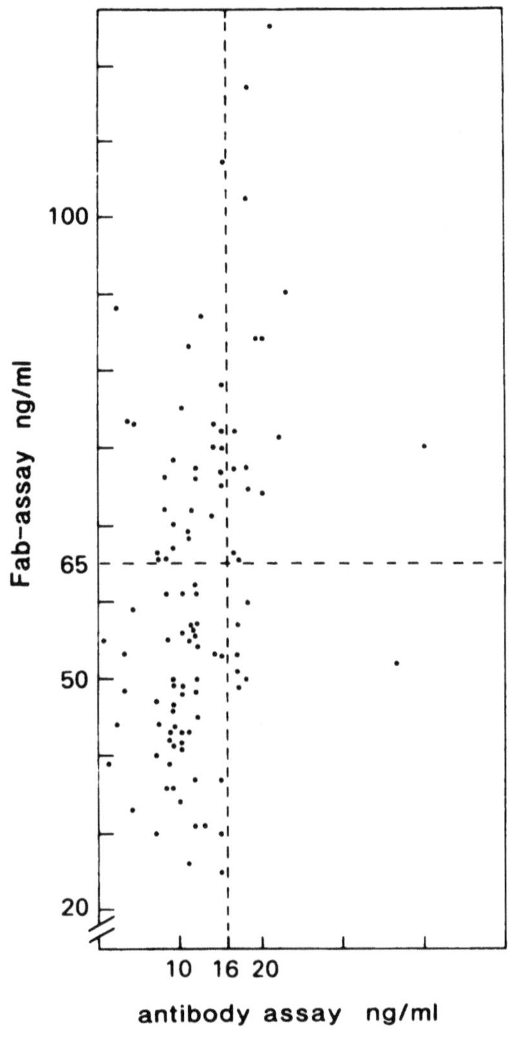

FIGURE 12. Serum concentration of the amino-terminal propeptide of type III procollagen in 101 scleroderma patients. PIIINP assays used were No. 1 (antibody assay) and No. 4 (Fab assay) in Table 3. The dashed lines indicate the upper limits of the reference intervals (mean ±2 SD). (Reprinted by permission of Elsevier Science Publishing Co., Inc. from Krieg, T., Langer, I., Gerstmeier, H., Keller, J., Mensing, H., Goerz, G., and Timpl, R., *J. Invest. Dermatol.*, 87, 788, 1986. Copyright 1986 by the Society for Investigative Dermatology, Inc.,)

serum concentration of the glycan; the concentration correlates with the clinically assessed degree of morning stiffness. These findings have been interpreted by assuming excessive accumulation of hyaluronan in the inflamed joints during the night.

F. DISEASES OF BONE AND MUSCLE

The main organic component of the bone matrix is type I collagen. In contrast to most other type I collagen-containing tissues, bone itself does not contain detectable amounts of type III collagen. On the other hand, the stroma of the bone marrow is rich in type III collagen.

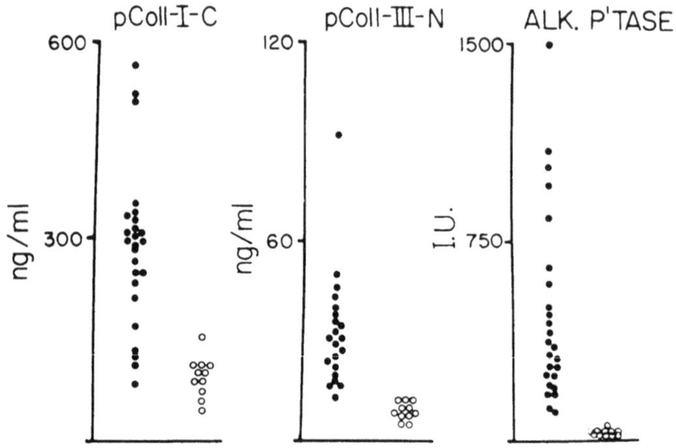

FIGURE 13. Serum concentration of the carboxy-terminal propeptide of type I procollagen, the amino-terminal propeptide of type III procollagen, and the levels of alkaline phosphatase in patients with Paget's disease (solid circles) compared to those in normal subjects (open circles). (From Simon, L. S., Krane, S. M., Wortman, P. D., Krane, I. M., and Kovitz, K. L., *J. Clin. Endocrinol. Metab.*, 58, 110, 1984. With permission.)

When the metabolic rate in bone is increased, as in Paget's disease, increased concentrations of both the carboxy-terminal propeptide of type I procollagen and of the amino-terminal propeptide of type III procollagen are found in serum (Figure 13).[49,51] In a heterogenous group of patients with metabolic bone diseases, the serum PICP concentration was found to correlate with the formation rate of cancellous bone.[53]

Several bone-specific proteins have been described during recent years.[156] Many of these are phosphorylated and/or can bind calcium. It has recently been shown that the amino-terminal propeptide of type I procollagen (PINP) is present in the matrix of developing bone and bears a phosphate residue attached to serine.[12] The possibility cannot be excluded at present that PINP has a function of its own in constituting the bone matrix. Specific immunologic assays have been developed for some of the noncollagenous bone proteins, such as osteocalcin (bone Gla-protein)[157] and bone morphogenetic protein.[158]

Fibrosis is a prominent feature of some muscular diseases. Modest increases in the serum concentration of PIIINP have been found in two such diseases, polymyositis and muscular dystrophy, but also in amyotrophic lateral sclerosis.[159]

G. MALIGNANT DISEASES

Interaction with the extracellular matrix is necessary both for the spread of a primary malignancy and for the establishment of metastases. This interaction usually results in the degradation of the matrix, but in certain tumors a fibrotic response, known as desmoplasia, also follows. The pathogenesis of the desmoplastic response is not known in detail, but some malignant cells are known to stimulate matrix formation in normal connective tissue cells. There is also evidence for the ability of some tumor cells to produce components of a fibrotic matrix. The prognostic significance of the desmoplastic reaction is not known in general.

There are still relatively few investigations of the serum connective tissue components in cancer patients. Furthermore, in most of these studies the patients have not been thoroughly characterized, e.g., in terms of the spread of the malignancy, ongoing or recent treatments, or other diseases present. Thus, it is not possible to trace the pathophysiological mechanisms responsible for the findings. However, malignant diseases are an area with great opportunities

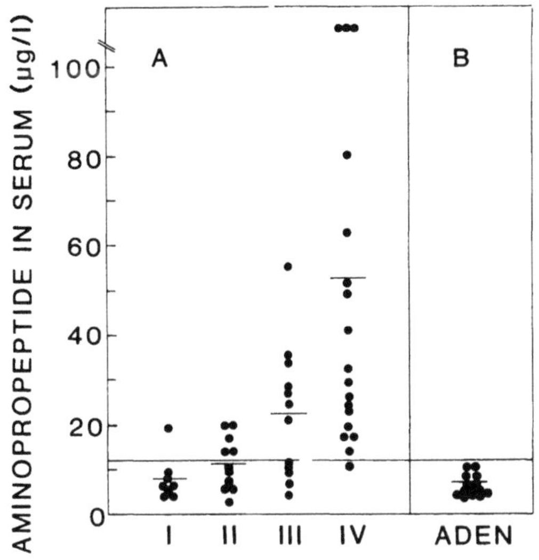

FIGURE 14. Serum concentration of the amino-terminal propeptide of type III procollagen (assay No. 1 in Table 3) in ovarian carcinoma. The numbers I to IV refer to the clinical stages I to IV, respectively. ADEN indicates a group of patients with benign ovarian cystadenomas. The short horizontal lines indicate the arithmetical means of the groups. (From Risteli, L., Kauppila, A., Mäkilä, U. -M., and Risteli, J., *Int. J. Cancer*, 41, 409, 1988. With permission.)

for applying new connective tissue methodology; further studies with clearly defined clinical questions and carefully characterized patient groups are warranted.

In hepatic malignancy, either primary or secondary, very high concentrations of the amino-terminal propeptide of type III procollagen can be found in serum.[127,128] This finding has been interpreted as indicative of collagen formation in the tumor, but turnover of an existing matrix could be contributing. It has been suggested that the PIIINP assay could be used in the follow-up of cirrhotic patients in order to detect the development of a hepatoma as early as possible.[128] This interesting hypothesis has not yet been tested in a prospective follow-up study.

Ascitic fluid from cancer patients contains high concentrations of the PIIINP antigens (up to 10 mg/l). This is true at least for ovarian[18] and pancreatic malignancies.[21] However, ascitic fluids from patients with malignant and nonmalignant diseases have not been compared. In a cross-sectional study of patients with ovarian cancer the serum PIIINP concentration correlated with the clinical stage of the disease (Figure 14);[18] the source of the serum antigens is most probably the ascitic fluid with its 50- to 100-fold higher concentration. In the same study, the follow-up of 13 patients for up to 2 years suggested that changes in serum PIIINP reflect the development of the disease. Thus, this propeptide could be regarded as a tumor-associated antigen in ovarian carcinoma. A recent prospective follow-up study has given further support to this idea.[160]

Mineralized bone tissue is devoid of type III collagen. This does not necessarily invalidate the assay for PIIINP in studies of malignancies involving bone, however, since these tumors also tend to affect the bone marrow. Bone metastases can be osteolytic, with degradation of the matrix predominating, or osteoblastic, with formation of new bone predominating, or have features of both. Such changes are not always easy to detect clinic. There seems to be a need for indicators of increased bone turnover, both for diagnostic and follow-up purposes.[161]

In patients with bone metastases from a breast carcinoma, the serum concentration of the amino-terminal propeptide of type III procollagen has been suggested to be as good an indicator of the clinical response as is urinary hydroxyproline, conventionally used for this purpose.[162] This may also be true for a primary bone malignancy, Ewing's sarcoma.[163]

There are cross-sectional findings on the serum PIIINP concentrations in several other malignant diseases. As no clinical information has been given on the patients studied, the contributions of new connective tissue synthesis and degradation of existing matrix to the serum antigens cannot be estimated.[127,128] Neither can the possible usefulness of the test, as an indicator of the spread or of the prognosis of the disease, be evaluated on the basis of the scattered findings available so far.

The serum concentration of the laminin P1 antigen was shown to be increased in a cross-sectional study of several malignant tumors including breast, colon, and lung carcinomas.[68] The observed changes were relatively small, however. Increases can be found also in patients with leukemias, but not in patients with sarcomas or lymphomas.[164] In patients with carcinoma of the colon, there was a close correlation between the serum concentrations of laminin and the carcinoembryonic antigen.[164] In other studies no such correlations have been detected between the serum laminin P1 level and the concentrations of tumor markers, such as carcinoembryonic antigen,[68,165] α-fetoprotein, or CA 19-9.[165] We found no relationship between the spread of gynecological malignancies and the serum laminin P1 or 7S domain of type IV collagen concentrations; in fact, almost all of the 104 patients studied had values within the reference range.[18]

H. ENDOCRINOLOGIC DISEASES

1. Growth Disorders and Acromegaly

The availability of recombinant, biosynthetically produced human growth hormone has made it possible to treat both growth hormone-deficient and otherwise short children. At the same time, the need for methods has increased that would help to predict the response to the expensive and tedious therapy as early as possible.

The concentration of the C-terminal propeptide of type I procollagen is abnormally low in growth hormone-deficient children and changes toward normal during treatment with the hormone.[50] The serum concentration of the amino-terminal propeptide of type III procollagen also increases during growth hormone treatment of children.[166-168] Most of the change takes place during the first 5 weeks of treatment; the magnitude of this early change seems to correlate with the growth rate to be observed 6 or 12 months later.[167,168] This finding suggests that PIIINP could be used as an early indicator of peripheral response to growth hormone treatment.

PICP has also been suggested to be useful as an indicator of the growth rate in children with inflammatory bowel diseases.[169] In fact, chronic disease in children often diminishes the growth rate, which will again increase after adequate treatment. Thus, when increased concentrations of PIIINP or PICP are found in fibrosing diseases in children, recovery from the disease can be associated with almost as high values, due to this catch-up growth.[131]

In acromegaly, the excessive production of growth hormone is associated with an elevated serum concentration of PIIINP. This changes toward normal upon medical or surgical treatment.[170]

2. Hyperthyroidism

Thyroid hormones are known to increase the catabolism of collagens. In fact, urinary excretion of hydroxyproline was used as a measure of the functional state of the thyroid before specific assays were available for the hormones of this gland. Specific immunologic methods for the thyroid hormones and their unbound fractions now form the basis of thyroid diagnostics. Quite recently it has become evident, however, that in many situations a parameter reflecting the peripheral response to thyroid hormones would be clinically useful. This is the case, e.g., when hypothyroidism or hyperthyroidism is being treated with thyroxine or with thyreostatic drugs, respectively.

The serum concentration of the amino-terminal propeptide of type III procollagen has been suggested to serve as this kind of peripheral marker, at least in hyperthyroidism.[171] In the only report published so far, a cross-sectional study of ten patients with hyperthyroidism was presented. The PIIINP concentrations were significantly higher than in the reference group, with no overlap between the two groups. It remains to be seen whether hyperthyroidism will become a new indication for the PIIINP assay or any other assay for a specific connective tissue protein.

3. Diabetes Mellitus

The development of diabetic microangiopathy is accompanied by thickening of basement membranes and thus involves accelerated synthesis or decreased turnover of basement membrane material or both. In experimentally diabetic animals, it can easily be shown that the serum concentrations of antigens derived from laminin and type IV collagen (the fragment P2 and the domains 7S and NC1, respectively) are increased.[64,66,172] The change can almost completely be prevented by insulin treatment,[64,66] whereas hypertension further increases at least the 7S domain concentration in serum.[173] The proportion of the large type IV collagen serum antigen, which bears both the 7S and the NC1 determinants, is preferentially increased in this condition, a phenomenon which the authors have interpreted in terms of an increased rate of basement membrane synthesis.[173] The 7S antigen is not found in urine, whereas increased urinary excretion of another antigenic fragment, related to one of the type IV collagen polypeptide chains, has been reported in genetically diabetic mice.[174]

In human diabetics, clinically evident microangiopathy is associated with a modestly elevated serum concentration of the laminin P1 antigen.[165,175,176] The difference between the serum laminin P1 concentrations in diabetics with and without microvascular complications has not been found statistically significant.[175]

The changes observed in the serum concentration of the 7S domain of type IV collagen in human diabetics depend on the assay used. With the assay for the long human 7S antigen, we have noted hardly any increases in diabetic patients. In contrast to this, with the assay involving the bovine short 7S antigen a significantly elevated concentration is seen in diabetics without complications, and a further significant difference between the diabetics with microangiopathy and those without complications.[62,176-178]

No follow-up studies with any of the basement membrane antigens are available on human diabetics. Thus, the usefulness of such assays in predicting the development of microangiopathic complications in individual patients awaits further study. In the case of the laminin P1 antigen, demonstration of such usefulness does not seem probable. Since the differences noted between the groups have been very small, any slow increase is likely to be masked by the interassay variation.

Scleroderma-like fibrotic changes of the skin and limited movement of small joints are not uncommon findings in diabetics. No investigations are currently available on the use of serum indicators of interstitial collagen metabolism in diabetes, although such a study seems indicated in these patients.

V. CONCLUSIONS

The methods that have been described during recent years for the assessment of connective tissue-related compounds in serum have mostly resulted as by-products of basic biochemical studies into the structures of these compounds. Thus, the methods have not been consciously chosen as best possible indicators of changes in connective tissue metabolism in tissues, neither have most of them been developed to meet the requirements of hospital laboratories. The clinical potentials of the assays have been evaluated to some extent, but at the same time it has become clear that in order to be adequately evaluated the method must be available to a great number of clinicians, which seems to require that the assay is distributed commercially. In addition to detecting fibrogenesis, which was the indication originally recognized for such assays, numerous other indications for connective tissue assays have been suggested lately, e.g., within endocrinology and oncology. On the other hand, the fact that changes have also been noted in disease states other than the fibrosing diseases has made the interpretation of the results complicated.

Taking the current knowledge of the biochemistry of the extracellular matrix into account, it seems worthwhile to aim at the assessment of such compounds that can be unequivocally traced back to one certain chemical component, e.g., collagen type. Thus, the recent emphasis on collagen type-specific methodology seems basically correct. Although this does not indicate tissue or organ specificity, the approach leads to better sensitivity than such general markers of collagen turnover as urinary hydroxyproline, and to the possibility of finding out time-courses of events.

As our objective is to make an estimate of the rate of connective tissue accumulation, at least if effective antifibrotic drugs become available that could slow this process down, it seems desirable to compare the rates of synthesis and degradation of connective tissue.[66,87] Although distinguishing between the synthesis and breakdown of, e.g., a certain collagen type is in principle clear-cut, this has turned out to be more problematic in practice. PIIINP in serum is the best example of a marker that was originally introduced as a parameter directly proportional to the rate of collagen synthesis, but that has later been shown to originate to some extent from the tissue form of type III collagen as well. If connective tissue assays will be more consciously developed in the future, different methods should be set up for a strictly precursor-specific sequence of a collagen type and another for the tissue form of the same type. Theoretically, there are a number of candidates for indicators of connective tissue degradation, since collagen molecules contain different domains and are chemically further modified in the mature state, e.g., cross-linked.

There must be profound differences in the basal rates of connective tissue turnover between different tissues and, thus, in their respective contributions to the connective tissue antigens in serum. The turnover rate in the lungs seems to be significantly higher than that in many other tissues.[9] Unfortunately, no corresponding data is available on the turnover rate of the connective tissue in the liver. Further investigations into these problems are clearly warranted.

The routes along which the connective tissue-related components reach the blood are not known. Hyaluronan is an exception, since its passage from loose connective tissue via the lymph into the circulation has been documented. At least in tissues where the endothelial lining of small blood vessels is discontinuous, the connective tissue components could possibly also gain access to the blood directly. Normal liver and bone marrow are examples of tissues with such sinusoidal structures. The large and small forms of certain connective tissue antigens, e.g., PIIINP, may also differ in this respect. Interestingly, the large form of 7S collagen seems to accumulate in blood in diseases of the liver and the bone marrow.

The clearance from circulation and further metabolism of the measured antigens naturally affect their steady-state concentration in blood. Again, hyaluronan is best known in this respect, but a considerable amount of information has recently been obtained on PIIINP as well. As demonstrated by both of these, different forms of the same component can have different clearing mechanisms, making the situation more complex to interpret.

The quality of the answers obtained, when these assays — and those that no doubt will be introduced in the future — are clinically evaluated, depends on the scientific quality of the questions asked. In general, the cross-sectional approach is valuable in the early detection of areas of potential interest, but longitudinal studies with careful analysis of the data from individual patients seem more promising in terms of evaluating the usefulness of an assay. The longitudinal approach is comparable to real clinical situations only, if the interassay variation is sufficiently small. In this respect, most of the assays presented clearly require further development. So far we have little knowledge of interindividual, probably genetically based differences in connective tissue metabolism. Such differences are likely to exist, however, and the serum parameters for fibrogenesis and fibrolysis should be useful in detecting them in carefully chosen clinical situations.

The use of animal experiments in finding answers to distinct clinical questions is complicated by the fact that every immunologic assay, also within one species, seems to have different characteristics. Thus, conclusions must be made on the basis of such animal experiments even more cautiously than otherwise.

Liver fibrosis and cirrhosis are a major health problem both in industrialized and in developing countries. Fibrosis also accompanies diseases of several other organs and is often responsible for the ensuing permanent functional deterioration. It has proved more difficult than perhaps originally expected to establish noninvasive methods that would give the clinician exact information on the rate of connective tissue accumulation in a patient. However, a lot has been learned about the physiology of extracellular matrix components in recent years and development of such noninvasive means still looks possible. Further work in the field is certainly needed, but this medical challenge still seems worth taking.

ACKNOWLEDGMENTS

The studies in the authors' laboratory were supported in part by grants from the Medical Research Council of the Academy of Finland, the Finnish Foundation for Alcohol Studies, and the Finnish Cancer Foundation.

REFERENCES

1. **Peters, J. H., Ginsberg, M. H., Bohl, B. P., Sklar, L. A., and Cochrane, C. G.,** Intravascular release of intact cellular fibronectin during oxidant-induced injury of the in vitro perfused rabbit lung, *J. Clin. Invest.*, 78, 1596, 1986.
2. **Hahn, E. G.,** Blood analysis for liver fibrosis, *J. Hepatol.*, 1, 67, 1984.
3. **Beaugrand, M.,** Quand et comment mesurer la fibrose hépatique?, *Gastroenterol. Clin. Biol.*, 9, 235, 1985.
4. **Hahn, E. G. and Schuppan, D.,** Ethanol and fibrogenesis in the liver, in *Alcohol Related Diseases in Gastroenterology*, Seitz, H. K. and Kommerell, B., Eds., Springer-Verlag, Berlin, 1985, 124.
5. **Gressner, A. M.,** Measurement of connective tissue parameters in serum for diagnosis and follow-up of liver fibrosis, *Ann. Clin. Biochem.*, 24, 283, 1987.
6. **Risteli, L. and Risteli, J.,** Radioimmunoassays for monitoring connective tissue metabolism, *Rheumatology*, 10, 216, 1986.
7. **Risteli, L. and Risteli, J.,** Assessment of type IV collagen and laminin metabolism by radioimmunoassays, in *Renal Basement Membranes in Health and Disease*, Price, R. G. and Hudson, B. G., Eds., Academic Press, London, 1987, 219.
8. **Risteli, L. and Risteli, J.,** Analysis of extracellular matrix proteins in biological fluids, *Methods Enzymol.*, 145, 391, 1987.
9. **Laurent, G. J.,** Dynamic state of collagen: pathways of collagen degradation in vivo and their possible role in regulation of collagen mass, *Am. J. Physiol.*, 252, C1, 1987.
10. **Miyahara, M., Njieha, F. K., and Prockop, D. J.,** Formation of collagen fibrils in vitro by cleavage of procollagen with procollagen proteinases, *J. Biol. Chem.*, 257, 8442, 1982.
11. **Davis, B. H. and Madri, J. A.,** An immunohistochemical and serum ELISA study of type I and III procollagen aminopropeptides in primary biliary cirrhosis, *Am. J. Pathol.*, 128, 265, 1987.

12. **Fisher, L. W., Robey, P. G., Tuross, N., Otsuka, A. S., Tepen, D. A., Esch, F. S., Shimasaki, S., and Termine, J. D.,** The M_r 24,000 phosphoprotein from developing bone is the NH_2-terminal propeptide of the α1 chain of type I collagen, *J. Biol. Chem.*, 262, 13457, 1987.
13. **Van der Rest, M., Rosenberg, L. C., Olsen, B. R., and Poole, A. R.,** Chondrocalcin is identical with the C-propeptide of type II procollagen, *Biochem. J.*, 237, 923, 1986.
14. **Hinek, A., Reiner, A., and Poole, A. R.,** The calcification of cartilage matrix in chondrocyte culture: studies of the C-propeptide of type II collagen (chondrocalcin), *J. Cell Biol.*, 104, 1435, 1987.
15. **Rohde, H., Vargas, L., Hahn, E., Kalbfleisch, H., Bruguera, M., and Timpl, R.,** Radioimmunoassay for type III procollagen peptide and its application to human liver disease, *Eur. J. Clin. Invest.*, 9, 451, 1979.
16. **Rojkind, M.,** The blue glass and the predictive value of serum amino-terminal propeptide of type III procollagen as a marker of liver fibrosis, *Hepatology*, 4, 977, 1984.
17. **Niemelä, O., Risteli, L., Sotaniemi, E. A., and Risteli, J.,** Aminoterminal propeptide of type III procollagen in serum in alcoholic liver disease, *Gastroenterology*, 85, 254, 1983.
18. **Risteli, L., Kauppila, A., Mäkilä, U. -M., and Risteli, J.,** Aminoterminal propeptide of type III procollagen in serum — an indicator of clinical behaviour of advanced ovarian carcinoma?, *Int. J. Cancer*, 41, 409, 1988.
19. **Risteli, J., Niemi, S., Trivedi, P., Mäentausta, O., Mowat, A. P., and Risteli, L.,** Rapid equilibrium radioimmunoassay for the amino-terminal propeptide of human type III procollagen, *Clin. Chem.*, 34, 715, 1988.
20. **Niemelä, O., Risteli, L., and Risteli, J.,** unpublished.
21. **Niemelä, O., Risteli, L., Parkkinen, J., and Risteli, J.,** Purification and characterization of the N-terminal propeptide of human type III procollagen, *Biochem. J.*, 232, 145, 1985.
22. **Bjermer, L., Thunell, M., and Hällgren, R.,** Procollagen III peptide in bronchoalveolar lavage fluid. A potential marker of altered collagen synthesis reflecting pulmonary disease in sarcoidosis, *Lab. Invest.*, 55, 654, 1986.
23. **Raedsch, R., Stiehl, A., Sieg, A., Walker, S., and Kommerell, B.,** Biliary excretion of procollagen type III peptide in healthy humans and in patients with alcoholic cirrhosis of the liver, *Gastroenterology*, 85, 1265, 1983.
24. **Gressner, A. M.,** Aminoterminal propeptides of type III procollagen in human cerebrospinal fluid, *J. Clin. Chem. Clin. Biochem.*, 22, 237, 1984.
25. **Gressner, A. M. and Neu, H. -H.,** Amino-terminal propeptides of type III procollagen in seminal fluid, *Clin. Chem.*, 30, 488, 1984.
26. **Gressner, A. M. and Neu, H. -H.,** N-terminal procollagen peptide and β_2-microglobulin in synovial fluids from inflammatory and non-inflammatory joint diseases, *Clin. Chim. Acta*, 141, 241, 1984.
27. **Rohde, H., Langer, I., Krieg, T., and Timpl, R.,** Serum and urine analysis of the aminoterminal procollagen peptide type III by radioimmunoassay with antibody Fab fragments, *Collagen Relat. Res.*, 3, 371, 1983.
28. **Haukipuro, K., Risteli, L., Kairaluoma, M. I., and Risteli, J.,** Aminoterminal propeptide of type III procollagen in healing wound in humans, *Ann. Surg.*, 206, 752, 1987.
29. **Rohde, H., Bruckner, P., and Timpl, R.,** Immunochemical properties of the aminopropeptide of procollagen type III, *Eur. J. Biochem.*, 135, 197, 1983.
30. **Galambos, M. R., Collins, D. C., and Galambos, J. T.,** A radioimmunoassay procedure for type III procollagen: its use in the detection of hepatic fibrosis, *Hepatology*, 5, 38, 1985.
31. **Niemelä, O.,** Radioimmunoassays for type III procollagen amino-terminal peptides in humans, *Clin. Chem.*, 31, 1301, 1985.
32. **Pierard, D., Nusgens, B. V., and Lapiere, C. M.,** Radioimmunoassay for the aminoterminal sequences of type III procollagen in human body fluids measuring fragmented precursor sequences, *Anal. Biochem.* 141, 127, 1984.
33. **Hartmann, D. et al.,** unpublished.
34. **Ackermann, W., Pott, G., Voss, B., Müller, K. -M., and Gerlach, U.,** Serum concentration of procollagen-III-peptide in comparison with the serum activity of *N*-acetyl-β-glucosaminidase for diagnosis of the activity of liver fibrosis in patients with chronic active liver disease, *Clin. Chim. Acta*, 112, 365, 1981.
35. **Niemelä, O., Risteli, L., Sotaniemi, E. A., and Risteli, J.,** Heterogeneity of the antigens related to the amino-terminal propeptide of type III procollagen in human serum, *Clin. Chim. Acta*, 124, 39, 1982.
36. **Schuppan, D., Dumont, J. M., Kim, K. Y., Hennings, G., and Hahn, E. G.,** Serum concentration of the aminoterminal procollagen type III peptide in the rat reflects early formation of connective tissue in experimental liver cirrhosis, *J. Hepatol.*, 3, 27, 1986.
37. **Savolainen, E.-R., Brocks, D., Ala-Kokko, L., and Kivirikko, K. I.,** Serum concentrations of the N-terminal propeptide of type III procollagen and two type IV collagen fragments and gene expression of the respective collagen types in liver in rats with dimethylnitrosamine-induced hepatic fibrosis, *Biochem. J.*, 249, 753, 1988.

38. **Davis, B. H. and Madri, J. A.,** Type I and type III procollagen peptides during hepatic fibrogenesis. An immunohistochemical and ELISA serum study in the CCl_4 rat model, *Am. J. Pathol.*, 127, 137, 1987.
39. **Wick, G., Brunner, H., Penner, E., and Timpl, R.,** The diagnostic application of specific antiprocollagen sera. II. Analysis of liver biopsies, *Int. Arch. Allergy Appl. Immunol.*, 56, 316, 1978.
40. **Fleischmajer, R., Perlish, J. S., and Timpl, R.,** Collagen fibrillogenesis in human skin, in *Biology, Chemistry, and Pathology of Collagen*, Fleischmajer, R., Olsen, B. R., and Kühn, K., Eds., New York Academy of Sciences, New York, 1985, 246.
41. **Karttunen, T., Sormunen, R., Risteli, L., Risteli, J., and Autio-Harmainen, H.,** Immunoelectron microscopic localization of laminin, type IV collagen and type III pN-collagen in reticular fibers of human lymph nodes, *J. Histochem. Cytochem.*, 37, 279, 1989.
42. **Bowness, J. M., Folks, J. E., and Timpl, R.,** Identification of a substrate site for liver transglutaminase on the aminopropeptide of type III collagen, *J. Biol. Chem.*, 262, 1022, 1987.
43. **Smedsrød, B.,** Aminoterminal propeptide of type III procollagen is cleared from the circulation by receptor-mediated endocytosis in liver endothelial cells, *Collagen Relat. Res.*, 8, 375, 1988.
44. **Bentsen, K. D., Boesby, S., Kirkegaard, P., Hansen, C. P., Jensen, S. L., Hørslev-Petersen, K., and Lorenzen, I.,** Is the aminoterminal propeptide of type III procollagen degraded in the liver? A study of type III procollagen peptide in serum during liver transplantation in pigs, *J. Hepatol.*, 6, 144, 1988.
45. **Gressner, A. M., Kropf, J., and Tittor, W.,** Estimation of the production rates of serum aminoterminal propeptide of type III procollagen and laminin in human fibrotic liver, *J. Clin. Chem. Clin. Biochem.*, 25, 553, 1987.
46. **Bentsen, K. D., Henriksen, J. H., Boesby, S., Hørslev-Petersen, K., and Lorenzen, I.,** Hepatic and renal extraction of circulating type III procollagen aminoterminal propeptide and hyaluronan in pig, *J. Hepatol.*, in press.
47. **Bentsen, K. D., Henriksen, J. H., Bendtsen, F., Hørslev-Petersen, K., and Lorenzen, I.,** Splanchnic and renal extraction of circulating type III procollagen aminoterminal propeptide in patients with normal liver function and in patients with alcoholic cirrhosis, *Hepatology*, in press.
48. **Taubman, M. B., Goldberg, B., and Sherr, C. J.,** Radioimmunoassay for human procollagen, *Science*, 186, 1115, 1974.
49. **Taubman, M. B., Kammerman, S., and Goldberg, B.,** Radioimmunoassay of procollagen in serum of patients with Paget's disease of bone, *Proc. Soc. Exp. Biol. Med.*, 152, 284, 1976.
50. **Carey, D. E., Goldberg, B., Ratzan, S. K., Rubin, K. R., and Rowe, D. W.,** Radioimmunoassay for type I procollagen in growth hormone-deficient children before and during treatment with growth hormone, *Pediatr. Res.*, 19, 8, 1985.
51. **Simon, L. S., Krane, S. M., Wortman, P. D., Krane, I. M., and Kovitz, K. L.,** Serum levels of type I and III procollagen fragments in Paget's disease of bone, *J. Clin. Endocrinol. Metab.*, 58, 110, 1984.
52. **Savolainen, E. -R., Goldberg, B., Leo, M. A., Velez, M., and Lieber, C. S.,** Diagnostic value of serum procollagen peptide measurements in alcoholic liver disease, *Alcohol Clin. Exp. Res.*, 8, 384, 1984.
53. **Parfitt, A. M., Simon, L. S., Villaneuva, A. R., and Krane, S. M.,** Procollagen type I carboxy-terminal extension peptide in serum as a marker of collagen biosynthesis in bone. Correlation with iliac bone formation rates and comparison with total alkaline phosphatase, *J. Bone Mineral Res.*, 2, 427, 1987.
54. **Melkko, J., Risteli, L., and Risteli, J.,** unpublished.
55. **Haukipuro, K., Melkko, J., Risteli, L., Kairaluoma, M., and Risteli, J.,** unpublished.
56. **McDonald, J. A., Broekelmann, T. J., Matheke, M. L., Crouch, E., Koo, M., and Kuhn, C., III,** A monoclonal antibody to the carboxyterminal domain of procollagen type I visualizes collagen-synthesizing fibroblasts. Detection of an altered fibroblast phenotype in lungs of patients with pulmonary fibrosis, *J. Clin. Invest.*, 78, 1237, 1986.
57. **Rohde, H. U., Nowack, H., Becker, U., and Timpl, R.,** Radioimmunoassay for the aminoterminal peptide of procollagen $p\alpha 1$(I)-chain, *J. Immunol. Methods*, 11, 135, 1976.
58. **Rojkind, M. and Pérez-Tamayo, R.,** Liver fibrosis, *Int. Rev. Connect. Tissue Res.*, 10, 333, 1983.
59. **Hahn, E. G., Wick, G., Pencev, D., and Timpl, R.,** Distribution of basement membrane proteins in normal and fibrotic human liver: collagen type IV, laminin and fibronectin, *Gut*, 21, 63, 1980.
60. **Timpl, R.,** Recent advances in the biochemistry of glomerular basement membrane, *Kidney Int.*, 30, 293, 1986.
61. **Risteli, J., Wick, G., and Timpl, R.,** Immunological characterization of the 7-S domain of type IV collagens, *Collagen Relat. Res.*, 5, 419, 1981.
62. **Högemann, B., Voss, B., Pott, G., Rauterberg, J., and Gerlach, U.,** 7S collagen: a method for the measurement of serum concentrations in man, *Clin. Chim. Acta*, 144, 1, 1984.
63. **Niemelä, O., Risteli, L., Sotaniemi, E. A., and Risteli, J.,** Type IV collagen and laminin-related antigens in human serum in alcoholic liver disease, *Eur. J. Clin. Invest.*, 15, 132, 1985.
64. **Risteli, J., Draeger, K. E., Regitz, G., and Neubauer, H. P.,** Increase in circulating basement membrane antigens in diabetic rats and effects of insulin treatment, *Diabetologia*, 23, 266, 1982.

65. **Risteli, J., Rohde, H., and Timpl, R.,** Sensitive assays for 7 S collagen and laminin: application to serum and tissue studies of basement membranes, *Anal. Biochem.*, 113, 372, 1981.
66. **Brocks, D. G., Neubauer, H. P., and Strecker, H.,** Type IV collagen antigens in serum of diabetic rats: a marker for basement membrane collagen biosynthesis, *Diabetologia,* 28, 928, 1985.
67. **Schuppan, D., Besser, M., Schwarting, R., and Hahn, E. G.,** Radioimmunoassay for the carboxy-terminal cross-linking domain of type IV (basement membrane) procollagen in body fluids, *J. Clin. Invest.*, 78, 241, 1986.
68. **Brocks, D. G., Strecker, H., Neubauer, H. P., and Timpl, R.,** Radioimmunoassay of laminin in serum and its application to cancer patients, *Clin. Chem.*, 32, 787, 1986.
69. **Pick-Kober, K. H., Negwer, A., and Gressner, A. M.,** Determination of laminin in serum by a competitive radioimmunoassay directed against the pepsin-resistant fragment P1, *J. Clin. Chem. Clin. Biochem.*, 23, 572, 1985.
70. **Jukkola, A., Risteli, J., Autio-Harmainen, H., and Risteli, L.,** Effects of experimental nephrosis on basement-membrane components and enzymes of collagen synthesis in rat kidney, *Biochem. J.*, 226, 243, 1985.
71. **Hardingham, T. E.,** Structure and biosynthesis of proteoglycans, *Rheumatology,* 10, 143, 1986.
72. **Engström-Laurent, A., Laurent, U. B. G., Lilja, K., and Laurent, T. C.,** Concentration of sodium hyaluronate in serum, *Scand. J. Clin. Lab. Invest.*, 45, 497, 1985.
73. **Frébourg, T., Delpech, B., Bercoff, E., Senant, J., Bertrand, P., Deugnier, Y., and Bourreille, J.,** Serum hyaluronate in liver diseases: study by enzymoimmunological assay, *Hepatology,* 6, 392, 1986.
74. **Tengblad, A., Laurent, U. B. G., Lilja, K., Cahill, R. N. P., Engström-Laurent, A., Fraser, J. R. E., Hansson, H. E., and Laurent, T. C.,** Concentration and relative molecular mass of hyaluronate in lymph and blood, *Biochem. J.*, 236, 5211, 1986.
75. **Laurent, T. C., Dahl, I. M. S., Dahl, L. B., Engström-Laurent, A., Eriksson, S., Fraser, J. R. E., Granath, K. A., Laurent, C., Laurent, U. B. G., Lilja, K., Pertoft, H., Smedsrød, B., Tengblad, A., and Wik, O.,** The catabolic fate of hyaluronic acid, *Connect. Tissue Res.*, 15, 33, 1986.
76. **Smedsrød, B., Pertoft, H., Eriksson, S., Fraser, J. R. E., and Laurent, T. C.,** Studies in vitro on the uptake and degradation of sodium hyaluronate in rat liver endothelial cells, *Biochem. J.*, 223, 617, 1984.
77. **Bentsen, K. D., Henriksen, J. H., and Laurent, T. C.,** Circulating hyaluronate: concentration in different vascular beds in man, *Clin. Sci.*, 71, 161, 1986.
78. **Hällgren, R., Engström-Laurent, A., and Nisbeth, U.,** Circulating hyaluronate. A potential marker of altered metabolism of the connective tissue in uremia, *Nephron,* 46, 150, 1987.
79. **Engström-Laurent, A. and Hällgren, R.,** Circulating hyaluronate in rheumatoid arthritis: relationship to inflammatory activity and the effect of corticosteroid therapy, *Ann. Rheum. Dis.*, 44, 83, 1985.
80. **Engström-Laurent, A., Feltelius, N., Hällgren, R., and Wasteson, Å.,** Elevated serum hyaluronate in scleroderma. An effect of growth factor-influenced activation of connective tissue cells?, *Ann. Rheum. Dis.*, 44, 614, 1985.
81. **Engström-Laurent, A., Lööf, L., Nyberg, A., and Schröder, T.,** Increased serum levels of hyaluronate in liver disease, *Hepatology,* 5, 638, 1985.
82. **Schuppan, D., Rühlmann, T., and Hahn, E. G.,** Radioimmunoassay for human type VI collagen and its application to tissue and body fluids, *Anal. Biochem.*, 149, 238, 1985.
83. **Darnule, T. V., McKee, M., Darnule, A. T., Turino, G. M., and Mandl, I.,** Solid-phase radioimmunoassay for estimation of elastin peptides in human sera, *Anal. Biochem.*, 122, 302, 1982.
84. **Koivu, J., Myllylä, R., Helaakoski, T., Pihlajaniemi, T., Tasanen, K., and Kivirikko, K. I.,** A single polypeptide acts both as the β subunit of prolyl 4-hydroxylase and as a protein disulfide-isomerase, *J. Biol. Chem.*, 262, 6447, 1987.
85. **Pihlajaniemi, T., Helaakoski, T., Tasanen, K., Myllylä, R., Huhtala, M.-L., Koivu, J., and Kivirikko, K. I.,** Molecular cloning of the β-subunit of human prolyl 4-hydroxylase. This subunit and protein disulphide isomerase are products of the same gene, *EMBO J.*, 6, 643, 1987.
86. **Cheng, S. -Y., Gong, Q. -H., Parkison, C., Robinson, E. A., Appella, E., Merlino, G. T., and Pastan, I.,** The nucleotide sequence of a human cellular thyroid hormone binding protein present in endoplasmic reticulum, *J. Biol. Chem.*, 262, 11221, 1987.
87. **Gressner, A. M. and Roebruck, P.,** Predictive values of serum *N*-acetyl-β-D-glucosaminidase for fibrotic liver disorders — correlation with monoamine oxidase activity, *Clin. Chim. Acta,* 124, 315, 1982.
88. **Morelli, A., Vedovelli, A., Fiorucci, S., Angelini, G. P., Palmerini, C. A., and Floridi, A.,** Type III procollagen peptide and PZ-peptidase serum levels in precirrhotic liver diseases, *Clin. Chim. Acta,* 148, 87, 1985.
89. **Kivirikko, K. I.,** Urinary excretion of hydroxyproline in health and disease, *Int. Rev. Connect. Tissue Res.*, 5, 93, 1970.
90. **Robins, S. P.,** Turnover and cross-linking of collagen, in *Collagen in Health and Disease,* Weiss, J. B. and Jayson, M. I. V., Eds., Churchill Livingstone, Edinburgh, 1982, 160.

91. **Niell, H. B., Palmieri, G. M., Neely, C. L., Jr., and McDonald, M. W.,** Postabsorptive urinary hydroxyproline test in patients with metastatic bone disease from breast cancer, *Arch. Intern. Med.*, 141, 1471, 1981.
92. **Elomaa, I., Blomqvist, C., Gröhn, P., Porkka, L., Kairento, A.-L., Selander, K., Lamberg-Allardt, C., and Holmström, T.,** Long-term controlled trial with diphosphonate in patients with osteolytic bone metastases, *Lancet*, 1, 146, 1983.
93. **Gunja-Smith, Z. and Boucek, R. J.,** Collagen cross-linking compounds in human urine, *Biochem. J.*, 197, 759, 1981.
94. **Fujimoto, D., Suzuki, M., Uchiyama, A., Miyamoto, S., and Inoue, T.,** Analysis of pyridinolone, a crosslinking compound of collagen fibres, in human urine, *J. Biochem.*, 94, 1133, 1983.
95. **Robins, S. P.,** An enzyme-linked immunoassay for the collagen crosslink pyridinoline, *Biochem. J.*, 207, 617, 1982.
96. **Robins, S. P., Stewart, P., Astbury, C., and Brid, H. A.,** Measurement of the cross linking compound, pyridinoline, in urine as an index of collagen degradation in joint disease, *Ann. Rheum. Dis.*, 45, 969, 1986.
97. **Bentsen, K. D., Lanng, C., Hørslev-Petersen, K., and Risteli, J.,** The aminoterminal propeptide of type III procollagen and basement membrane components in serum during wound healing in man, *Acta Chir. Scand.*, 154, 97, 1988.
98. **Haukipuro, K., Risteli, L., Kairaluoma, M., and Risteli, J.,** unpublished.
99. **Trivedi, P., Cheeseman, P., Portmann, B., Hegarty, J., and Mowat, A. P.,** Variation in serum type III procollagen peptide with age in healthy subjects and its comparative value in the assessment of disease activity in children and adults with chronic active hepatitis, *Eur. J. Clin. Invest.*, 15, 69, 1985.
100. **Trivedi, P., Risteli, J., Risteli, L., Tanner, M. S., Bhave, S., Pandit, A. N., and Mowat, A. P.,** Serum type III procollagen and basement membrane proteins as noninvasive markers of hepatic pathology in Indian childhood cirrhosis, *Hepatology*, 7, 1249, 1987.
101. **Risteli, L., Puistola, U., Hohtari, H., Kauppila, A., and Risteli, J.,** Collagen metabolism in normal and complicated pregnancy: changes in the aminoterminal propeptide of type III procollagen in serum, *Eur. J. Clin. Invest.*, 17, 81, 1987.
102. **Bieglmayer, C., Feiks, A., and Rudelstorfer, R.,** Laminin in pregnancy, *Gynecol. Obstet. Invest.*, 22, 7, 1986.
103. **Raedsch, R., Stiehl, A., Waldherr, R., Mall, G., Gmelin, K., Götz, R., Walker, S., Czygan, P., and Kommerell, B.,** Procollagen-type III-peptide serum concentrations in chronic persistent and chronic active hepatis and in cirrhosis of the liver and their diagnostic value, *Z. Gastroenterol.*, 20, 738, 1982.
104. **Colombo, M., Annoni, G., Donato, M. F., Conte, D., Martines, D., Zaramella, M. G., Bianchi, P. A., Piperno, A., and Tiribelli, C.,** Serum type III procollagen peptide in alcoholic liver disease and idiopathic hemochromatosis: its relationship to hepatic fibrosis, activity of the disease and iron overload, *Hepatology*, 5, 475, 1985.
105. **Monarca, A., Petrini, C., Perolini, S., Pozzi, F., Adelasco, L., Natangelo, R., and Crode, G.,** Procollagen-type III peptide serum concentrations in alcoholic and non-alcoholic liver disease, *Ric. Clin. Lab.*, 15, 167, 1985.
106. **Hartmann, D. J., Trinchet, J. C., Galet, B., Callard, P., Nusgens, B., Lapiere, C. M., Beaugrand, M., and Ville, G.,** Measurement of serum procollagen type III N terminal propeptide in patients with alcoholic liver disease, in *Marker Proteins in Inflammation*, Vol. 3, Bienvenu, Grimaud, and Laurent, Eds., Walter de Gruyter & Co., Berlin, 1986, 443.
107. **Torres-Salinas, M., Parés, A., Caballeria J., Jiménez, W., Heredia, D., Bruguera, M., and Rodés, J.,** Serum procollagen type III peptide as a marker of hepatic fibrogenesis in alcoholic hepatitis, *Gastroenterology*, 90, 1241, 1986.
108. **Nouchi, T., Worner, T. M., Sato, S., and Lieber, C. S.,** Serum procollagen type III N-terminal peptides and laminin P1 peptide in alcoholic liver disease, *Alcohol Clin. Exp. Res.*, 11, 287, 1987.
109. **Surrenti, C., Casini, A., Milani, S., Ambu, S., Ceccatelli, P., and D'Agata, A.,** Is determination of serum N-terminal procollagen type III peptide (sPIIIP) a marker of hepatic fibrosis?, *Dig. Dis. Sci.*, 32, 705, 1987.
110. **Lu, W., Bantok, I., Desai, S., Lloyd, B., Hinks, L., and Tanner, A. R.,** Aminoterminal procollagen III peptide elevation in alcoholics who are selenium and vitamin E deficient, *Clin. Chim. Acta*, 154, 165, 1986.
111. **Frei, A., Zimmermann, A., and Weigand, K.,** The N-terminal propeptide of collagen type III in serum reflects activity and degree of fibrosis in patients with chronic liver disease, *Hepatology*, 4, 830, 1984.
112. **Sato, S., Nouchi, T., Worner, T. M., and Lieber, C. S.,** Liver fibrosis in alcoholics. Detection by Fab radioimmunoassay of serum procollagen III peptides, *JAMA*, 256, 1471, 1986.
113. **Tanaka, Y., Minato, Y., Hasumura, Y., and Takeuchi, J.,** Evaluation of hepatic fibrosis by serum proline and amino-terminal type III procollagen peptide levels in alcoholic patients, *Dig. Dis. Sci.*, 31, 712, 1986.

114. **Bentsen, K. D., Horn, T., Risteli, J., Risteli, L., Engström-Laurent, A., Hørslev-Petersen, K., Lorenzen, I., and the Copenhagen Study Group on Fibrotic Diseases,** Serum aminoterminal type III procollagen peptide and the 7S domain of type IV collagen in patients with alcohol abuse. Relation to ultrastructural fibrosis in the acinar zone 3 and to serum hyaluronan, *Liver,* 7, 339, 1987.
115. **Colombo, M., Annoni, G., Donato, M. F., Fargion, S., Tiribelli, C., and Dioguardi, N.,** Serum marker of type III procollagen in patients with idiopathic hemochromatosis and its relationship to hepatic fibrosis, *Am. J. Clin. Pathol.,* 80, 499, 1983.
116. **Bolarin, D. M., Savolainen, E. -R., and Kivirikko, K. I.,** Three serum markers of collagen biosynthesis in Nigerians with cirrhosis and various infectious diseases, *Eur. J. Clin. Invest.,* 14, 90, 1984.
117. **Annoni, G., Cargnel, A., Colombo, M., and Hahn, E. G.,** Persistent elevation of the aminoterminal peptide of procollagen type III in serum of patients with acute viral hepatitis distinguishes chronic active hepatitis from resolving or chronic persistent hepatitis, *J. Hepatol.,* 2, 379, 1986.
118. **Bentsen, K. D., Hørslev-Petersen, K., Junker, P., Juhl, E., Lorenzen, I., and the Copenhagen Hepatitis Acuta Programme,** Serum aminoterminal procollagen type III peptide in acute viral hepatitis. A long-term follow-up study, *Liver,* 7, 96, 1987.
119. **Ballardini, G., Faccami, A., Bianchi, F. B., Fallani, M., Patrono, D., Capelli, M., and Pisi, E.,** Steroid treatment lowers hepatic fibroplasia, as explored by serum aminoterminal procollagen type III peptide, in chronic liver disease, *Liver,* 4, 348, 1984.
120. **Weigand, K., Zaugg, P.-Y., Frei, A., and Zimmermann, A.,** Long-term follow-up of serum N-terminal propeptide of collagen type III levels in patients with chronic liver disease, *Hepatology,* 4, 835, 1984.
121. **McCullough, A. J., Stassen, W. N., Wiesner, R. H., and Czaja, A. J.,** Serum type III procollagen peptide concentrations in severe chronic active hepatitis: relationship to cirrhosis and disease activity, *Hepatology,* 7, 49, 1987.
122. **McCullough, A. J., Stassen, W. N., Wiesner, R. H., and Czaja, A. J.,** Serial determinations of the amino-terminal peptide of type III procollagen in severe chronic active hepatitis, *J. Lab. Clin. Med.,* 109, 55, 1987.
123. **Savolainen, E. -R., Miettinen, T. A., Pikkarainen, P., Salaspuro, M. P., and Kivirikko, K. I.,** Enzymes of collagen synthesis and type III procollagen aminopropeptide in the evaluation of D-penicillamine and medroxyprogesterone treatments of primary biliary cirrhosis, *Gut,* 24, 136, 1983.
124. **Smith, A., Haboubi, N. Y., and Warnes, T. W.,** Serum procollagen-III-peptide concentration in primary biliary cirrhosis (PBC), *Gut,* 24, A503, 1983.
125. **Eriksson, S. and Zettervall, O.,** The N-terminal propeptide of collagen type III in serum as a prognostic indicator in primary biliary cirrhosis, *J. Hepatol.,* 2, 370, 1986.
126. **Niemelä, O., Risteli, L., Sotaniemi, E. A., Stenbäck, F., and Risteli, J.,** Serum basement membrane and type III procollagen related antigens in primary biliary cirrhosis, *J. Hepatol.,* 6, 307, 1988.
127. **Bolarin, D. M., Savolainen, E.-R., and Kivirikko, K. I.,** Enzymes of collagen synthesis and type III procollagen amino-propeptide in serum from Nigerians with hepatocellular carcinoma and other malignant diseases, *Int. J. Cancer.,* 29, 401, 1982.
128. **Hatahara, T., Igarashi, S., and Funaki, N.,** High concentrations of N-terminal peptide of type III procollagen in the sera of patients with various cancers, with special reference to liver cancer, *Gann,* 75, 130, 1984.
129. **El-Mohandes, M., Hassanein, H., El-Badwary, N., Voss, B., and Gerlach, U.,** Serum concentration of N-terminal procollagen peptide of collagen type III in schistosomal liver fibrosis, *Exp. Mol. Pathol.,* 46, 383, 1987.
130. **Risteli, J., Søgaard, H., Oikarinen, A., Risteli, L., Karvonen, J., and Zachariae, H.,** Aminoterminal propeptide of type III procollagen in methotrexate-induced liver fibrosis and cirrhosis, *Br. J. Dermatol.,* 119, 321, 1988.
131. **Trivedi, P., Cheeseman, P., Portmann, B., and Mowat, A. P.,** Serum type III procollagen peptide as a noninvasive marker of liver damage during infancy and childhood in extrahepatic biliary atresia, idiopathic hepatitis of infancy and alfal antitrypsin deficiency, *Clin. Chim. Acta,* 161, 137, 1986.
132. **Gressner, A. M. and Tittor, W.,** Serum laminin — its concentration increases with portal hypertension in cirrhotic liver disease, *Klin. Wochenschr.,* 64, 1242, 1986.
133. **Gressner, A. M., Tittor, W., and Negwer, A.,** Serum concentrations of N-terminal propeptide of type III procollagen and laminin in the outflow of fibrotic livers compared with liver-distal regions, *Hepato-gastroenterology,* 33, 191, 1986.
134. **Sotaniemi, E. A., Niemelä, O., Risteli, L., Stenbäck, F., Pelkonen, R. O., Lahtela, J. T., and Risteli, J.,** Fibrotic process and drug metabolism in alcoholic liver disease, *Clin. Pharmacol. Ther.,* 40, 46, 1986.
135. **Brocks, D. G., Bickel, M., and Engelbart, K.,** Type IV collagen antigens in serum of rats with experimental fibrosis of the liver, *Alcohol Alcoholism Suppl.,* 1, 497, 1987.
136. **Clark, J. G., Kuhn, C., III, McDonald, J. A., and Mecham, R. P.,** Lung connective tissue, *Int. Rev. Connect. Tissue Res.,* 10, 249, 1983.

137. **Low, R. B., Cutroneo, K. R., Davis, G. S., and Gianicola, M. S.,** Lavage type III procollagen N-terminal peptides in human pulmonary fibrosis and sarcoidosis, *Lab. Invest.*, 48, 755, 1983.
138. **Aresu, G., Pascalis, L., Pia, G., Rosetti, L., and Giglio, S.,** Procollagen III peptide in the blood and in the fluid of bronchoalveolar lavage of subjects affected by pulmonary fibrosis, *IRCS Med. Sci.*, 14, 871, 1986.
139. **Kirk, J. M. E., Bateman, E. D., Haslam, P. L., Laurent, G. J., and Turner-Warwick, M.,** Serum type III procollagen peptide concentration in cryptogenic fibrosing alveolitis and its clinical relevance, *Thorax*, 39, 726, 1984.
140. **Anttinen, H., Terho, E. O., Myllylä, R., and Savolainen, E. -R.,** Two serum markers of collagen biosynthesis as possible indicators of irreversible pulmonary impairment in farmer's lung, *Am. Rev. Respir. Dis.*, 133, 88, 1986.
141. **Bjermer, L., Engström-Laurent, A., Lundgren, R., Rosenhall, L., and Hällgren, R.,** Hyaluronate and type III procollagen peptide concentrations in bronchoalveolar lavage fluid as markers of disease activity in farmer's lung, *Br. Med. J.*, 295, 803, 1987.
142. **Bégin, R., Martel, M., Desmarais, Y., Drapeau, G., Boileau, R., Rola-Pleszczynski, M., and Massé, S.,** Fibronectin and procollagen 3 levels in bronchoalveolar lavage of asbestos-exposed human subjects and sheep, *Chest*, 89, 237, 1986.
143. **Hällgren, R., Eklund, A., Engström-Laurent, A., and Schmekel, B.,** Hyaluronate in bronchoalveolar lavage fluid: a new marker in sarcoidosis reflecting pulmonary disease, *Br. Med. J.*, 290, 1778, 1985.
144. **Watanabe, Y., Yamaki, K., Yamakawa, I., Takagi, K., and Satake, T.,** Type III procollagen N-terminal peptides in experimental pulmonary fibrosis and human respiratory diseases, *Eur. J. Respir. Dis.*, 67, 10, 1985.
145. **Apaja-Sarkkinen, M., Autio-Harmainen, H., Alavaikko, M., Risteli, J., and Risteli, L.,** Immunohistochemical study of basement membrane proteins and type III procollagen in myelofibrosis, *Br. J. Haematol.*, 63, 571, 1986.
146. **Hochweiss, S., Fruchtman, S., Hahn, E. G., Gilbert, H., Donovan, P. B., Johnson, J., Goldberg, J. D., and Berk, P. D.,** Increased serum procollagen III aminoterminal peptide in myelofibrosis, *Am. J. Hematol.*, 15, 343, 1983.
147. **Vellenga, E., Mulder, N. H., van Zanten, A. K., Nieweg, H. O., and Woldring, M. G.,** The significance of the aminoterminal propeptide of type III procollagen in paroxysmal nocturnal haemoglobinuria and myelofibrosis, *Eur. J. Nucl. Med.*, 8, 499, 1983.
148. **Arrago, J. P., Poirier, O., and Najean, Y.,** Evolution des polyglobulies primitives vers la myélofibrose. Surveillance par le dosage de l'amino-propeptide du procollagène III, *Presse Med.*, 13, 2429, 1984.
149. **Hasselbalch, H., Junker, P., Lisse, I., and Bentsen, K. D.,** Serum procollagen III peptide in chronic myeloproliferative disorders, *Scand. J. Haematol.*, 35, 550, 1985.
150. **Arrago, J. P., Poirier, O., Chomienne, C., D'Agay, M. F. D., and Najean, Y.,** Type III aminoterminal propeptide of procollagen in some haematological malignancies, *Scand. J. Haematol.*, 36, 288, 1986.
151. **Najean, Y., Poirier, O., and Arrago, J. P.,** Interet clinique des dosages du fragment aminoterminal du procollagene III dans les desordres myeloprolifetifs, *Nouv. Rev. Fr. Hematol.*, 29, 123, 1987.
152. **Hasselbalch, H., Junker, P., Lisse, I., Bentsen, K. D., Risteli, L., and Risteli, J.,** Serum markers for type IV collagen and type III procollagen in the myelofibrosis-osteomyelosclerosis syndrome and other chronic myeloproliferative disorders, *Am. J. Hematol.*, 23, 101, 1986.
153. **Hørslev-Petersen, K., Bentsen, K. D., Junker, P., and Lorenzen, I.,** Serum amino-terminal type III procollagen peptide in rheumatoid arthritis, *Arthritis Rheum.*, 29, 592, 1986.
154. **Hørslev-Petersen, K., Bentsen, K. D., Halberg, P., Junker, P., Kivirikko, K. I., Majamaa, K., Risteli, L., Risteli, J., and Lorenzen, I.,** Connective tissue metabolites in serum as markers of disease activity in patients with rheumatoid arthritis, *Clin. Exp. Rheumatol.*, 6, 129, 1988.
155. **Krieg, T., Langer, I., Gerstmeier, H., Keller, J., Mensing, H., Goerz, G., and Timpl, R.,** Type III collagen aminopropeptide levels in serum of patients with progressive systemic scleroderma, *J. Invest. Dermatol.*, 87, 788, 1986.
156. **Fisher, L. W., Gehron Robey, P., Young, M. F., and Termine, J.,** Bone glycoproteins, *Methods Enzymol.*, 145, 269, 1987.
157. **Deftos, L. J., Parthemore, J. G., and Price, P. A.,** Changes in plasma bone Gla protein during treatment of bone disease, *Calcif. Tissue Int.*, 34, 121, 1982.
158. **Urist, M. R. and Hudak, R. T.,** Radioimmunoassay for bone morphogenetic protein in serum: a tissue-specific parameter of bone metabolism, *Proc. Soc. Exp. Biol. Med.*, 176, 472, 1984.
159. **Myllylä, R., Myllylä, V. V., Tolonen, U., and Kivirikko, K. I.,** Changes in collagen metabolism in diseased muscle. I. Biochemical studies., *Arch. Neurol.*, 39, 752, 1982.
160. **Kauppila, A., Puistola, U., Risteli, J., and Risteli, L.,** Aminoterminal propeptide of type III procollagen: a new prognosis indicator in human ovarian cancer, *Cancer Res.*, 49, 1885, 1989.
161. **Scher, H. I. and Yagoda, A.,** Bone metastases: pathogenesis, treatment, and rationale for use of resorption inhibitors, *Am. J. Med.*, 82 (Suppl. 2A), 6, 1987.

162. **Blomqvist, C., Elomaa, I., Virkkunen, P., Porkka, L., Karonen, S. -L., Risteli, L., and Risteli, J.,** Response evaluation of bone metastases in mammary carcinoma. The value of radiology, scintigraphy and biochemical markers of bone metabolism, *Cancer,* 60, 2907, 1987.
163. **Elomaa, I., Risteli, L., and Risteli, J., et al.,** unpublished.
164. **Rochlitz, C., Hasslacher, C., Brocks, D. G., and Herrmann, R.,** Serum concentration of laminin, and course of the disease in patients with various malignancies, *J. Clin. Oncol.,* 5, 1424, 1987.
165. **Hartmann, D. J. and Ville, G.,** Dosage sérique du fragment P1 de la laminine humaine, *Ann. Biol. Clin.,* 45, 166, 1987.
166. **Danne, T., Grüters, A., Schnabel, K., Burger, W., l'Allemand, D., Enders, I., Helge, H., and Weber, B.,** Longterm monitoring of treatment with recombinant human growth hormone by serial determination of type III procollagen related antigens in serum, *Pediatr. Res.,* 23, 167, 1988.
167. **Tapanainen, P., Risteli, L., Knip, M., Käär, M. -L., and Risteli, J.,** Serum aminoterminal propeptide of type III procollagen: a potential predictor of the response to growth hormone therapy, *J. Clin. Endocrinol. Metab.,* 67, 1244, 1988.
168. **Trivedi, P., Hindmarsh, P., Risteli, J., Risteli, L., Mowat, A. P., and Brook, C. G. D.,** Growth velocity, growth hormone treatment and serum concentrations of the aminoterminal propeptide of type III procollagen, *J. Pediatr.,* 114, 225, 1989.
169. **Hyams, J. S., Carey, D. E., Leichtner, A. M., and Goldberg, B. D.,** Type I procollagen as a biochemical marker of growth in children with inflammatory bowel disease, *J. Pediatr.,* 109, 619, 1986.
170. **Verde, G. G., Santi, I., Chiodini, P., Cozzi, R., Dallabonzana, D., Oppizzi, G., and Liuzzi, A.,** Serum type III procollagen propeptide levels in acromegalic patients, *J. Clin. Endorinol. Metab.,* 63, 1406, 1986.
171. **Charrie, A. M., Fleury-Goyon, M. C., Dutey, P., and Tourniaire, J.,** Procollagen-III-peptide and hyperthyroidism, *Lancet,* 2, 1221, 1986.
172. **Hasslacher, C. H., Reichenbacher, R., Gechter, F., and Timpl, R.,** Glomerular basement membrane synthesis and serum concentration of type IV collagen in streptozotocin-diabetic rats, *Diabetologia,* 26, 150, 1984.
173. **Hasslacher, C., Brocks, D., Mann, J., Mall, G., and Waldherr, R.,** Influence of hypertension on serum concentration of type IV collagen antigens in streptozotocin-diabetic and non-diabetic rats, *Diabetologia,* 30, 344, 1987.
174. **Mellon, C. D., Sasser, B. W., and Bowman, B. H.,** Increased concentrations of basement membrane collagen fragment in urine of diabetic mice, *Biochem. Gen.,* 22, 631, 1985.
175. **Hasslacher, C. and Brocks, D. G.,** Serum concentration of laminin in type I diabetic patients with and without microangiopathy, *Transplant. Proc.,* 18, 1534, 1986.
176. **Högemann, B., Voss, B., Altenwerth, F. J., Schneider, M., Rauterberg, J., and Gerlach, U.,** Concentrations of 7S collagen and laminin P1 in sera of patients with diabetes mellitus, *Klin. Wochenschr.,* 64, 382, 1986.
177. **Högemann, B., Voss, B., and Rauterberg, J.,** 7 S-Kollagen — Serumkonzentrationen eines Basalmembranfragments bei Patienten mit Diabetes mellitus, *Med. Welt.,* 36, 809, 1985.
178. **Högemann, B., Balleisen, L., Rauterberg, J., Voss, B., and Gerlach, U.,** Basement membrane components (7S collagen, laminin P1) are increased in sera of diabetics and activate platelets in vitro, *Haemostasis,* 16, 428, 1986.

Chapter 4

THE USE OF MOLECULAR HYBRIDIZATION TECHNIQUES AS TOOLS TO EVALUATE HEPATIC FIBROGENESIS

Mark A. Zern, Mark J. Czaja, and Francis R. Weiner

TABLE OF CONTENTS

I. INTRODUCTION

Collagens are a heterogeneous group of proteins that make up a major portion of the extracellular matrix in addition to proteoglycans and noncollagenous glycoproteins. There are at least ten types of collagens, for which constituent chains are coded by a minimum of 17 genes.[1-5] The collagens are briefly described in Tables 1 through 3 as they have been grouped by Miller.[6] This chapter will attempt to delineate two elements which are crucial to the understanding of collagen synthesis: the level of gene regulation responsible for changes in collagen content and factors that affect collagen gene expression.

Extensive work has illustrated changes in the content of the collagen proteins in normal and pathological states. More recently, molecular hybridization techniques have been employed to evaluate the bases for these changes in collagen expression. Molecular studies of collagen regulation have been done in several organs systems including lung,[7-9] skin,[10-12] and bone.[13-15] This chapter will focus on hepatic fibrosis as a representative system of collagen regulation that lends itself to molecular studies. An increase in collagen content plays a direct role in the pathophysiology of chronic liver disease through its effect on both hepatic structure and function. Classic studies have demonstrated that normal human liver contains approximately equal amounts of types I and III and lesser amounts of type IV (basement membrane-like) and type V collagen.[16] There is recent evidence that type VI collagen is also present in liver.[17] In early human cirrhosis, all of these types of collagen are increased; however, with more advanced liver disease there is a striking increase in type I collagen.[16] There is evidence that at least in some models, this increase in collagen content is due to increased collagen synthesis.[18]

We have evaluated *in vitro* systems, animal models, and human liver samples in an attempt to better understand the molecular bases of hepatic fibrogenesis as it applies to man. The benefit of an *in vitro* study is that it allows for precise manipulation of a system. However, analysis of an *in vitro* system is imperfect since the interactions of cells in culture cannot match the true complexity of the human *in vivo* system. The human studies obviously portray the perfect model, but they do not allow for a clear delineation of individual factors, and experimental manipulation is usually impossible. For example, it is relatively easy to investigate the effects of a steroid on collagen gene expression in a cell line, but this immortal cell clone does not realistically portray the *in vivo* situation, and it is not subject to the enormous complexity of the entire organism. In man we cannot readily sort out the effects of exogenous steroid therapy from other multiple and complex endogenous factors that may also be affecting hepatic collagen regulation *in vivo*. With these disclaimers in mind, this chapter will provide a short overview of the values and pitfalls in using a few molecular techniques in the evaluation of collagen gene expression.

II. LEVELS OF GENE REGULATION

A. STEADY-STATE mRNA LEVELS

Initial investigations of gene regulation often begin with a determination of whether the steady-state mRNA levels of a specific gene correlate with the amount of a specific protein present in a pathophysiological state. For example, since type I collagen protein content is markedly increased in cirrhosis,[16] it is informative to determine whether the level of type I procollagen mRNA is increased. Since DNA must be transcribed into mRNA in order to make the protein, answering such a question is then helpful in making preliminary statements about the expression of a gene product. The increased level of procollagen mRNAs that is associated with an increase in hepatic collagen content in fibrosis[19,20] is strong evidence that an enhanced synthesis of the protein is the primary mechanism for the pathological state, rather than a decrease in the degradation of the protein. Furthermore, evidence for an increase

TABLE 1
Collagens of Group 1

Type	Chain composition	General comments
I	$[\alpha_1(I)]_2\alpha_2(I)$	Most abundant in skin and bone; present in many organs; it forms large banded fibrils; synthesized as procollagen and deposited as collagen
II	$[\alpha_1(II)]_3$	Abundant in cartilage and vitreous humor; it forms small 67-nm banded fibrils; synthesized as procollagen and deposited as collagen
III	$[\alpha_1(III)]_3$	Most abundant and codistributed with type I collagen (except for bone and teeth); it corresponds to some reticulin fibrils; synthesized as procollagen and deposited as procollagen or as a partially processed product that contains part or all of the pN-domain
V	$\alpha_1(V)$, $\alpha_2(V)$, $\alpha_3(V)$	Present in most organs; less abundant than types I, II, and III collagens; small fibrils in close association with smooth muscle cells; its tissue arrangement is unknown; synthesized as procollagen and probably deposited as such; extracellular type V is 1.5 to 2.0 times larger than pepsin-treated type V

Reprinted from Rojkind, M., in *The Liver: Biology and Pathobiology,* 2nd ed., Arias, I. M., Jakoby, W. B., Popper, H., Schachter, D., and Shafritz, D. A., Eds., Raven Press, New York, 1988, 707. With permission.

TABLE 2
Collagens of Group 2

Type	Chain composition	General comments
IV	$\alpha_1(IV)$, $\alpha_2(IV)$	Found only in basement membranes; interrupted helix susceptible to proteolytic attack; molecule synthesized and deposited as procollagen; it has no homology with interstitial collagens; high 3-Hydroxyproline content (gly-3-Hypro-4-Hypro); 7S-domain represents the 60-nm-long NH_2-terminal end of four molecules; it is collagenase resistant; NCl-domain contains 229 amino acid residues; it is a globule with a M_r of 170 kDa; hexameric structure containing two type IV collagen molecules cross-linked at the COOH-terminal end; it may function to align the three α-chains
VI	$\alpha_1(VI)$, $\alpha_2(VI)$, $\alpha_3(VI)$	Small 100-nm-banded fibrils; present in many organs forming part of the loose connective tissue; helical domain $^1/_3$ of the molecule; nonhelical, globular regions ($^2/_3$ of the molecule) at both sides of the helix; it forms dimers and tetramers via end-to-end aggregation; α_3 has a M_r of 200 kDa and originates from a 260-kDa precursor; it is collagenase resistant (?); α_1 and α_2 have a M_r of 140 kDa; no precursor chain detected; they are collagenase sensitive; probably a hybrid molecule containing a collagenous and a noncollagenous domain
VII	Unknown	Major structural component of anchoring fibrils; 450 nm long; present in the subbasal lamina underlying squamous epithelia of ectodermal origin; present in basement membranes of chorioamion and skin but not of placenta and muscle; helical domain 1.0 to 1.5 times larger than type I; pepsin-solubilized shows antiparallel dimeric structures with 60-nm overlap
VIII	$\alpha_1(VIII)$	Produced by some endothelial cells; (60% of total collagens produced); α-chains with a M_r of 61 kDa, β-chains with M_r of 124 kDa, and γ-chains with M_r of 194 kDa; the helical domain occupies $^2/_3$ of the molecule; no S-S cross-links; very sensitive to proteolytic attack

Reprinted from Rojkind, M., in *The Liver: Biology and Pathobiology,* 2nd ed., Arias, I. M., Jakoby, W. B., Popper, H., Schachter, D., and Shafritz, D. A., Eds., Raven Press, New York, 1988, 707. With permission.

TABLE 3
Collagens of Group 3

Type	Chain composition	General comments
IX	Unknown	Known as HMW (high molecular weight) or M_1, and LMW (low molecular weight) or M_2; minor components of cartilage; several nonhelical domains susceptible to proteolytic attack; it is assembled from three genetically distinct chains; the three HMW chains correspond to the NH_2-terminal domain, and the three LMW chains correspond to the COOH-terminal domain; it contains S-S cross-links; the NH_2-terminal end is a knob-like structure; it contains chondroitin sulfate side chains
X	Unknown	Known as type G or short chain; short helix 150 nm long; globular domain at one end of the molecule; the arrangement in the matrix is unknown; major product of cultured chondrocytes and of mineralizing cartilage

Reprinted from Rojkind, M., in *The Liver: Biology and Pathobiology*, 2nd ed., Arias, I. M., Jakoby, W. B., Popper, H., Schachter, D., and Shafritz, D. A., Eds., Raven Press, New York, 1988, 707. With permission.

in the steady-state levels of procollagen mRNA would indicate that an increase in transcription of the gene or a decreased degradation of the mRNA was crucial in the enhanced expression of the collagen gene in fibrosis. However, determining the changes in steady-state mRNA levels of a specific gene does not provide information as to whether other levels of gene expression are also affected. It does not investigate variation in translational efficiency, posttranslational modification, or protein transport and degradation. A steady-state mRNA determination does, however, provide a basis for further studies of the mechanisms of collagen regulation in a pathophysiological state.

During the past 2 decades, progress in our understanding of the basic principles of nucleic acid structure and interaction has made it possible to use recombinant DNA probes to study gene organization and function. Molecular hybridization is a reaction between two molecules of DNA or one molecule of DNA and RNA, based on the ability of complementary nucleotides in these molecules to interact and form stable base pairs by hydrogen bonding (adenine [A] recognizes thymidine [T] or uracil [U], and guanine [G] recognizes cytosine [C]). This bonding is the same principle which forms the basis of the Watson-Crick structure of DNA, the "double helix". Molecular hybridization can precisely quantitate the content of a specific mRNA sequence and is not dependent on an assay requiring biological activity. This represents an advantage over assays utilizing cell-free protein synthesis, since the latter may be highly variable and is not quantitative.[21,22]

If a radiolabeled cDNA (complementary DNA) probe is prepared from a purified unique mRNA species, hybridization of this cDNA to a total RNA fraction quantitates sequences in the total RNA that are in turn complementary to the DNA. When this recombinant, double-stranded cDNA is to be utilized as a probe, it must be labeled with a radioactive compound. One such reaction is termed nick-translation,[23] in which labeled dXTPs are substituted enzymatically for nucleotides in the DNA by bacterial DNA polymerase I. This method produces DNA probes with high radioactive activity (more than 10^8 cpm/μg DNA). Such probes can detect less than 1 pg of a specific nucleic acid sequence (DNA or RNA). Thus, molecular hybridization can be performed with [^{32}P]-labeled probes under conditions which are 1000 times more sensitive on a weight basis than standard radioimmunoassays for the detection of specific proteins.

The technique which is most frequently employed to detect and quantitate the mRNA content of a specific gene is called "Northern" blot hybridization (Figure 1). Total RNA is extracted from the cells or tissue of interest by using materials and procedures which limit

SIZE CHARACTERIZATION OF SPECIFIC mRNA BY MOLECULAR HYBRIDIZATION

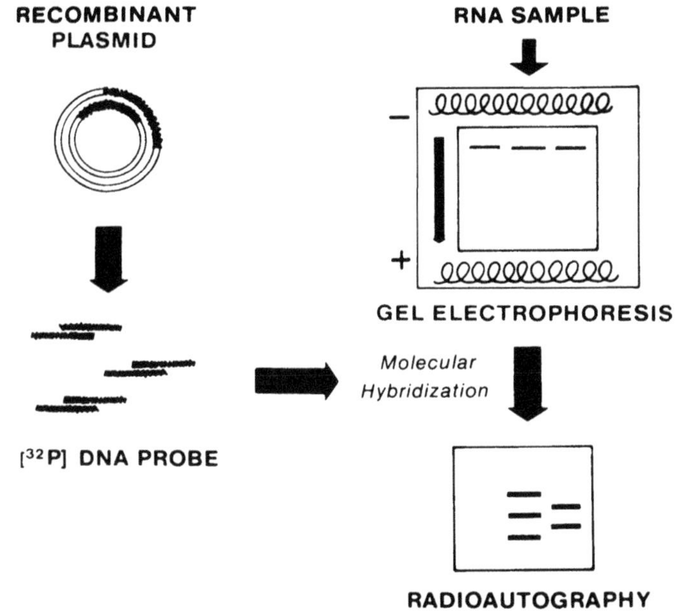

FIGURE 1. The Northern blot hybridization technique that assays steady-state mRNA levels of specific genes. Total RNA is extracted from tissue or cells, separated by size employing agarose gel electrophoresis, then transferred (or blotted) to a solid matrix. This blot is then hybridized with a radiolabeled chimeric bacterial plasmid which contains the cDNA of interest. The radiolabeled bands can then be detected by autoradiography.

the presence and activity of RNase, a ubiquitous enzyme which can degrade the RNA sample. A frequently employed approach is homogenizing the tissue in guanidine thiocyanate (a strong denaturing agent) and then separating the components of the homogenate by centrifuging the homogenate through a CsCl gradient.[24] This technique yields high quantities of relatively undegraded RNA which is then denatured and electrophoresed in an agarose gel to separate the RNA by size. The RNA from the gel is then "blotted" or transferred to a solid matrix, baked onto the matrix, hybridized with the labeled, denatured DNA probe, and washed under stringent conditions to remove nonspecific binding. Because the radiolabeled DNA probe binds only to the specific mRNA to which it is complementary, the blot is then exposed to X-ray film to detect the amount of the particular mRNA. The hybridization signal can be assessed in a semiquantitative manner by densitometry scanning of the developed X-ray film.

Another approach to quantitating specific mRNA species is to employ the "dot" hybridization assay. The extracted RNA is serially diluted in the denaturing agent formaldehyde before being dotted directly onto the solid matrix. The blot containing the dotted RNA is then baked, hybridized, washed, and autoradiography performed exactly as are the Northern blots. The advantage of the dot blot is that it provides a somewhat better means of quantitating mRNA levels but is somewhat less specific because the size of the mRNA species is not determined. We have employed both the Northern and dot blot methods in our studies of collagen gene regulation.[19,20,25]

A major problem in assessing the molecular events associated with the development of cirrhosis in man is determining an appropriate and feasible model system to study. Murine schistosomiasis is an acceptable model from several perspectives. It is inexpensive, fibrosis develops after only 8 weeks of infection, and schistosomiasis may be the most common worldwide cause of chronic liver disease in man.[26] The eggs of *Schistosoma* cause an inflammatory response when trapped in liver sinusoids, leading to granuloma formation and ultimately to hepatic fibrosis. When mice are infected with *S. mansoni,* they consistently develop hypoalbuminemia[27] and a massive increase in liver collagen.[28] Therefore, murine schistosomiasis is an appropriate model to study gene regulation of collagen, albumin, and other hepatic proteins in the presence of this fibrogenic stimulus. Studies of schistosomiasis by Takahashi et al. suggest that the major change in hepatic fibrosis is increased collagen synthesis and not diminished collagenase activity.[18] In our initial studies, we demonstrated that the increase in hepatic collagen content in this model is associated with an increase in type I procollagen mRNA content.[20] Our use of this model to further characterize the fibrogenic process is described in later portions of this chapter. The immunologic injury caused by schistosomiasis, however, differs significantly from the hepatotoxic effects of chronic ethanol exposure, which is the predominant form of liver disease seen in developed countries. Therefore, in an attempt to further delineate the molecular mechanisms responsible for the increased collagen content in human liver, we evaluated the changes in the type I procollagen mRNA content in the livers of ethanol-fed baboons compared with their pair-fed controls.

Another advantage of using the baboon model[29] of chronic ethanol administration is that the animals develop a wide spectrum of liver disease from fatty liver to cirrhosis. We investigated animals both at the fatty liver stage and those with significant fibrosis in order to evaluate the events that lead to the development of cirrhosis. We used percutaneous liver biopsies as the method for obtaining liver samples for histology and molecular studies because this approach could then be applied to human investigations. Using percutaneous biopsy specimens makes sampling error a possibility, but alcoholic liver disease is known to affect the liver in a relatively uniform manner, thus minimizing the possibility of error.

Percutaneous liver biopsy specimens were obtained from 11 ethanol-fed baboons and their pair-fed controls.[19] The specimens were then examined histologically, and RNA was extracted. Six of the ethanol-fed animals developed fatty liver, and no significant differences were found when the RNA from the control and fatty livers was analyzed with specific cDNA probes. The other five baboons given ethanol developed significant fibrosis, and the type I procollagen mRNA content from their livers was significantly higher per microgram of liver RNA as determined by hybridization analysis. A representative Northern and dot blot showing increased pro αl(I) collagen mRNA levels in the fibrotic liver of an ethanol-fed animal is shown in Figure 2. Control genes were also analyzed. There were higher levels of albumin mRNA in the livers of ethanol-fed baboons that developed fibrosis. There was no change, however, in the mRNA levels of β-actin, a representative constitutive protein. These findings indicate that in fibrotic animals, ethanol consumption increases type I procollagen mRNA which may account for the increase in fibrosis. These studies provide further evidence that chronic ethanol ingestion does not injure the protein synthetic machinery. In addition, these results demonstrated that the percutaneous needle biopsy procedure can supply sufficient tissue to evaluate molecular changes in the collagen content of human liver disease.

Utilizing the techniques refined in the baboon study, we extracted total RNA from unused portions of percutaneous liver biopsies from patients with a variety of liver abnormalities.[25] Figure 3 is an autoradiograph of a dot blot which shows a variation in the hybridization signal when RNA from different human biopsies was serially diluted and hybridized with a type I procollagen cDNA probe. It demonstrates an increased hybridization signal for procollagen mRNA in those patients with active liver disease when compared to the control. The quantitative representation of these data appears in Table 4. This table

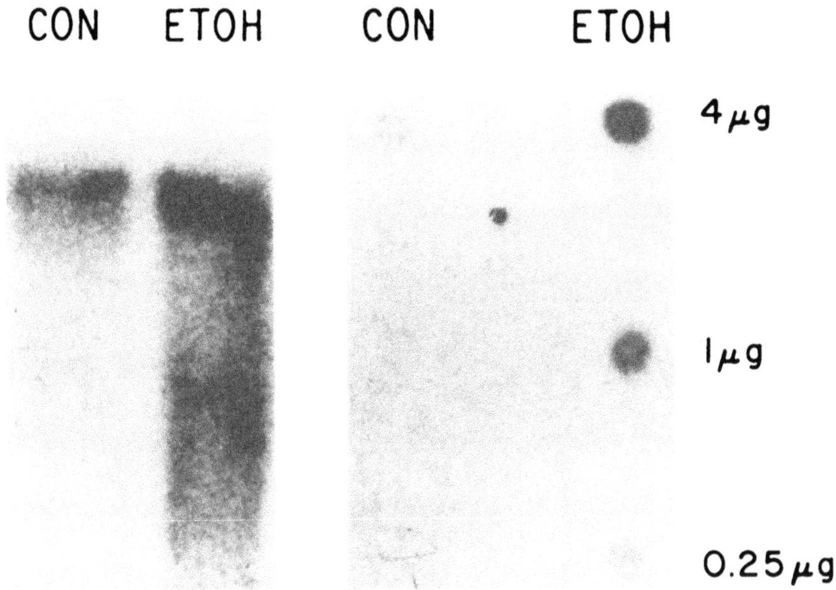

FIGURE 2. "Northern" hybridization and "dot" blot hybridization that identify pro αl(I) collagen mRNA in total RNA isolated from livers of control (CON) and ethanol (ETOH)-fed baboons that developed fibrotic liver. Total RNA was extracted from liver samples using guanidine thiocyanate. The RNA was then used in a "Northern" transfer or spotted (in the amount indicated) on Gene-Screen matrix and hybridized with a human pro αl(I) collagen cDNA probe. Both hybridization techniques demonstrate more procollagen mRNA in the RNA from the ethanol-fed animals. (Reprinted from Zern, M. A., Leo, M. A., Giambrone, M. -A., and Lieber, C. S., *Gastroenterology*, 89, 1123, 1985. With permission from Elsevier Science Publishing, Inc., Copyright 1985 by The American Gastroenterological Association.)

compares the clinical data in five of the patients with the molecular findings for procollagen mRNA. The patient (J.N.) with an essentially normal biopsy is considered the control. In general, the procollagen mRNA levels tended to correlate with the activity of chronic liver disease and the active formation of fibrosis. For example, the one patient (J.A.) with "inactive" cirrhosis had levels of procollagen mRNA only slightly higher than normal, whereas the patients with chronic active hepatitis had substantially more procollagen mRNA. The increased levels of type I procollagen mRNA content found in the livers of baboons and humans with active liver fibrosis are not surprising. Nor is it surprising that a patient with inactive cirrhosis had procollagen mRNA levels similar to patients with essentially normal histology. The lack of active inflammation suggests limited hepatocyte injury and subsequent diminution of procollagen mRNA content and new collagen deposition.

Many studies of hepatic protein synthesis and fibrogenesis have been performed in cell culture or in isolated perfused organ systems. The advantage of evaluating human liver tissue directly is clear. The use of RNA-DNA hybridization in man provides a useful tool for evaluating changes caused in specific genes by pathophysiological states and the future likelihood for developing fibrosis. Finally, the techniques we have described have great potential application in the evaluation of small human specimens from other organ systems.

B. *IN SITU* HYBRIDIZATION: LOCALIZATION OF mRNA

The information obtained from RNA-DNA hybridization studies is limited in that the RNA is extracted from the entire tissue, and it is difficult to determine which cell type is synthesizing the protein of interest. Therefore, besides employing RNA extraction and RNA-

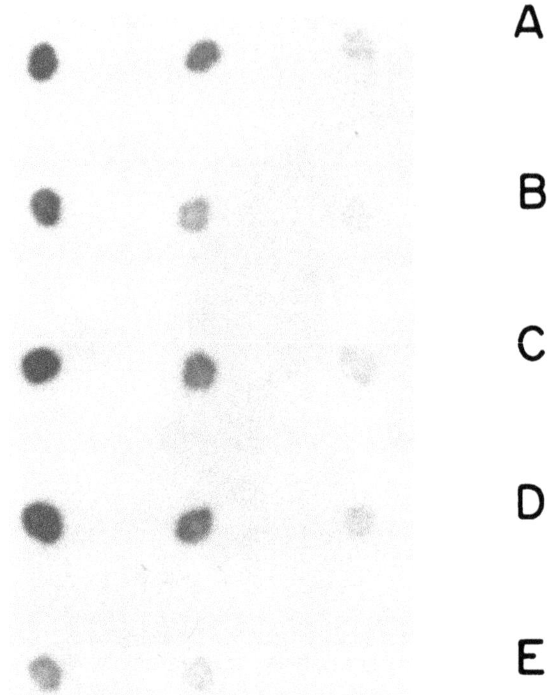

FIGURE 3. A "dot" blot assay of type I procollagen mRNA content in total RNA from liver biopsy specimens of humans with liver dysfunction. Total RNA was extracted, spotted in decreasing concentrations on Gene Screen filters and hybridized with a human pro αl(I) collagen cDNA probe. Each row represents RNA from a different patient.

TABLE 4
Clinical Information Correlated with Densitometry Tracings of Total RNA Hybridized with Collagen cDNA Probe

Patient	Clinical data	Liver histology	Serum albumin (mg/dl)	Collagen mRNA (% of control)[a]
J. N.	Psoriasis, methotrexate, ETOH^{-b}	Mild fatty changes	4.4	100
J. A.	HBcAb^{+c} ETOH^{+}	Fatty changes, inactive cirrhosis	3.3	149
R. F.	Sarcoidosis, ETOH^{-}	Granulomatous hepatitis, slight fibrosis	4.3	195
B. H.	HBsAg^{-d}, ETOH^{-}, chronic active hepatitis	Chronic active hepatitis, fibrosis	3.7	267
R. J	HBsAg^{+}, ETOH^{-}	Chronic active hepatitis, fibrosis	4.2	698

[a] Control considered to be J. N. with essentially normal biopsy.
[b] ETOH = ethanol.
[c] HBcAb = hepatitis B core antibody.
[d] HBsAg = hepatitis B surface antigen.

Reprinted from Weiner, F. R., Czaja, M. J., Wu, C. H., Wu, G. Y., Giambrone, M. -A., and Zern, M. A., *Hepatology*, 7, 19s, 1987. With permission from the American Association for the Study of Liver Diseases.

DNA hybridization, other molecular techniques are presently being utilized to better determine the effects on the liver of various pathophysiological states. One such methodology is *in situ* hybridization.

Investigations of which cell(s) is primarily responsible for the production of collagen in normal and fibrotic liver have yielded conflicting results.[30-41] Fibroblasts,[30] adipocytes,[31,32] myofibroblasts,[33] smooth muscle-like cells,[34] Kupffer cells,[35] and hepatocytes[36-41] have all been shown to produce collagen or be associated with increased fibrogenesis in the liver. *In situ* hybridization is a useful technique to resolve this issue because it can localize mRNAs for collagens or other unique proteins in a highly sensitive and specific manner. It detects the presence of mRNA in intact tissue or in isolated cells prior to culture, at a time before alterations in subcellular mechanisms or dedifferentiation can occur.

In situ hybridization involves the same base-pair interactions as employed in the Northern blot hybridization studies. Radiolabeled, denatured cDNA probes are hybridized directly with the mRNA in the cell. Thus, the base-pairing occurs in the cytoplasm of the cell, rather than on a solid matrix as with Northern blots. Following stringent washes and dehydration, the cells or tissue sections fixed on microscope slides are directly coated with a radiation-sensitive emulsion and developed following days to weeks of exposure. The sections are then histologically stained, and the autoradiographic grains can be counted in standardized fields as a means of semiquantitating specific mRNA content. Stringent washing of the hybridized cells as well as the use of appropriate controls are essential to insure the elimination of nonspecific hybridization. The advantage of the technique is that individual cells can be evaluated for the presence of a specific mRNA. The disadvantage is that the cellular architecture must be somewhat disrupted to allow for the penetration of the probe into the cell.

In situ hybridization studies have been done with isolated hepatocytes as well as with tissue samples from mice infected with *S. mansoni* and controls.[40,41] Hepatocytes were isolated from infected or normal liver and then used for *in situ* hybridization analysis. As shown in Figure 4, individual cells vary significantly in the amount of hybridization signal they exhibit (i.e., they differ significantly in the amount of the specific mRNA they possess). Whereas hepatocytes from infected mice showed significantly less albumin mRNA than control hepatocytes, there were more grains of procollagen mRNA in hepatocytes from infected as compared to control cells.[41] Hybridization of infected liver tissue sections with the collagen probe showed more grains per field in granulomas than in regions of normal liver (see Figure 5), whereas with the albumin probe, again useful as a control gene, there was more hybridization in liver tissue than in granulomas.[41] These results suggest that although the granuloma seems to be the primary source of type I collagen synthesis, hepatocytes are also capable of synthesizing collagen, especially under fibrogenic stimulation.

These studies which were done originally in a mouse model can readily be extended to man. *In situ* hybridization can allow for the localization of procollagen mRNA content in the cells of liver biopsy specimens taken directly from patients. Therefore, this technique can provide direct evidence for the cell type most responsible for the synthesis of specific collagen types in the human liver.

C. TRANSCRIPTIONAL RATES OF GENES

Another limitation of Northern hybridization analysis is that although it is useful for obtaining steady-state levels of specific mRNAs in the liver, a more precise delineation of the level of gene regulation requires further analysis. Steady-state mRNA levels can be affected by differences in transcriptional rates (the rate at which polymerase II makes a complementary RNA copy of the gene), processing and transport of the primary RNA transcripts (heterologous RNA or hnRNA), or mRNA degradation. Frequently, the further characterization of changes induced in steady-state mRNA levels focuses on variations in

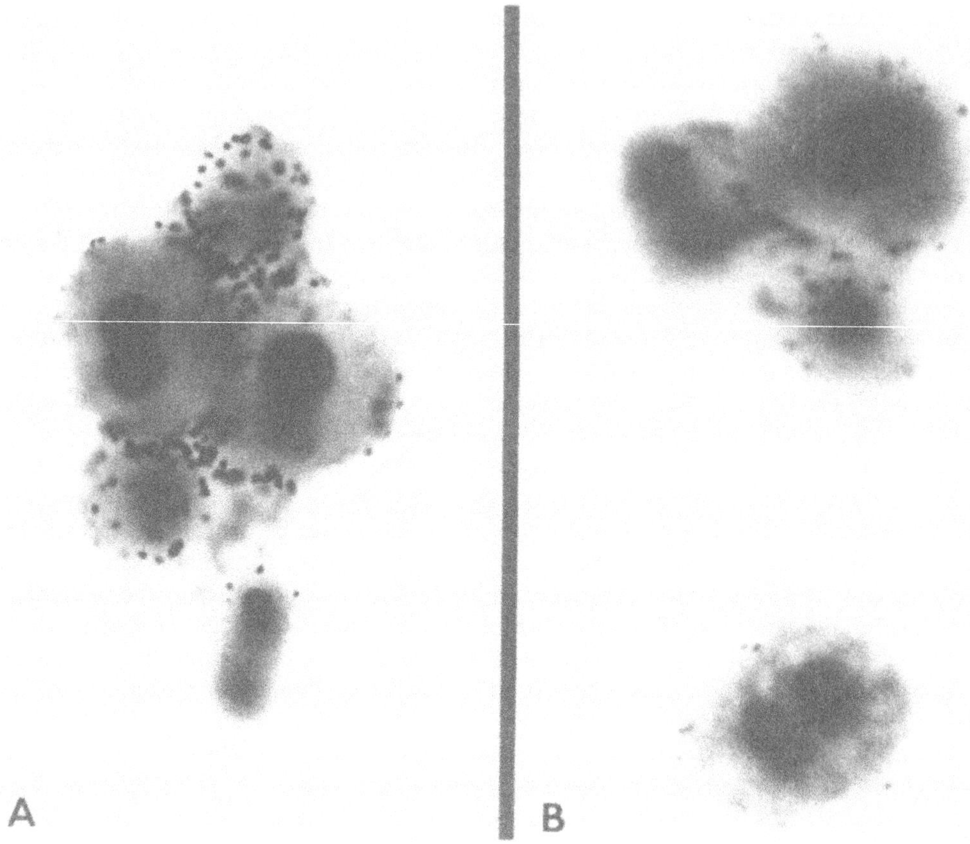

FIGURE 4. *In situ* hybridization of isolated mouse hepatocytes with [^{3}H]-labeled cloned albumin cDNA. Hepatocytes were isolated from mouse livers after *in situ* perfusion by collagenase digestion, sedimentation, and resuspension. In microcentrifuge tubes, cells were fixed, treated with hydrochloric acid, and dehydrated prior to hybridization with a [^{3}H]albumin cDNA probe. After hybridization the cells were washed extensively and mounted on a microscope slide prior to dehydration and evaluation by autoradiography. (A) Hepatocytes from liver of normal mouse; (B) hepatocytes from liver of schistosoma-infected mouse. (Original magnification × 400.) Note that hepatocytes from the normal liver contain significantly more autoradiographic grains, indicating more albumin mRNA, than hepatocytes from an infected liver. (Reprinted from Saber, M. A., Shafritz, D. A., and Zern, M. A., *J. Cell Biol.*, 97, 986, 1983. With permission from The Rockefeller University Press.)

transcriptional rates. This is because current evidence suggests that transcription is the most common level of gene regulation in eukaryotes.[42] In previous studies, for example, cell culture systems have shown transcriptional regulation of type I collagen.[43] It is important to note, however, that evidence for transcriptional regulation does not preclude other levels of regulation.

The nuclear run-on assay has been employed to investigate transcriptional rates of the collagens and other hepatic genes.[44-48] This technique actually measures elongation of previously initiated hnRNA transcripts, but it remains that most widely used and reproducible means of evaluating transcriptional rates. The methodology is delineated in Figure 6. First, nuclei are isolated from liver samples (or from cultured cells). Transcription within these nuclei is then allowed to take place in the presence of [^{32}P]-UTP. The nuclei are lysed, and the labeled RNA transcripts are isolated and hybridized to an excess of the cDNAs of interest

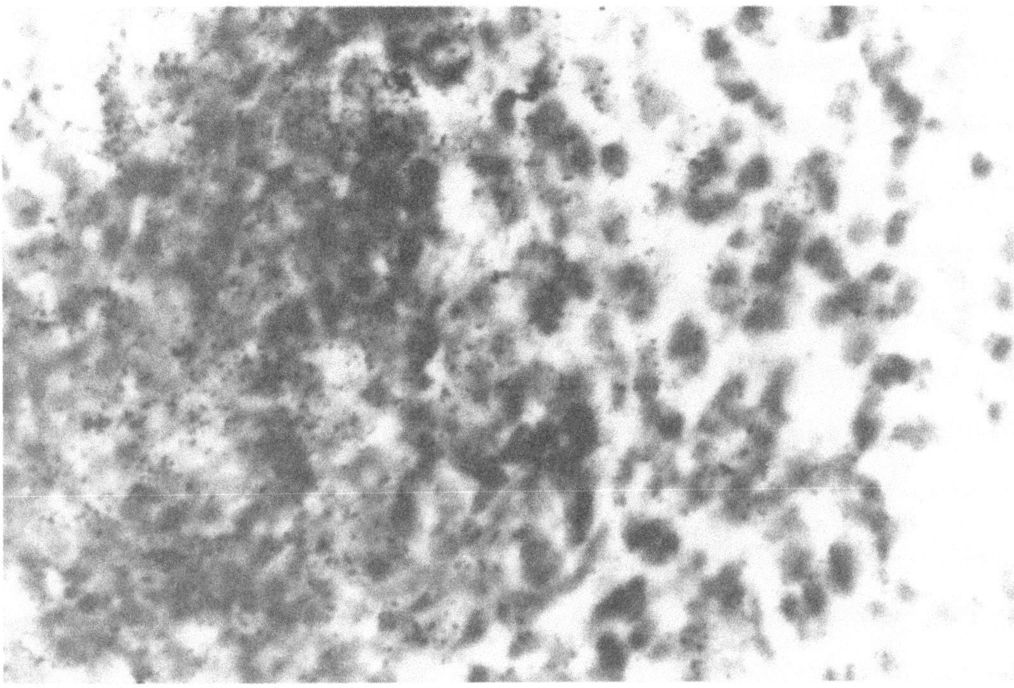

FIGURE 5. *In situ* hybridization of liver tissue from infected mouse liver with a[^{3}H]-labeled pro α2(I) collagen probe. Sections were prepared by *in situ* perfusion of the liver and cryostat sectioning. The left side of the picture shows the granuloma tissue with more silver radiographic grains than are found in normal liver tissue on the right. (Original magnification × 100.) (Reprinted from Saber, M. A., Shafritz, D. A., and Zern, M. A., *J. Cell Biol.*, 97, 986, 1983. With permission from The Rockefeller University Press.)

which have been bound to a nitrocellulose membrane. Finally, the membranes containing the hybridized, radiolabeled transcripts are washed and exposed to X-ray film.

In order to evaluate the effects of schistosomiasis on the transcriptional rates of several hepatic genes,[45] control mice and littermates infected with *S. mansoni* were evaluated at 8 weeks postinfection, the time of maximum collagen synthesis.[18] First, RNA was extracted and hybridized with cDNA probes. Dot blot hybridization assays indicated that there were threefold increases in types I and IV procollagen and β-actin steady-state mRNA levels in infected liver, while albumin mRNA content was only 24% of control levels. When transcriptional rates were then evaluated by the nuclear run-on assay, it appeared that all the changes in steady-state levels could be accounted for by a transcriptional level of regulation. Figure 7 is a representative transcription assay.

We have adapted the nuclear run-on assay and have begun to apply it to the evaluation of human liver samples. While we initially employed this technique to evaluate the regulation of the ceruloplasmin gene in Wilson's disease,[49] the technique can readily be used to study matrix gene regulation. In an attempt to better delineate the level of gene expression responsible for a decrease in the steady-state levels of ceruloplasmin mRNA in Wilson's disease patients, the nuclear run-on assay was used to analyze transcriptional rates by modifying previously described techniques.[50] Nuclei were extracted from portions of human livers that had been harvested prior to transplantation, then frozen at −70°C for periods of 1 year or more. Despite this long period of freezing, the nuclei were still able to actively transcribe hnRNA (an example of an assay is shown in Figure 8). The amount of ceruloplasmin gene transcription in four Wilson's patients was decreased to 44% that of three controls. These results indicate that nuclear run-on assays can be readily done in livers that

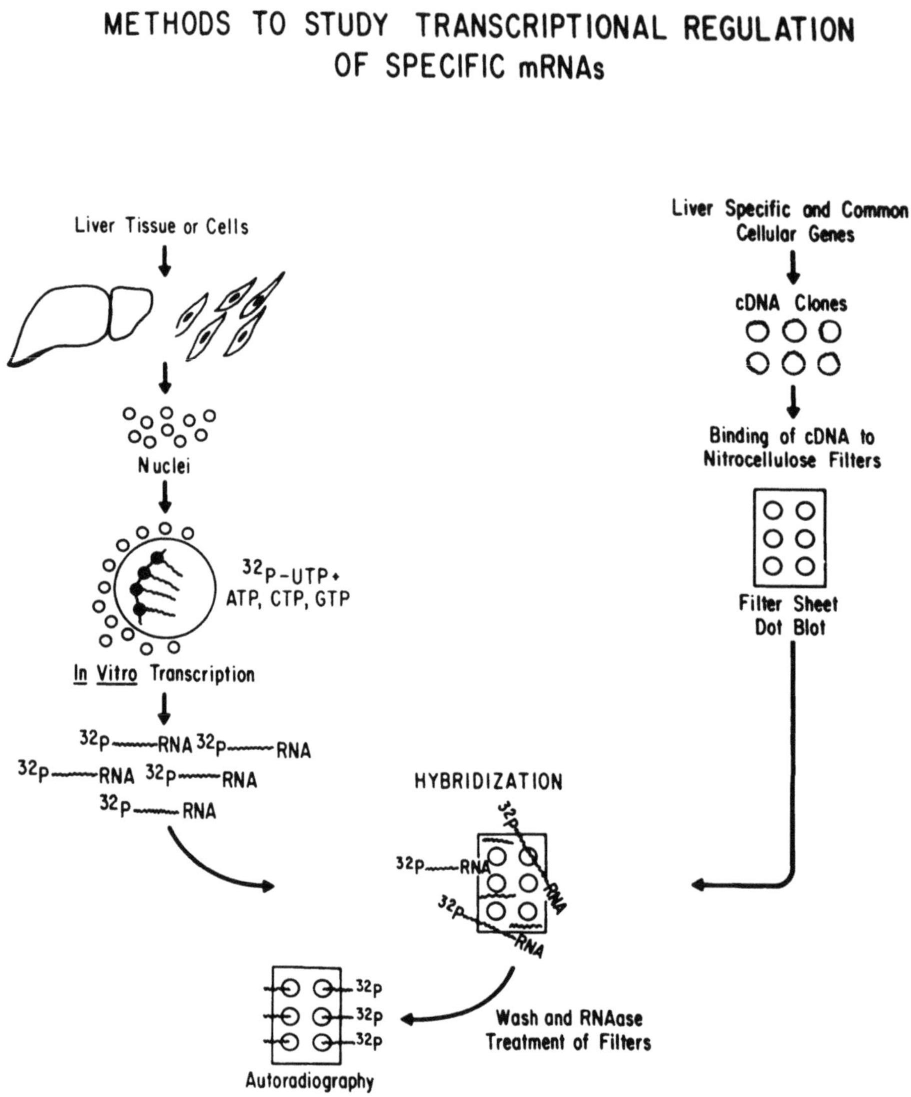

FIGURE 6. Nuclear run-on assay that evaluates transcriptional rates of specific genes. Nuclei are isolated from liver tissue or cells. RNA transcription within these nuclei is then allowed to take place in the presence of radiolabeled-UTP. The labeled RNA transcripts are then isolated and hybridized with the specific cDNA probes which are bound to a nitrocellulose membrane. After washing and RNase treatment, autoradiography is performed on the membrane containing the hybridized, radiolabeled transcripts.

have been frozen for prolonged periods, thus making analysis of human hepatic collagen gene transcription a realistic possibility.

III. FACTORS THAT REGULATE HEPATIC COLLAGEN SYNTHESIS

The liver is composed of a variety of cells organized on a scaffold of extracellular matrix. A large number of factors help maintain the proper equilibrium of matrix deposition and remodeling. The purpose of the next portion of this chapter is to assess how molecular techniques may be used to evaluate the factors or signals that disrupt this homeostatic system

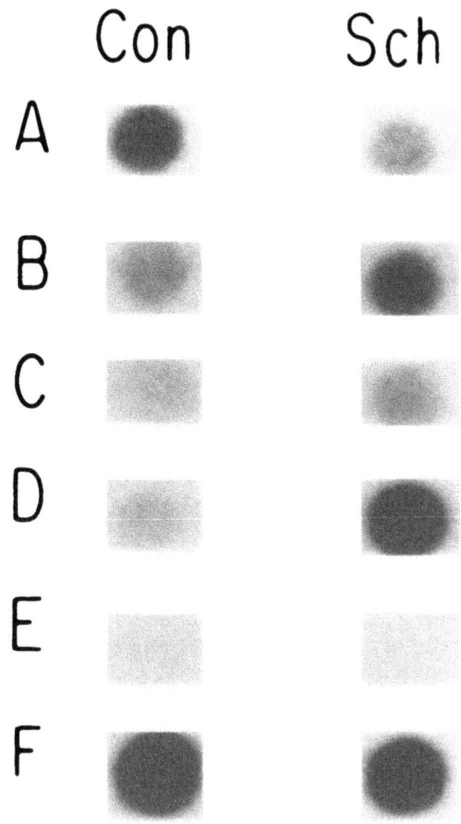

FIGURE 7. Autoradiograph of specific gene transcriptional rate analysis in mouse liver (nuclear run-on assay). Nascent labeled nuclear RNA transcripts from whole liver of control (Con) and infected mice (Sch) were isolated, hybridized with 5 μg of specific cDNAs on nitrocellulose, and autoradiographed. Analysis of transcription of the following genes was performed: (A) albumin; (B) type I procollagen; (C) type IV procollagen; and (D) β-actin. *Escherichia coli* plasmid pBR-322 DNA (E) was used to determine background activity, and rat t-arginine (F) was used to determine that total transcription between conditions was equal. Livers of mice infected with *S. mansoni had increased transcriptional rates of types I and III procollagen and β-actin; they had diminished rates of albumin transcription. (Reprinted from Weiner, F. R., Czaja, M. J., Giambrone, M. -A., Takahashi, S., Biempca, L., and Zern, M. A., Biochemistry,* 26, 1557, 1987. With permisson from the American Chemical Society.)

during the development of cirrhosis. We hypothesize that (1) there are several endogenous effector molecules that normally enhance the deposition of the matrix (such as the cytokine transforming growth factor-beta 1 [TGF-β1]), and other endogenous factors that normally inhibit this process (such as γ-interferon and corticosteroids); and (2) during hepatic injury, the balance is shifted so that there is a relative or absolute superabundance of the fibrogenic factors. Subsequently, fibrosis and cirrhosis intervene.

A. FACTORS THAT ENHANCE HEPATIC FIBROGENESIS AND COLLAGEN REGULATION

Despite extensive investigation, the underlying mechanisms responsible for the initiation of increased matrix synthesis in cirrhosis are largely unknown. Suggestions have been made that cytokines released by macrophages or lymphocytes may be significant in controlling the fibrogenic process,[51-53] but it has not been determined which factors are actually responsible. Seyer and co-workers have isolated a fraction from liver homogenates of fibrotic animals that stimulates types I and IV procollagen mRNA at a transcriptional level.[54,55] Armendariz-Borunda and Rojkind have begun to investigate a factor produced by Kupffer

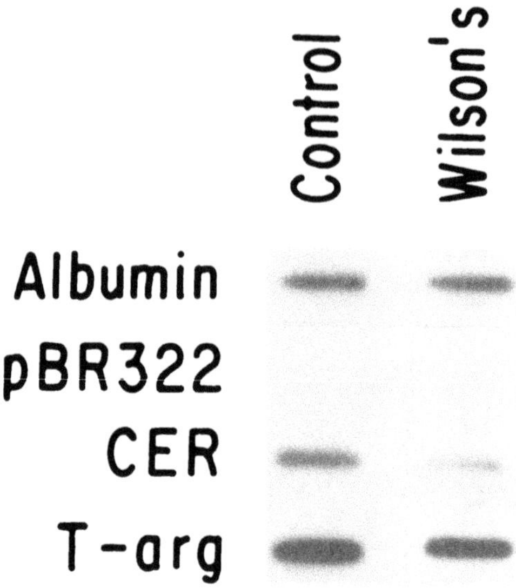

FIGURE 8. Autoradiogram of a representative nuclear run-on assay analyzing the transcriptional rates of genes in a Wilson's disease patient and in a control with primary biliary cirrhosis. Nascent labeled nuclear RNA transcripts from liver nuclei of the two patients were hybridized with 5 μg of specific cDNA probes bound to a nitrocellulose filter. Autoradiography was then performed. The patient with Wilson's disease had lower levels of ceruloplasmin gene transcription. (Reprinted from Czaja, M. J., Weiner, F. R., Schwarzenberg, S. J., Sternlieb, I., Scheinberg, I. H., Van Thiel, D. H., Giambrone, M. -A., Kirschner, R., LaRusso, N., Koschinsky, M. L., MacGillivray, R. T. A., and Zern, M. A., *J. Clin. Invest.* 80, 1200, 1987. With permission from The American Society for Clinical Investigation, Inc.)

cells from CCl_4-treated rats.[56] This polypeptide appears to stimulate fibroblast proliferation but does not enhance collagen synthesis directly. However, these and other putative "fibrogenic factors" have not been fully characterized or purified.

More recently, molecular techniques have allowed for the cloning and production of relatively large quantities of several growth factors or cytokines that may directly affect the fibrogenic process. The advantage of these recombinant proteins is their ready availability and absolute purity. With the use of these recombinant proteins, there is no concern about contamination of column fractions or partially purified proteins by small amounts of an extraneous factor. Such minor contamination can otherwise be a major problem because these cytokines are highly active in picomolar or nanomolar concentrations. Although several cytokines may eventually prove to be crucial in the development of hepatic fibrosis, a likely agent is transforming growth factor-β1 (TGF-β1).

1. Transforming Growth Factor-β1

TGF-β1 is a 25-kDa homodimer. Many cell types including lymphocytes, platelets, and macrophages synthesize this polypeptide,[57,58] and essentially all cells have a specific, high-affinity TGF-β1 receptor.[59] TGF-β1 has been implicated as having a major role in the production of extracellular matrix proteins. This cytokine has been shown to induce collagen formation in myoblasts, fibroblasts, and osteoblasts,[60-64] and it appears to do this by enhancing the procollagen mRNA steady-state levels.[65] TGF-β1 is also known to stimulate the production of fibronectin and proteoglycans.[60,66] When administered to mice, TGF-β1 enhances the production of collagen in granulation tissue[67] and hastens wound healing.[68] However, despite much speculation, no previous study has demonstrated a role for this factor in a pathophysiological state.

The extremely low concentration at which TGF-β1 exerts its effects on hepatocytes in culture[69] and its possible regulatory role in matrix formation have led us to investigate this cytokine for its possible role in hepatic fibrogenesis. We have demonstrated that primary cultures of hepatic cells treated with TGF-β have markedly increased levels of proα2(I) collagen mRNA due to a posttranscriptional mechanism. In addition, when livers of mice infected with *S. mansoni* were evaluated, they exhibited increased levels of TGF-β1 gene expression at times of maximal collagen synthesis. These findings suggest a role for TGF-β1 in hepatic fibrogenesis. Further investigation is clearly indicated to examine the precise mechanism by which TGF-β1 may initiate fibrogenesis *in vivo*.

B. FACTORS THAT INHIBIT HEPATIC FIBROGENESIS AND COLLAGEN REGULATION

1. Gamma-Interferon

The interferons are a group of endogenous proteins that exhibit a variety of biological functions. Many studies have evaluated that antiviral effects of α-interferon on chronic hepatitis B virus infection.[71,72] However, γ-interferon has never been studied as an antifibrogenic agent in chronic liver disease despite evidence that it profoundly inhibits collagen synthesis *in vitro*.[73,74] In order to evaluate the antifibrogenic actions of interferon, we have delineated the level of regulation responsible for γ-interferon-induced changes in collagen and fibronectin gene expression in cultured fibroblasts.[46] Confluent human skin fibroblasts were exposed to 500 antiviral units per milliliter of γ-interferon. This did not cause a significant change in total protein or DNA content, but interferon exposure did lead to a 60% decrease in collagen formation in the treated cells. RNA was extracted from the cells, and steady-state mRNA levels were determined by Northern and dot blot hybridization studies. Cells exposed to interferon had type I procollagen mRNA levels that were 23% of control and type III procollagen mRNA levels only 7% of control. The interferon-treated cells also had β-actin mRNA levels that were decreased to 51% that of untreated cells, but had fibronectin steady-state mRNA levels that were 560% of control levels.[46] Nuclear run-on assays revealed that interferon did not affect the transcriptional rates of types I and III collagen or β-actin, but it did increase the transcriptional rate of fibronectin to 670% of control levels (see Figure 9). These findings demonstrate that γ-interferon causes a marked decrease in types I and III procollagen mRNA levels *in vitro* by a posttranscriptional mechanism while inducing fibronectin expression at a transcriptional level.

Becase we had analyzed the levels of gene expression by which γ-interferon affected fibrogenesis *in vitro*, our next project was to test the effectiveness of γ-interferon as an antifibrogenic agent *in vivo*. Control CF-1 mice and mice 4 weeks postinfection with *S. mansoni* were treated with 100,000 antiviral units of γ-interferon IM daily[75]. At the end of 4 weeks, the mice were sacrificed for histological and molecular studies. Liver histology and high performance liquid chromatography (HPLC) analysis showed a marked decrease in the amount of collagen in the infected animals treated with the interferon. Northern and dot blot hybridizations demonstrated decreases in types I and III procollagen mRNA in the interferon-treated animals. These findings suggest that γ-interferon may prove to be a valuable antifibrotic agent in chronic liver disease.

2. Corticosteroids

As opposed to γ-interferon, corticosteroids have been studied extensively as antifibrogenic agents in a number of systems including the liver.[76-80] It is felt that endogenous or exogenous glucocorticoids may have a beneficial effect on the development of at least some forms of chronic liver disease. We have recently shown that detoxified alcoholics with high serum cortisol levels have a trend toward better liver function.[81] In addition, corticosteroids are given to patients with a variety of chronic liver diseases in an attempt to improve their

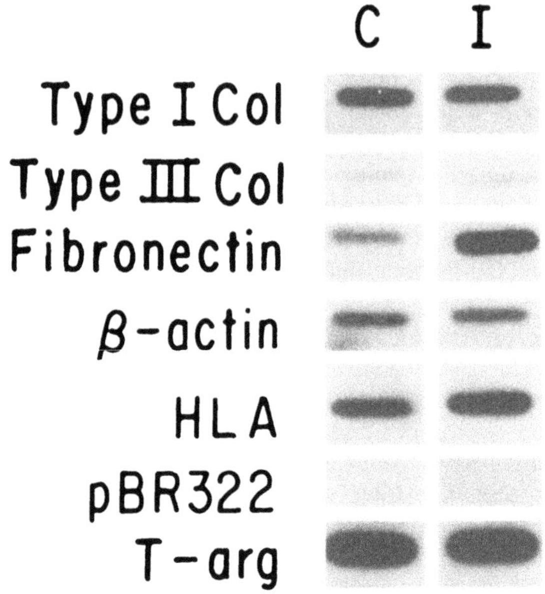

FIGURE 9. Autoradiogram of a nuclear run-on assay demonstrating the effects of interferon treatment (I) on the transcriptional rate of various genes in human skin fibroblasts as compared to untreated controls (C). Labeled nuclear RNA transcripts were hybridized with cDNA probes bound to nitrocellulose membrane. Interferon treatment had no effect on the transcriptional rates of the types I and III collagen and the β-actin genes. However, there was a large increase in fibronectin transcription with γ-interferon. Note also the expected increase in HLA-A2 gene transcription with interferon. (Reprinted from Czaja, M. J., Weiner, F. R., Eghbali, M., Giambrone, M. -A., Eghbali, M. A., Blumenfeld, O. O., and Zern, M. A., *J. Biol. Chem.*, 262, 13348, 1987. With permission from The American Society for Biochemistry and Molecular Biology, Inc.)

prognosis and retard the development of cirrhosis.[76,82] There are several possible mechanisms by which steroid therapy may improve chronic liver disease. Experimental evidence exists that corticosteroids may help maintain the differentiated function of hepatocytes,[83,84] and we have shown that dexamethasone increases albumin mRNA content in cultured hepatocytes[85] and both albumin mRNA content and serum albumin levels in animals.[45] Corticosteroids have also been shown to decrease collagen synthesis and fibrosis formation in a variety of ways.[77-80] In addition to their generalized antiinflammatory action,[77] steroids have been found to decrease the levels of posttranslational enzymes associated with collagen synthesis.[78,79] Studies of the molecular effects of dexamethasone on collagen production have shown that corticosteroids decrease type I procollagen mRNA content in chicken,[86] rat,[87] and human[88] skin fibroblasts. The level of regulation responsible for this inhibition of steady-state procollagen mRNA levels by corticosteroids has been a subject of considerable study. Cockayne and co-workers[86] have suggested that transcriptional regulation in fibroblasts may be responsible for this inhibition while Raghow et al.[87] and Hämäläinen et al.[88] have indicated that steroid-induced reduction of type I procollagen mRNA stability may be significant. We have done a series of studies using molecular biology techniques in an attempt to better delineate the antifibrogenic mechanisms of steroids.[44,45]

When 4 μg/ml of dexamethasone was added to the drinking water of mice infected with *S. mansoni*, there was a 75% decrease in the liver collagen content as determined by HPLC. RNA-DNA hybridization analysis revealed that the corticosteroids suppressed the steady-state levels of types I and IV procollagen mRNA by 40%, had no effect on β-actin mRNA, and enhanced the level of albumin mRNA nearly fourfold. The level of gene expression

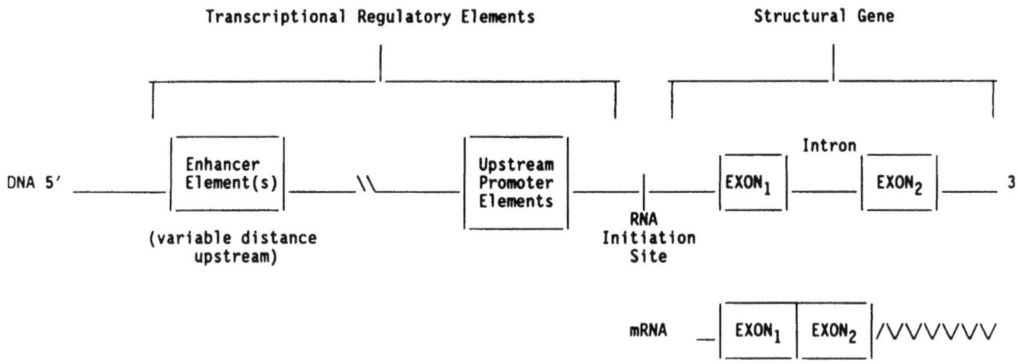

FIGURE 10. Elements involved in the transcription of a eukaryotic gene. The structural gene is composed of exons (coding segments) and intervening sequences or introns (noncoding regions). The RNA initiation site is the location where polymerase II initiates transcription of the hnRNA. The regulatory sequences reside in the region which is upstream in the 5′ direction from the initiation start site and include the gene promoter elements and enhancers. These sequences are crucial in regulating the qualitative and quantitative transcription of the structural gene.

responsible for these steady-state changes was evaluated by nuclear run-on analysis. Dexamethasone exerted its effect on different genes in the injured liver by diverse mechanisms, i.e., decreasing collagen synthesis at a transcriptional level and increasing albumin by a posttranslational mechanism.[45] Some of this decrease in procollagen mRNA content could be secondary to a diminution of inflammatory infiltrate in the fibrotic liver caused by the steroids and a subsequent loss of mesenchymal cell RNA. To address this issue, we evaluated the effects of dexamethasone on *in vitro* collagen gene expression, thereby eliminating the antiinflammatory effects of the hormone. We examined the level of regulation responsible for the dexamethasone-induced changes in gene expression of two cultured cell types implicated in hepatic fibrosis. We focused specifically on transcriptional regulation of the pro α2(I) collagen gene using the nuclear run-on assay and the chloramphenicol acetyltransferase (CAT) transient expression vector system described by Gorman et al.[89] Primary cultures of adult rat hepatocytes treated with and without dexamethasone under classical cell culture conditions or using serum-free hormonally defined media were evaluated for their synthesis and abundance of procollagen and β-actin mRNA.[44] Cells treated with dexamethasone had decreased steady-state levels of types I and IV procollagen mRNA. On the other hand, β-actin mRNA levels were unaffected by dexamethasone. Nuclear run-on assays revealed that the decrease in steady-state levels of both procollagen mRNAs in the presence of dexamethasone was associated with decreased transcription of the collagen genes.[44] However, the change in transcription was not always of sufficient magnitude to account for the marked decrease in the cytoplasmic abundance of the mRNA. Therefore, both transcriptional and posttranscriptional mechanisms appear to affect collagen synthesis. There was no apparent effect of dexamethasone on β-actin gene transcription.

To confirm the effect of dexamethasone on type I procollagen gene transcription as well as to define more precisely the mechanism by which dexamethasone inhibits transcription of this gene, transient expression vector experiments with the mouse pro α2(I) collagen gene promoter were performed. The purpose of transient expression vector experiments is to evaluate the activity of the regulatory region of the gene of interest removed from the structural region of that gene. As shown in Figure 10, the regulatory region of the gene is 5′ or upstream from the structural gene. The regulatory region contains several areas which are necessary for the appropriate qualitative and quantitative regulation of gene transcription. Thus, alterations in the regulatory region or variation in the way that soluble factors interact with this region can markedly affect gene expression. The activity of promoters can be determined if the regulatory region is isolated from its own gene and placed in the appropriate position upstream from a marker gene which lacks its own promoter. Such constructs are then ligated to regions from a virus in a bacterial plasmid. This enables the chimeric plasmid to be reproduced in bulk in bacteria or to be introduced into and expressed by eukaryotic cells. When introduced into a eukaryotic cell, expression of the marker gene is controlled by the activity of the promoter of interest. This plasmid is called an expression vector because after the RNA of the marker gene is transcribed, it is then translated by the protein synthetic machinery of the cells into which it has been introduced. The amount of the protein marker thus synthesized can then be quantitated as a measure of promoter activity. For our analysis of the effect of dexamethasone on the pro α2(I) collagen promoter, we employed the recombinant plasmid pAZ1009.[90] This chimeric plasmid contains the mouse α2(I) procollagen gene promoter linked to the structural CAT gene (the marker), the SV40 viral enhancer, and the bacterial plasmid pBR322.[90] The bacterial gene (CAT) is a useful marker because there is no CAT activity in eukaryotic cells, and its expression must therefore be a result of the transfection of this marker gene into the eukaryotic cell and subsequent synthesis of its protein.

A schema of the CAT transient expression assay is shown in Figure 11. The pAZ1009 vector is transfected into the fibroblast cell line (either NIH-3T3 or L-cells) using the standard calcium phosphate precipitation method.[89] Half of the cells are treated with 10^{-6} *M* dexamethasone for 48 h, and extracts of the cells are prepared by sonification and centrifugation. The cell extracts are then assayed for CAT activity by evaluating the acetylation of [^{14}C]-labeled chloramphenicol by silica gel thin-layer chromatography and autoradiography.[89] An example of such an assay is shown in Figure 12. It demonstrates that cells transfected with the collagen-containing promoter and treated with dexamethasone have significantly less CAT activity than transfected, untreated cells. The results of six such experiments showed that the activity of the collagen promoter was reduced by greater than 50% by dexamethasone.[44] Therefore, there appears to be strong evidence that steroids affect collagen synthesis by several mechanisms, including a direct inhibition of collagen gene transcription.

IV. NEW DIRECTIONS

As described in previous sections of this chapter, there is convincing evidence that several changes in collagen gene expression are due to transcriptional regulation. For example, the increased collagen content in models of hepatic fibrosis appears to be regulated, at least in part, at the transcriptional level, while the inhibition of collagen synthesis caused by steroids appears to have a transcriptional component. However, most of the previous studies have focused on determining the level of gene regulation; fewer studies have evaluated the mechanisms responsible for these changes in collagen gene expression. The analyses of these mechanisms will be a major thrust of future investigations.

Modulation of specific gene transcription can be caused by two basic mechanisms: differences in the structure of the gene itself and of its regulatory region (*cis* elements), or variation in proteins or other factors that act on the promoter (*trans*-acting factors). In both cases the promoter of the gene is the region most crucial in the gene's regulation. The study of *cis* elements includes the analysis of variations in the nucleic acid sequences of specific genes, the study of methylation patterns and DNase hypersensitivity sites, as well as analyses of restriction site polymorphisms. In addition to the use of restriction site analysis and DNA sequencing, a powerful technique has been developed recently and used to evaluate the promoter region of the type I collagen gene. deCrombrugghe and co-workers have studied the promoter function of the mouse pro α2 type I collagen gene in a series of studies using the CAT expression vector system.[90,91] Studies employing the CAT vector to analyze promoter function in alcoholics with and without cirrhosis might be a valuable use of this

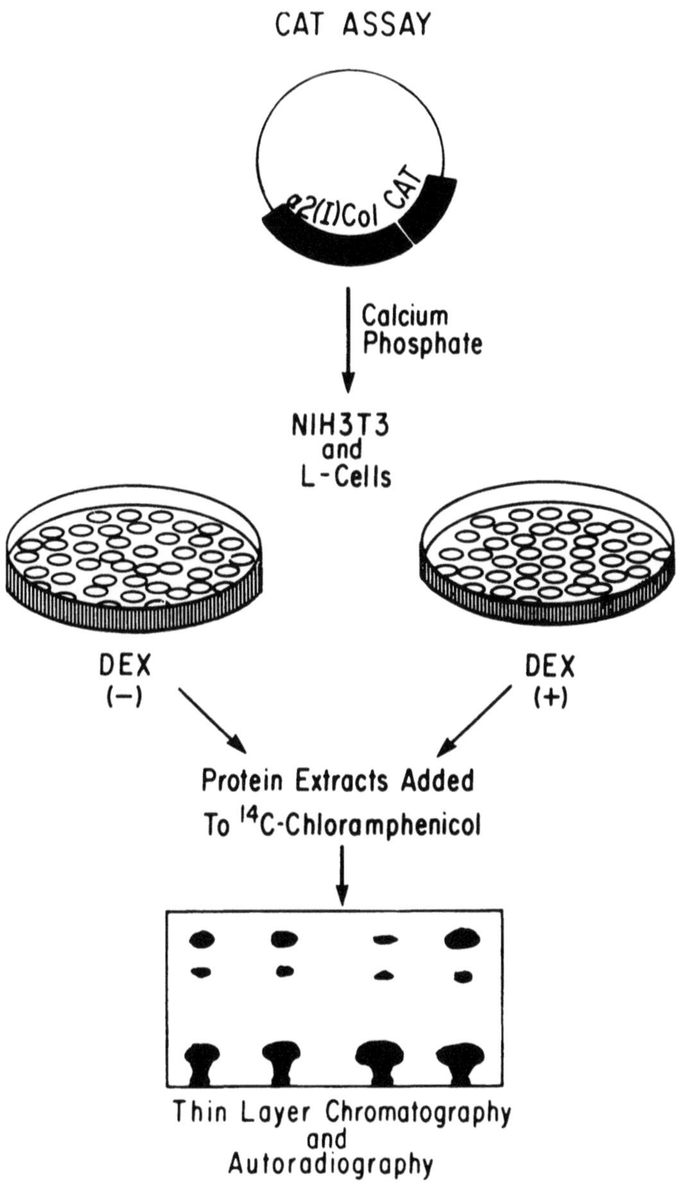

FIGURE 11. Chloramphenicol acetyltransferase (CAT) transient expression vector assay. The chimeric plasmid which contains the pro α2(I) collagen gene promoter immediately upstream to the CAT gene, as well as regions from the SV40 virus and bacterial plasmid pBR-322, is introduced into a mouse fibroblast cell line by the calcium phosphate precipitation method. Cells are grown in the absence or presence of dexamethasone, harvested, and their proteins are extracted. CAT activity in cell extracts are then determined by analyzing the amount of ^{14}C-labeled chloramphenicol which is acetylated. This assessment of CAT activity is then representative of the collagen gene promoter activity because the collagen promoter directs the transcription of the CAT gene.

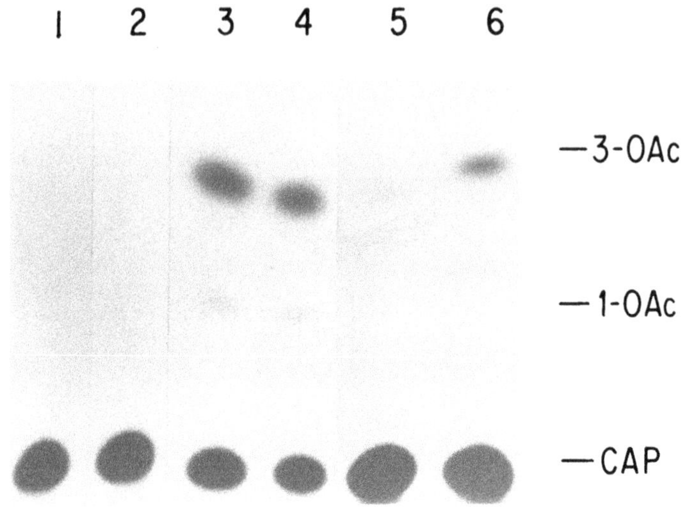

FIGURE 12. Chloramphenicol acetyltransferase (CAT) activity in NIH-3T3 cells transfected with pAZ1009, $pA_{10}CAT$, and pSV_2CAT. DNA was introduced into the cells by calcium phosphate precipitation; then half the cells were treated with 10^{-6} *M* dexamethasone for 48 h. Cells were harvested, and extracts were prepared by sonification and centrifugation. Equal amounts of protein were assayed for CAT activity as described by Gorman et al.[89] Ethyl acetate-extracted samples were then spotted on silica gel thin-layer plates, and [^{14}C]chloramphenicol (CAP) as well as its acetylated forms were detected by autoradiography. The bottom band represents unacetylated product. The upper two bands represent acetylated chloramphenicol (3-OAc and 1-OAc). Lane 1, represents cells infected with the negative control, $pA_{10}CAT$, and treated with dexamethasone; lane 2, represents the $pA_{10}CAT$, without dexamethasone; lane 3, the positive control pSV_2CAT (lacking a collagen promoter) treated with dexamethasone; lane 4, pSV_2CAT without dexamethasone; lane 5, the construct containing the collagen promoter, pAZ1009, treated with dexamethasone; lane 6, pAZ1009 without dexamethasone. Note that dexamethasone treatment inhibits CAT activity in the cells transfected with the collagen promoter but does not affect CAT activity in cells transfected with control vectors. (Reprinted from Weiner, F. R., Czaja, M. J., Jefferson, D. M., Giambrone, M. -A., Tur-Kaspa, R., Reid, L. M., and Zern, M. A., *J. Biol. Chem.*, 262, 6955, 1987. With permission from the American Society for Biochemistry and Molecular Biology, Inc.)

technology. It would provide a new method for investigating the variation in the fibrotic response to alcohol-induced hepatic injury. The foundation for such studies has been established by the cloning and analysis of the promoter regions of a human type I collagen gene.[1]

Most recently, considerable attention has been focused on the delineation of *trans*-acting factors which may be essential for accurate gene transcription. A series of proteins have been found which bind to specific regions of the eukaryotic promoter and seem to regulate transcription.[92,93] Alterations in such proteins have been shown to regulate the expression of several genes that are active in the liver including metallothionein.[94] Much of the work which explores the interactions of *trans*-acting factors and the collagen genes has been done by deCrombrugghe and co-workers. They have used a series of techniques, including the exonuclease III assay, DNase I footprinting, and the DNA gel retardation assay to delineate the relationship of nuclear proteins to the promoter region of the collagen genes.[95,96] These techniques can certainly be applied to the regulation of the collagen genes in hepatic fibrosis.

V. CONCLUSION

There has been enormous progress in the last 10 to 15 years in the use of molecular techniques to evaluate gene expression in eukaryotic systems. However, much of the work has been done in basic systems not directly applicable to pathophysiological states. It remains the task of "disease-oriented" investigators to adopt and adapt these approaches to answer questions that are pertinent to specific diseases. The use of molecular techniques to evaluate human pathophysiology provides an opportunity to delineate clinically relevant information while at the same time offering the additional advantage of providing fundamental knowledge about the disease process.

REFERENCES

1. **Chu, M. -L., Myers, J., Bernard, M. P., Ding, J. -F., and Ramirez, J.,** Cloning and characterization of five overlapping cDNAs specific for the human pro α 1(I) collagen chain, *Nucleic Acids Res.*, 10, 5925, 1982.
2. **Schmidt, A., Yamada, Y., and de Crombrugghe, B.,** DNA sequence comparison of the regulatory signals at the 5′ end of the mouse and chick 2 type I collagen genes, *J. Biol. Chem.*, 259, 7411, 1984.
3. **Prockop, D. J. and Kivirikko, K. I.,** Heritable diseases of collagen, *N. Engl. J. Med.*, 311, 376, 1984.
4. **Martin, G. R., Timple, R., Muller, P., and Kuhn, K.,** The genetically distinct collagens, *Trends Biochem. Sci.*, 19, 285, 1985.
5. **Burgeson, R. E.,** Genetic heterogeneity of collagens, *J. Invest. Dermatol.*, 79, 25, 1982.
6. **Miller, E. J.,** The structure of fibril-forming collagens, *Ann. N.Y. Acad. Sci.*, 460, 1, 1985.
7. **Raghow, R., Lurie, S., Seyer, J. M., and Kang, A. H.,** Profile of steady state levels of messenger RNAs coding for type I procollagen, elastin, and fibronectin in hamster lungs undergoing bleomycin-induced interstitial pulmonary fibrosis, *J. Clin. Invest.*, 76, 1733, 1985.
8. **Sterling, K. M., Harris, M. J., Mitchell, J. J., and Cutroneo, K. R.,** Bleomycin treatment of chick fibroblasts causes an increase of polysomal type I procollagen mRNA, *J. Biol. Chem.*, 258, 14438, 1983.
9. **Tolstoshev, P., Berg, R. A., Rennard, S. I., Bradley, K. H., Trapnell, B. C., and Crystal, R. G.,** Procollagen production and procollagen messenger RNA levels and activity in human lung fibroblasts during periods of rapid and stationary growth, *J. Biol. Chem.*, 256, 3135, 1981.
10. **Vitto, J., Perejda, A. J., Abergel, R. P., Chu, M. -L., and Ramirez, F.,** Altered steady-state ratio of type I/III procollagen mRNAs correlates with selectively increased type I procollagen biosynthesis in cultured keloid fibroblasts, *Proc. Natl. Acad. Sci. U.S.A.*, 82, 5935, 1985.
11. **Jimenez, S. A., Williams, C. J., Myers, J. C., and Bashey, R. I.,** Increased collagen biosynthesis and increased expression of type I and type III procollagen genes in tight skin (TSK) mouse fibroblasts, *J. Biol. Chem.*, 261, 657, 1986.
12. **Jimenez, S. A., Feldman, G., Bashey, R. I., Bienkowski, R., and Rosenbloom, J.,** Co-ordinate increase in the expression of type I and type III collagen genes in progressive systemic sclerosis fibroblasts, *Biochem. J.*, 237, 837, 1986.
13. **Kream, B. E., Rowe, D. W., Gworek, S. C., and Raisz, L. G.,** Parathyroid hormone alters collagen synthesis and procollagen mRNA levels in fetal rat calvaria, *Proc. Natl. Acad. Sci. U.S.A.*, 77, 5654, 1980.
14. **Rowe, D. W. and Kream, B. E.,** Regulation of collagen synthesis in fetal rat calvaria by 1,25-hydroxy-vitamin D_3, *J. Biol. Chem.*, 257, 8009, 1982.
15. **Shapiro, J. R. and Rowe, D. W.,** Collagen genes and brittle bones, *Ann. Intern. Med.*, 99, 700, 1983.
16. **Rojkind, M., Giambrone, M. -A., and Biempica, L.,** Collagen types in normal and cirrhotic liver, *Gastroenterology*, 76, 710, 1979.
17. **Martinez-Hernandez, A.,** personal communication, 1987.
18. **Takahashi, S., Dunn, M., and Seifter, S.,** Liver collagenase in murine schistosomiasis, *Gastroenterology*, 78, 1425, 1980.
19. **Zern, M. A., Leo, M. A., Giambrone, M. -A., and Lieber, C. S.,** Increased type I procollagen mRNA levels and in vitro protein synthesis in the baboon model of chronic alcoholic liver disease, *Gastroenterology*, 89, 1123, 1985.
20. **Zern, M. A., Saber, M. A., and Shafritz, D. A.,** Changes in the molecular mechanisms of hepatic protein synthesis induced by schistosome infection in mice, *Biochemistry*, 22, 6072, 1983.
21. **Shafritz, D. A.,** Messenger RNA and its translation, in *Molecular Mechanisms of Protein Biosynthesis*, Weissbach, H. and Pestka, S., Eds., Academic Press, New York, 1977, 556.
22. **Tse, T. P. H. and Taylor, J. M.,** Translation of albumin messenger RNA in a cell-free protein synthesizing system derived from wheat germ, *J. Biol. Chem.*, 252, 1272, 1977.
23. **Rigby, P. W. J., Dieckmann, M., Rhodes, C., and Berg, P.,** Labelling deoxyribonucleic acid to high specific activity *in vitro* by nick translation with DNA polymerase I, *J. Mol. Biol.*, 113, 237, 1977.
24. **Chirgwin, J. M., Przbyla, A. E., Macdonald, R. J., and Rutter, W. J.,** Isolation of biologically active ribonucleic acid from sources enriched in ribonucleases, *Biochemistry*, 18, 5294, 1979.

25. **Weiner, F. R., Czaja, M. J., Wu, C. H., Wu, G. Y., Giambrone, M. -A., and Zern, M. A.,** Use of molecular hybridization technology to evaluate albumin and collagen mRNA content in baboons and man, *Hepatology,* 7, 19S, 1987.
26. **Warren, K. S.,** Regulation of the prevalence and intensity of schistosomiasis in man. Immunology or ecology?, *J. Infect. Dis.,* 127, 595, 1973.
27. **Page, C. R., Etges, F. J., and Ogle, J. D.,** Experiment prepatent *Schistosomiasis mansoni:* quantitative analysis of proteins, enzyme activity and free amino acids in mouse serum, *Exp. Parasitol.,* 31, 241, 1972.
28. **Dunn, M. A., Rojkind, M., Warren, S., Hait, P. K., Rifas, L., and Seifter, S.,** Liver collagen synthesis in murine schistosomiasis, *J. Clin. Invest.,* 59, 666, 1977.
29. **Lieber, C. S., DeCarli, L. M., and Rubin, E.,** Sequential production of fatty liver, hepatitis and cirrhosis in subhuman primates fed ethanol with adequate diets, *Proc. Natl. Acad. Sci. U.S.A.,* 72, 437, 1973.
30. **Galambos, J. T., Hollingsworth, M. A., Falek, A., and Warren, D.,** The rate of synthesis of glycosaminoglycans and collagen by fibroblasts cultured from adult human liver biopsies, *J. Clin. Invest.,* 60, 170, 1977.
31. **Kent, G., Gay, S., Inouye, T., Bahu, R., Minick, T., and Popper, H.,** Vitamin A-containing lipocytes and formation of type III collagen in liver injury, *Proc. Natl. Acad. Sci. U.S.A.,* 75, 3719, 1976.
32. **Friedman, S. L., Roll, F. J., Boyles, J., and Bissell, D. M.,** Hepatic lipocytes: the principal collagen-producing cells of normal rat liver, *Proc. Natl. Acad. Sci. U.S.A.,* 82, 8681, 1985.
33. **Nakano, M., Worner, T. M., and Lieber, C. S.,** Perivenular fibrosis in alcoholic liver injury: ultrastructure and histologic progression, *Gastroenterology,* 83, 777, 1982.
34. **Voss, R., Rauterberg, J., Pott, G., Bredmer, V., Allam, S., Tedmann, R., and Bassewitz, D. B. V.,** Nonparenchymal cells cultivated from explants of fibrotic liver resemble endothelial and smooth muscle cells from blood vessels, *Hepatology,* 2, 19, 1982.
35. **Voss, B., Allam, S., Rauterberg, J., Pott, G., and Bredmer, V.,** Liver derived macrophages synthesize collagen and fibronectin, in *Connective Tissue of the Normal and Fibrotic Human Liver,* Gerlach, U., Rauterberg, J., and Voss, B., Eds., Georg Thieme Verlag, Stuttgart, 1982, 97.
36. **Hata, R., Ninomiya, Y., and Tsukada, Y.,** Biosynthesis of interstitial types of collagen by albumin-producing rat liver parenchymal cell (hepatocyte) clones in culture, *Biochemistry,* 19, 1980, 169.
37. **Tseng, S. C. G., Lee, P. C., Ells, P. F., Bissell, D. M., Smuckler, A., and Stern, R.,** Collagen production by rat hepatocytes and sinusoidal cells in primary monolayer culture, *Hepatology,* 2, 12, 1982.
38. **Diegelmann, R. F., Guzelian, P. S., Gay, R., and Gay, S.,** Collagen formation by the hepatocyte in primary monolayer culture and *in vivo, Science,* 219, 1343, 1983.
39. **Chojkier, M.,** Hepatocyte collagen production *in vivo* in normal rats, *J. Clin. Invest.,* 78, 333, 1986.
40. **Saber, M. A., Zern, M. A., and Shafritz, D. A.,** Use of *in situ* hybridization to identify the presence of collagen and albumin mRNAs in isolated mouse hepatocytes, *Proc. Natl. Acad. Sci. U.S.A.,* 80, 4017, 1983.
41. **Saber, M. A., Shafritz, D. A., and Zern, M. A.,** Changes in collagen and albumin mRNA in liver tissue of mice infected with *Schistosoma mansoni* as determined by *in situ* hybridization, *J. Cell Biol.,* 97, 986, 1983.
42. **Darnell, J. E.,** Variety in the level of gene control in eukaryotic cells, *Nature (London),* 197, 365, 1982.
43. **Sandmeyer, S., Gallis, B., and Bornstein, P.,** Coordinate transcriptional regulation of type I procollagen genes by Rous sarcoma virus, *J. Biol. Chem.,* 256, 5022, 1981.
44. **Weiner, F. R., Czaja, M. J., Jefferson, D. M., Giambrone, M. -A., Tur-Kaspa, R., Reid, L. M., and Zern, M. A.,** The effects of dexamethasone on *in vitro* collagen gene expression, *J. Biol. Chem.,* 262, 6955, 1987.
45. **Weiner, F. R., Czaja, M. J., Giambrone, M. -A., Takahashi, S., Biempica, L., and Zern, M. A.,** Transcriptional and post-transcriptional effects of dexamethasone in murine schistosomiasis, *Biochemistry,* 26, 1557, 1987.
46. **Czaja, M. J., Weiner, F. R., Eghbali, M., Giambrone, M. -A., Eghbali, M. A., Blumenfeld, O. O., and Zern, M. A.,** Differential effects of gamma-interferon on collagen and fibronectin gene expression, *J. Biol. Chem.,* 262, 13348, 1987.
47. **Panduro-Cerda, A., Shalaby, A., Biempica, L. F., Weiner, F. R., Zern, M. A., and Shafritz, D. A.,** Transcriptional switch from albumin to alpha-fetoprotein and changes in transcription of other genes during CCl_4-induced liver regeneration, *Biochemistry,* 25, 1414, 1986.
48. **Aycock, R. S., Raghow, R., Stricklin, G. P., Seyer, J. M., and Kang, A. H.,** Post-transcriptional inhibition of collagen and fibronectin synthesis by a synthetic homolog of a portion of the carboxyl-terminal propeptide of human type I collagen, *J. Biol. Chem.,* 261, 14355, 1986.
49. **Czaja, M. J., Weiner, F. R., Schwarzenberg, S. J., Sternlieb, I., Scheinberg, I. H., Van Thiel, D. H., Giambrone, M. -A., Kirschner, R., LaRusso, N., Koschinsky, M. L., MacGillivray, R. T. A., and Zern, M. A.,** Molecular studies of ceruloplasmin deficiency in Wilson's disease, *J. Clin. Invest.,* 80, 1200, 1987.

50. **Clayton, D. F., and Darnell, J. E.,** Changes in liver specific compared to common gene transcription during primary cultures of mouse hepatocytes, *Mol. Cell. Biol.*, 3, 1552, 1983.
51. **Postlethwaite, A. E., Smith, G. N., Mainardi, C. L., Seyer, J. M., and Kang, A. H.,** Lymphocyte modulation of fibroblast function *in vitro*: stimulation and inhibition of collagen production by different effector molecules, *J. Immunol.*, 132, 2470, 1984.
52. **Wahl, S. M., Wahl, L. M., and McCarthy, J. B.,** Lymphocyte-mediated activation of fibroblast proliferation and collagen production, *J. Immunol.*, 121, 942, 1978.
53. **Fallon, A., Bradley, J. F., Burns, J., and McGee, J. O. D.,** Collagen stimulating factors in hepatic fibrogenesis, *J. Clin. Pathol.*, 37, 542, 1984.
54. **Hatahara, T. and Seyer, J. M.,** Isolation and characterization of a fibrogenic factor from CCl_4-damaged rat liver, *Biochim. Biophys. Acta*, 716, 377, 1982.
55. **Raghow, R., Gossage, D., Seyer, J. M., and Kang, A. H.,** Transcriptional regulation of type I collagen genes in cultured fibroblasts by a factor isolated from thioacetamine-induced fibrotic rat liver, *J. Biol. Chem.*, 259, 12718, 1984.
56. **Armendariz-Borunda, J. and Rojkind, M.,** Mediators of fibroblast proliferation in experimental liver cirrhosis, *Hepatology*, 5 (Abstr.) 1037, 1985.
57. **Derynck, R., Jarrett, J. A., Chen, E. Y., Eaton, D. H., Bell, J. R., Assoian, R. K., Roberts, A. B., Sporn, M. B., and Goeddel, D. V.,** Human transforming growth factor-beta complementary DNA sequence and expression in normal and transformed cells, *Nature (London)*, 316, 701, 1985.
58. **Childs, C. B., Proper, J. A., Tucker, R. F., and Moses, H. L.,** Serum contains a platelet-derived transforming growth factor, *Proc. Natl. Acad. Sci. U.S.A.*, 79, 5312, 1982.
59. **Sporn, M. B., Roberts, A. B., Wakefield, L. M., and Assoian, R. K.,** Transforming growth factor-beta: biological function and chemical structure, *Science*, 232, 532, 1986.
60. **Ignatz, R. A. and Massagué, J.,** Transforming growth factor-beta stimulates the expression of fibronectin and collagen and their incorporation into the extracellular matrix, *J. Biol. Chem.*, 261, 4337, 1986.
61. **Fine, A. and Goldstein, R. H.,** The effect of transforming growth factor-beta on cell proliferation and collagen formation by lung fibroblasts, *J. Biol. Chem.*, 262, 3897, 1987.
62. **Varga, J. and Jimenez, S. A.,** Stimulation of normal human fibroblast collagen production and processing by transforming growth factor-beta, *Biochem. Biophys. Res. Commun.*, 138, 974, 1987.
63. **Massagué, J., Cheifetz, S., Endo, T., and Nadal-Girard, B.,** Type β transforming growth factor is an inhibitor of myogenic differentiation, *Proc. Natl. Acad. Sci. U.S.A.*, 83, 8206, 1986.
64. **Centrella, M., McCarthy, T. L., and Canalis, E.,** Transforming growth factor-beta is a bifunctional regulator of replication and collagen synthesis in osteoblast-enriched cell cultures from fetal rat bone, *J. Biol. Chem.*, 262, 2869, 1987.
65. **Raghow, R., Postlethwaite, A. E., Keski-Oja, J., Moses, H. L., and Kang, A. H.,** Transforming growth factor-beta increases steady state levels of type I procollagen and fibronectin messenger RNAs posttranscriptionally in cultured human dermal fibroblasts, *J. Clin. Invest.*, 79, 1285, 1987.
66. **Chen, J. -K., Hoshi, H., and McKeehan, W. L.,** Transforming growth factor type β specifically stimulates synthesis of proteoglycan in human adult arterial smooth muscle cells, *Proc. Natl. Acad. Sci. U.S.A.*, 84, 5287, 1987.
67. **Roberts, A. B., Sporn, M. B., Assoian, R. K., Smith, J. M., Roche, N. S., Wakefield, L. M., Heine, V. I., Liotta, L. A., Falanga, V., Kehrl, J. H., and Fauci, A. S.,** Transforming growth factor type-beta: rapid induction of fibrosis and angiogenesis *in vivo* and stimulation of collagen formation *in vitro*, *Proc. Natl. Acad. Sci. U.S.A.*, 83, 4167, 1986.
68. **Mustoc, T. A., Pierce, G. F., Thomason, A., Gramates, P., Sporn, M. B., and Deuel, T. F.,** Accelerated healing of incisional wounds in rats induced by transforming growth factor-beta, *Science*, 237, 1333, 1987.
69. **Hayashi, I. and Carr, B. I.,** DNA synthesis in rat hepatocytes: inhibition by a platelet factor and stimulation by an endogenous factor, *J. Cell. Physiol.*, 125, 82, 1985.
70. **Czaja, M. J., Weiner, F. R., Giambrone, M. -A., Wind, R., Biempica, L., and Zern, M. A.,** Transforming growth factor-beta stimulates collagen synthesis *in vitro* and is elevated in an *in vivo* model of hepatic fibrosis, *J. Cell Biol.*, 108, 2477, 1989.
71. **Davis, G. L. and Hoofnagle, J. H.,** Interferon in viral hepatitis: role in pathogenesis and treatment, *Hepatology*, 6, 1038, 1986.
72. **Dusheiko, G., Dibisceglie, A., Bourjer, S., Sachs, E., Ritchie, M., Schoub, B., and Kew, M.,** Recombinant leukocyte interferon treatment of chronic hepatitis B, *Hepatology*, 5, 556, 1985.
73. **Jimenez, S. A., Freundlich, B., and Rosenbloom, J.,** Selective inhibition of human diploid fibroblast collagen synthesis by interferons, *J. Clin. Invest.*, 74, 1112, 1984.
74. **Rosenbloom, J., Feldman, G., Freundlich, B., and Jimenez, S. A.,** Transcriptional control of human diploid fibroblast collagen synthesis by γ-interferon, *Biochem. Biophys. Res. Commun.*, 122, 365, 1984.
75. **Czaja M. J., Weiner, F. R., Giambrone, M. -A., Takahashi, S., Biempica, L., and Zern, M. A.,** Antifibrogenic effects of gamma-interferon in murine schistosomiasis, *Hepatology*, in press.

76. **Wands, J. R., Koff, R. S., and Isselbacher, K. J.**, Chronic active hepatitis, in *Harrison's Principles of Internal Medicine*, 10th ed., Petersdorf, R., Adams, R. D., and Braunwald, E., Eds., McGraw-Hill, New York, 1983, 1801.
77. **James, S. P., Hoofnagle, J. H., and Strober, W.**, Primary biliary cirrhosis: a model autoimmune disease, *Ann. Intern.Med.*, 99, 500, 1983.
78. **Newman, R. A. and Cutroneo, K. R.**, Glucocorticoids selectively decrease the synthesis of hydroxylated collagen peptides, *Mol. Pharmacol.*, 14, 185, 1977.
79. **Risteli, J.**, Effect of prednisolone on the activities of the intracellular enzymes of collagen biosynthesis in rat liver and skin, *Biochem. Pharmacol.*, 26, 185, 1977.
80. **Sterling, K. M., Harris, M. J., Mitchell, J. J., and Cutroneo, K. R.**, Dexamethasone decreases the amounts of type I procollagen mRNA *in vivo* and in fibroblast cell cultures, *J. Biol. Chem.*, 258, 7644, 1983.
81. **Zern, M. A., Halbreich, V., Bacon, K., Galanter, M., Kang, B. Y., and Gasparini, F.**, Relationship between serum cortisol, liver function and depression in detoxified alcoholics, *Alcoholism Clin. Exp. Res.*, 10, 320, 1986.
82. **Czaja, A. J., Beaver, S. J., and Shiels, M. T.**, Sustained remission after corticosteroid therapy of severe hepatitis B surface antigen-negative chronic active hepatitis, *Gastroenterology*, 92, 215, 1987.
83. **Guguen-Guillouzo, C. and Guillouzo, A.**, Modulation of functional activities in cultured rat hepatocytes, *Mol. Cell. Biochem.*, 53/54, 35, 1983.
84. **Laishes, B. A. and Williams, G. M.**, Conditions affecting primary cell cultures of functional adult rat hepatocytes. II. Dexamethasone enhanced longevity and maintenance of morphology, *In Vitro*, 12, 821, 1976.
85. **Jefferson, D. M., Giambrone, M. -A., Shafritz, D. A., Reid, L. M., and Zern, M. A.**, Effects of dexamethasone on albumin and type I procollagen mRNA levels in primary rat hepatocyte cultures, *Hepatology*, 5, 14, 1985.
86. **Cockayne, D., Sterling, K. M., Shull, S., Mintz, K. P., Illeyne, S., and Cutroneo, K. R.**, Glucocorticoids decrease the synthesis of type I procollagen mRNAs, *Biochemistry*, 25, 3202, 1986.
87. **Raghow, R., Gossage, D., and Kang, A. H.**, Pretranslational regulation of type I collagen, fibronectin, and a 50-kilodalton noncollagenous extracellular protein by dexamethasone in rat fibroblasts, *J. Biol. Chem.*, 261, 4677, 1986.
88. **Hämäläinen, L., Oikarinen, J., and Kivirikko, K. I.**, Synthesis and degradation of type I procollagen mRNAs in cultured human skin fibroblasts and the effect of cortisol, *J. Biol. Chem.*, 260, 720, 1985.
89. **Gorman, C. M., Moffat, L. F., and Howard, B. H.**, Recombinant genomes which express chloramphenicol acetyltransferase in mammalian cells, *Mol. Cell. Biol.*, 2, 1044, 1982.
90. **Schmidt, A., Rossi, P., and de Crombrugghe, B.**, Transcriptional control of the mouse α2(I) collagen gene: functional deletion analysis of the promoter and evidence for cell-specific expression, *Mol. Cell. Biol.*, 6, 347, 1986.
91. **Khillan, J. S., Schmidt, A., Overbeek, P. A., de Crombrugghe, B., and Westphal, H.**, Developmental and tissue-specific expression directed by the α2 type I collagen promoter in transgenic mice, *Proc. Natl. Acad. Sci. U.S.A.*, 83, 725, 1986.
92. **Dynan, W. S. and Tjian, R.**, Control of eukaryotic messenger RNA synthesis by sequence-specific DNA-binding proteins, *Nature (London)*, 316, 774, 1985.
93. **Briggs, M. R., Kadonaga, J. T., Bell, S. P., and Tjian, R.**, Purification and biochemical characterization of the promoter-specific transcription factor, Spl, *Science*, 234, 47, 1986.
94. **Karin, M., Haslinger,A., Hotlgreve, H., Cathala, G., Slater, E., and Baxter, J. D.**, Activation of a heterologous promoter in response to dexamethasone and cadmium by metallothionein gene 5′-flanking DNA, *Cell*, 36, 371, 1984.
95. **Hatamochi, A., Paterson, B., and de Crombrugghe, B.**, Differential binding of a CCAAT DNA binding factor to the promoters of the mouse α2(I) and α1(III) collagen genes, *J. Biol. Chem.*, 261, 11310, 1986.
96. **Oikarinen, J., Hatamochi, A., and de Crombrugghe, B.**, Separate binding sites for nuclear factor 1 and a CCAAT DNA binding factor in a mouse α2(I) collagen promoter, *J. Biol. Chem.*, 262, 11064, 1987.
97. **Rojkind, M.**, Extracellular matrix, in *The Liver: Biology and Pathobiology*, 2nd ed., Arias, I. M., Jakoby, W. B., Popper, H., Schachter, D., and Shafritz, D. A., Eds., Raven Press, New York, 1988, 707.

Chapter 5

PULMONARY FIBROSIS: HUMAN AND EXPERIMENTAL DISEASE

Moises Selman Lama

TABLE OF CONTENTS

I. INTRODUCTION

Diffuse pulmonary fibrosis is the final common pathway of a heterogenous and numerous group of lung disorders, known as interstitial lung diseases (ILD), which may vary in their form and time of progression in some pathogenic mechanisms and in the response to therapy, but which may be classified together because they share common clinical, radiographic, and physiological features.[1,2] In addition, in spite of their different etiologies, the ILD present some similar histological changes characterized by alveolitis, that is, an interstitial and occasionally alveolar inflammation, and alterations in connective tissue, mainly of the collagens in the lung parenchyma.

On the other hand, since almost all ILD may sooner or later lead to diffuse interstitial fibrosis, they may also be called fibrotic lung disorders.

The ILD constitute more than 130 different diseases, and the most commonly used classification is according to their etiology.[1-3] Perhaps a more simplified way to classify them is as "primary" or idiopathic pulmonary fibrosis (IPF) and "secondary to" different diseases or groups of diseases of known or unknown etiology (Table 1). In this sense, although most ILD of known etiology affect the lungs exclusively, knowing their cause allows them to be distinguished easily from IPF. In addition, most of the ILD of unknown etiology, other than IPF, present characteristic morphological changes or affect other organs, and in this way can also be clearly differentiated from this disorder.

The ILD are a group of diseases more common than most physicians believe, and in the U.S. they have a prevalence of 5 to 10 cases per 100,000 inhabitants.[3] In Mexico, as well as in many other Third World countries, there are no reliable statistics for these diseases, but in our institute we have come across approximately 1000 patients in the last 5 years with some type of ILD. The most common were IPF, hypersensitivity pneumonitis (HP), and those associated with collagen-vascular diseases (CVD). However, albeit with lesser frequency, almost all of the diseases shown in Table 1 have been diagnosed, and a similar picture seems to rise from other local pneumological centers.[4]

TABLE 1
Classification of the Fibrotic Lung Disorders

Primary
Idiopathic pulmonary fibrosis (cryptogenic fibrosing alveolitis; usual interstitial pneumonia)

Secondary to
A. Known injurious agent
  1. Inhalation of inorganic particles (pneumonoconiosis)
  2. Inhalation of organic and chemical particles (HP)
  3. Inhalation of gases, fumes, vapors, and aerosols
  4. Use of drugs
  5. Poisons
  6. Chronic aspiration from gastroesophageal reflux
  7. Infectious agents
  8. Chronic pulmonary edema
  9. Chronic uremia
  10. Neoplasms
B. Unknown injurious agent
  1. Collagen-vascular diseases
  2. Sarcoidosis
  3. Histiocytosis X
  4. Pulmonary hemorrhagic syndromes
  5. Inherited disorders
  6. Ankylosing spondylitis
  7. Lymphocytic infiltrative disorders
  8. Giant cell interstitial pneumonia
  9. Lymphangioleiomyomatosis

With regard to IPF which is one of the most frequent ILD, some confusion exists about its terminology, mainly derived from the evolution of its description and from the difficulty inherent in characterizing an entity of unknown etiology based only on histopathological observations.

For many years after Hamman and Rich reported the first patients with diffuse fibrosis of the alveolar walls,[5] this disease was recognized and designated the "Hamman-Rich syndrome". However, this term has been progressively abandoned.

In 1965, Liebow et al.[6] described 18 cases of a lung disorder which they considered to be a different type of reaction to injury than IPF and named it "desquamative interstitial pneumonia" (DIP). In their opinion this lesion was clearly different from those fibrotic cases "usually" described, so they called these latter cases "usual interstitial pneumonia" (UIP).

On the other hand, Scadding,[7] a year before, had proposed the term cryptogenic fibrosing alveolitis (CFS) ("cryptogenic" for unknown etiology), and this term is normally used by our British colleagues. Moreover, Scadding and Hinson[8] proposed that both types of lung reactions are part of the range of pathological changes in the evolution of the same disease and added "desquamative type" and "mural type" to their morphological descriptions. This concept agrees with most authors and allows a more flexible interpretation of the pulmonary tissue alterations. In fact, as idiopathic and secondary forms of interstitial fibrosis have been recognized and biopsied more frequently, it is evident that a spectrum of changes and severity may be observed in the lungs.

Finally, many physicians, including us, use the name IPF, adding "of inflammatory or fibrotic predominance" according to the morphological modifications.

Nevertheless, IPF, CFA, and UIP are synonyms of one and the same disease, although unfortunately most British authors include the interstitial fibrosis associated with CVD in the term CFA.[9]

Irrespective of the controversy, these designations generally imply an interstitial lung disease without systemic commitment, with variable degrees of inflammation and fibrosis, and without any clinical or morphological finding which may suggest an etiology or histological changes diagnostic of other ILD of unknown cause, such as sarcoidosis or eosinophilic granuloma, for example. On the other hand, although IPF is a relatively well-characterized entity, it is probably not a single disease; the term refers to complex pathological changes which may result from several unknown etiologic agents. For instance, some evidence suggests that at least in some patients with IPF, the initial mechanism of injury is viral,[10] and recently Vergnon et al.[11] have found more evidence supporting this possibility. In the same way, other etiologic agents, apart from virus, may induce an initial damage which subsequently develops inflammatory and fibrotic lesions in the lungs. When the patients are biopsied, normally in an advanced stage of the disease, only pathological findings compatible with IPF are observed, and the track of the triggering injurious agent has disappeared.

Finally, notwithstanding therapy, many of the ILD do not resolve. They gradually progress to interstitial fibrosis and in the late phases may be, from a pathological point of view, undistinguishable from IPF.

This chapter will focus on the complex mechanisms of the pathogenesis of the fibrotic lung disorders, considering mainly

1. The linkage between inflammation and fibrosis and the dynamics of the abnormal accumulation of collagens in interstitium
2. The main clinical, radiographic, and physiological abnormalities common to all ILD
3. Why, how, and where the lung biopsy may be performed
4. The possible usefulness of some methods in the staging and follow-up of the patients, such as gallium index, collagen markers, and bronchoalveolar fluid analysis
5. The evaluation of the different attempts for inducing experimental pulmonary fibrosis

And finally, it will also give some insight as to their treatment.

In most of these topics, I will try to analyze the controversies and agreements among different authors, emphasizing the limitations of some methods and techniques used, both in the clinical approach and in research studies.

II. THE PATHOGENESIS OF DIFFUSE PULMONARY FIBROSIS

Regardless of the etiology, the ILD commonly present a sequence of steps which lead to the development of pulmonary fibrosis. These are

1. Initial lung damage
2. Alveolitis
3. Linkage between inflammation and fibrosis
4. Abnormal accumulation of collagen until the end-stage lung

The first two stages are closely associated with the particular disease in question. Since more than 130 diffuse interstitial lung disorders may evolve into fibrosis, it would be impossible to detail in this review what happens in each of them. Therefore, only a short and generalized description of these phases will be given. The two latter stages, however, are probably shared to a greater or lesser extent by all these lung disorders and constitute a common pathway which will be commented upon more extensively.

A. INITIAL LUNG DAMAGE

The initial phases in the pathogenesis of interstitial pulmonary fibrosis are difficult to analyze in humans because these diseases have an insidious onset and are diagnosed in advanced stages. Thus, much information has been obtained from animal models, although these do not always resemble the human disease, as will be seen further on.

Depending on the injurious agent, any parenchymal cell can be affected or stimulated, and the response will represent a balance between the type of cells affected, the magnitude of the injury, and the defense mechanisms initiated by the damage. For instance, some toxic agents such as high concentrations of oxygen (O_2), bleomycin, and butylated hydroxytoluene (BHT)[12-14] provoke early damage and a decrease in the number of capillary endothelial cells and/or type I pneumocytes. This is not surprising since both cells are quite sensitive to injury, whereas type II pneumocytes are relatively resistant.

On the other hand, the inhalation of inorganic dusts, mainly silica, appears to initiate the mechanisms of cell injury and repair through the alveolar macrophage as the target cell. The ingestion of silica has a cytotoxic effect on macrophages and induces the release of a great number of factors which amplify the damage and the inflammatory response.[15] Finally, the exposure to organic particles triggers an immunopathological response mediated mainly through a hyperreactivity of T-lymphocytes.[16] A similar mechanism seems to operate in some granulomatous diseases like sarcoidosis.[17]

B. ALVEOLITIS

Interstitial and alveolar inflammation are very important steps in the pathogenesis of these lung disorders and always precede the lung fibrosis, as demonstrated in biopsies of patients with early interstitial pulmonary disease[2] and in experimental animal models.[18] (See Figure 1.)

As previously mentioned, the characteristics of alveolitis depend on the disease or group of diseases analyzed, and there may be remarkable differences among them. However, all of them share two basic features: the increase in the number of inflammatory and immune effector cells within the alveolar structures as well as shifts in the relative proportions of these cells. In order to simplify this problem, Crystal et al.[3] have classified the inflammation into two types: lymphocyte alveolitis and neutrophil alveolitis. The former type is characterized by an increase of lymphocytes in lung tissue and bronchoalveolar lavage (BAL), and is generally found in granulomatous diseases such as sarcoidosis and HP. The latter type is dominated by the macrophage, but its most remarkable feature is the chronic presence of neutrophils, as occurs in IPF and asbestosis. However, it has been reported that eosinophils are frequently increased in some neutrophil alveolitis,[9,19] and I believe that perhaps granulocyte alveolitis is a more appropriate term. Unfortunately, only a few ILD are well characterized, and this concept may not be representative of them all.

Two mechanisms may participate in the generation and perpetuation of alveolitis: (1) the influx of inflammatory and immune cells from the bloodstream due to the local release of chemoattractant agents and (2) the *in situ* replication of these cells. For instance, in IPF, probably one of the most studied interstitial diseases, neutrophils are attracted by activated alveolar macrophages through the release of a chemotactic factor which seems to be the leukotriene B_4.[20] In addition, collagen peptide fragments show a chemotactic effect on human neutrophils,[21] and this may be another mechanism of recruitment of these cells to the lung. On the other hand, although the mediators responsible for the accumulation of eosinophils are unknown, mast cells, which are frequently increased in these disorders,[22] and macrophages probably release selective chemotactic factors for these cells. It is important to remark that, at least for neutrophils, once in tissues they have a very short half-life, which implies that chemotactic factors must be continually operating in granulocyte alveolitis.

Alternatively, although macrophages may be recruited as precursor monocytes from blood by lymphokines, it has been demonstrated with a combination of autoradiographic, flow cytometry, morphological, ultrastructural, and *in vitro* culture data that macrophages from patients with chronic lung inflammation replicate *in situ* in a proportion of 2- to 15-

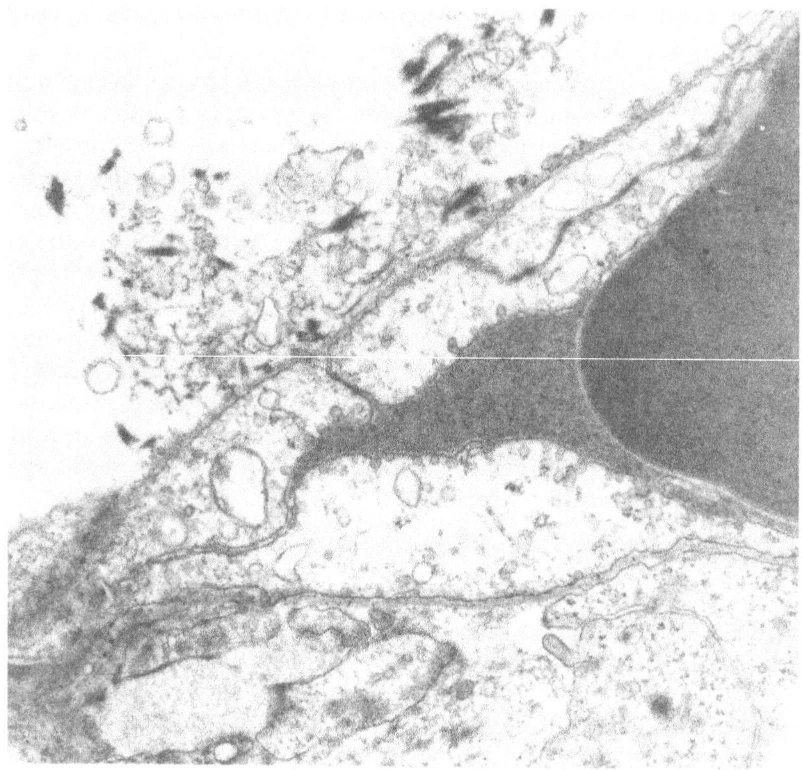

A

FIGURE 1. Sequential events of pulmonary fibrosis in the paraquat plus oxygen model studied by electron microscopy.[18] (A) Portion of an alveolar-capillary structure in which there is severe endothelial and epithelial damage, with swelling of the endothelial cell and destruction of the alveolar type I cell, 12 h after aggression. (Magnification × 10,000.) (B) Presence of interstitial and intraalveolar inflammatory cells 7 d after aggression. (Magnification × 2500.) (C) Increase of collagen fibers in the interstitium, 4 weeks after aggression. (Magnification × 10,000.)

fold compared with normal cells.[23] In the case of sarcoidosis (a lymphocyte alveolitis), for example, it has been suggested that the T-lymphocyte exerts a critical influence on cell traffic through the lung parenchyma. Furthermore, T-cells are actively proliferating in pulmonary sarcoid tissue.[17]

Once in the lung tissue, both inflammatory and immunocompetent cells can produce a great amount of mediators which may seriously damage the alveolar structures. With regard to this, neutrophils and macrophages release oxygen-derived radicals which produce, among other effects, a cytotoxic injury to epithelial cells.[24] In addition, they have a broad armamentarium of enzymes which, through a proteolytic effect, may exert a profound derangement to the lung structure. Eosinophils also have the capacity to injury lung cells and the extracellular matrix. Studies performed with unstimulated guinea pig eosinophils and those from BAL fluid of patients with ILD, have demonstrated a cytotoxic effect mainly for epithelial cells.[19] The role of eosinophils, nowadays poorly understood, has taken on great relevance with the recent evidence that their increase, at least in BAL fluid, seems to be associated with a bad prognosis in these disorders.[9,25]

Regarding immunocompetent cells, they can deteriorate the lung structures through an excessive helper function, through a subsequent abnormal T-lymphocyte production of lym-

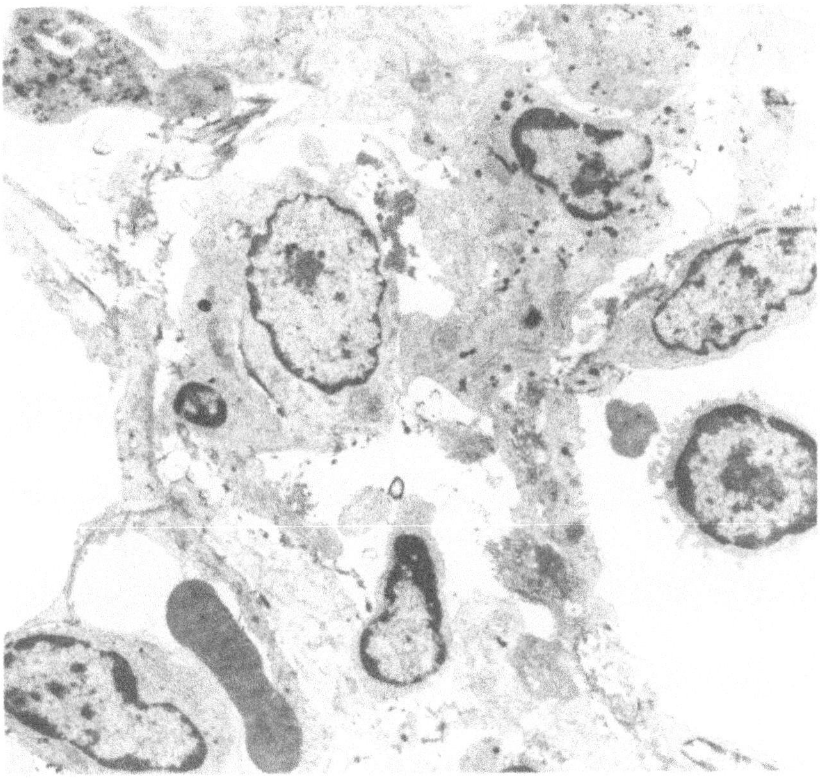

FIGURE 1B

phokines and cytotoxic effects,[17] and/or through hyperactive B-cell populations with the release of immunoglobulins and the formation and deposit of immune complexes.[26] Another pathological mechanism mediated by T-cells may be an autoimmune reaction to some macromolecules of the alveolar interstitium, mainly collagen, as has been suggested both in human and experimental lung fibrosis.[27,28] This phenomenon may explain the persistence of the inflammation in some disorders, even in cases when the initial injurious agent has disappeared.

Finally, I would like to emphasize that in absolute quantities, all these cells are increased and probably activated in almost all ILD. In this context, all of them may be involved to some degree in a series of interconnected events which cause the derangement of the alveolar-capillary structures during the stage of alveolitis. Thus, the inflammatory reaction is excessive, inappropriate, and very complex. Obviously high local concentrations of proteolytic enzymes, oxidant agents, and immune reactants overwhelm the inhibitory defenses.

Nevertheless, up to this phase of the disease the lung architecture and the alveolar-capillary structures have been relatively respected, and the pulmonary damage may still be reversible.

C. THE LINKAGE BETWEEN INFLAMMATION AND FIBROSIS

One of the most fascinating questions concerning the pathogenesis of pulmonary fibrosis is the linkage between alveolitis and fibrosis. In fact, there are two very different fates in the natural history or treated course of interstitial lung disorders. One of them consists of the resolution of the inflammation with the partial or total restitution of the parenchyma, and the other leads to progressive, irreversible, and usually lethal pulmonary fibrosis.

Although the presence of chronic alveolitis has been postulated as necessary to produce

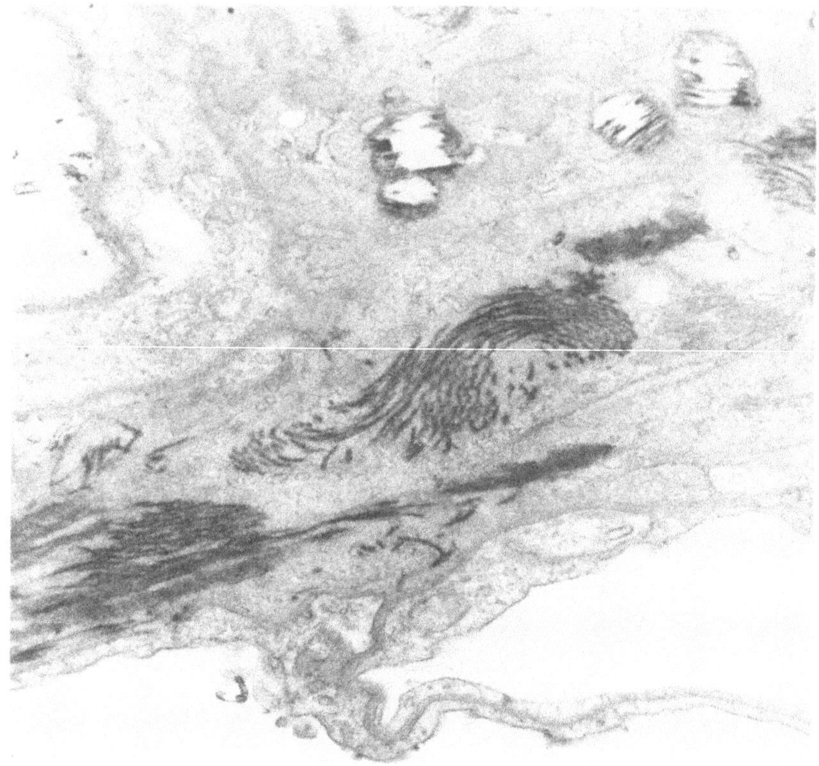

FIGURE 1C

the derangement of the alveolar structures which ultimately leads to fibrosis,[3] the permanence and autoperpetuation of the inflammation does not answer this question completely. Some subacute or chronic interstitial diseases such as DIP sarcoidosis, and HP[29-31] can resolve, whereas some acute lung disorders such as the adult respiratory distress syndrome (ARDS)[32] may lead to fibrosis within days or weeks. In this way, independent of the duration of the inflammation, some specific conditions must occur in order to trigger the fibrotic process.

This part of the chapter will focus on some events, both structural and cellular, which may culminate in this abnormality.

1. The Integrity of the Epithelial and Endothelial Basement Membranes

Basement membranes (BMs) are involved in a number of biologically critical processes. Among other properties they have positional information and spatially orient the attached cells. In addition, they play a role in the relationship between the cells and the interstitial matrix. Although the biochemical characterization of their components is not completely known, they seem to be formed at least by type IV collagen, proteoglycans, laminin, and fibronectin, among other molecules. (For review see Reference 33.)

The importance of BMs in the orderly regeneration of the damaged lung has been recognized for a long time.

Vracko[34] in 1971 studied by electron microsocopy the sequential changes produced in lung tissue of dogs injured with an intravenous injection of oleic acid. The early alterations consisted of necrosis of all types of lung cells. However, the epithelial and endothelial BMs remained unaltered and could be identified by their relation to interstitial collagen and elastic fibers. Afterward, regeneration of the epithelial and endothelial cells from the adjacent viable portions of the lung was observed. Apparently guided by their BMs, these cells proliferated and in 3 weeks almost completely reestablished the structure and function of the lung. The new cells grew directly upon the original BMs and did not deposit a new layer.

More recently, Raghu et al.,[32] using indirect immunofluorescence with polyclonal affinity-purified antibodies to human type IV collagen and laminin, found that there is a disruption of the alveolar BMs in patients with lung fibrotic disorders. This alteration was observed in 12 out of 17 fibrotic lungs and was demonstrated by the presence of "gaps" in the distribution of these components in the BM. Moreover, this break occurred much more frequently in initial stages of the disease, suggesting that the damage of this structure is an early event in pulmonary fibrosis.

Taken together, these observations support the notion that integrity or disruption of BMs may play a major role in the signals responsible for the regeneration or abnormal repair of the lung tissue.

On the other hand, two types of cells, the mast cells and the polymorphonuclear leukocytes, have enzymes that may destroy some components of the BM. Mast cells have several neutral proteases which can degrade fibronectin, collagen type IV, and proteoglycans.[35] Human leukocytes present proteolyic activity against type IV collagen, with an enzyme different from the collagenases which destroy types I and III collagens.[36]

2. The Role of Epithelial Cells

Haschek and Witschi[37] have hypothesized that fibrosis can be touched off if reepithelialization following acute lung damage is compromised by a second toxic agent. They demonstrated that the administration of BHT to mice produces necrosis of type I epithelial cells, and afterward an active proliferation of type II pneumocytes can be observed. If mice are exposed to high concentrations of O_2 during this period, type II cells are severely affected in contrast to interstitial cells which continue proliferating. The result is that BHT alone produces slight lung fibrosis, whereas BHT plus O_2 causes extensive interstitial fibrosis determined by collagen measurements and histopathological examination.

Similar results were obtained by us.[38] We demonstrated with paraquat and O_2 that the timing of the second aggression is critical. The pulmonary paraquat toxicity-enhancement effect of O_2 in rats was found to last approximately 24 to 48 h. During this period, exposure to 74% O_2 results in the development of severe pulmonary lesions and early death. Two days after paraquat administration, exposure to the same high O_2 concentration has no effect at all on the survival and on the normal histology and collagen content of the lungs, although in this study reepithelialization was not analyzed.

These data suggest that exposure to two agents with a specific timing has a synergistic, damaging effect, and normal proliferation of type II pneumocytes may in some way control fibroblast multiplication and, thus, probably help to avoid the development of fibrosis.

3. Factors Which Influence Fibroblast Behavior

Independent of their etiology, an increased number of fibroblasts is always observed in pulmonary fibrosis, both in humans and animals,[18,39] and the expansion of the lung fibroblast population seems to play a major role in the development of fibrosis.

In the last few years, a plethora of soluble mediators capable of regulating several fibroblast functions has been described,[40] and they may act in three ways: as chemoattractants, as proliferation factors, and as effectors of collagen production. However, it is important to note that most of the studies have been performed *in vitro,* and different variables including cell-culture conditions, fibroblast source, and methods to identify these molecules have been used. On the other hand, the same cells may produce both stimulating and inhibiting factors.[41] All this may explain some of the conflicting results reported in the literature. In pulmonary fibrosis several of these mediators have been found and partially characterized.

The early experiments were performed by Heppleston and Styles.[42] They exposed rat macrophages to silica, transferred the supernatant to independently grown cultures of fibroblast, and found enhanced cell proliferation and collagen production. More recently, Schmidt et al.[43] have suggested that the factor responsible for fibroblast duplication may be the interleukin-1. Nevertheless, and emphasizing the contradictions that may arise in *in vitro* studies, Gritter et al.[44] have demonstrated that the response of fibroblasts remarkably changes with the culture conditions. In this sense, supernatants of silica-treated alveolar macrophages inhibited division of rapidly growing fibroblasts, whereas the same material stimulated growth of slowly dividing cells. On the other hand, supernatants increased collagen production by confluent fibroblasts for 3 d, whereas the same supernatants inhibited collagen synthesis by confluent cells for at least 8 d. Thus, the macrophage-fibroblast interaction *in vivo* must be much more complex than we thought and will depend on a series of local conditions which at this time we do not know with certainty.

In human pulmonary fibrosis several factors that affect the function of fibroblasts have been described, all of them proceeding from mononuclear phagocytes.

Rennard et al.[45] studied the production of fibronectin from alveolar macrophages obtained from 48 patients with ILD and 7 normal volunteers. They found that macrophages from patients with IPF produced fibronectin at a rate 20 times higher than that of normal macrophages. In the sarcoidosis cases, the results showed 10 times the normal rate, and macrophages from 6 out of 10 patients with other interstitial disorders produced fibronectin at rates greater than the highest normal control. On the other hand, this fibronectin was chemotactic for human lung fibroblasts, suggesting that this molecule may be one of the chemoattractant factors for these cells in lung fibrosis. Their source is the macrophage.

Bitterman et al.[46] studied alveolar macrophages obtained from normal subjects in the search for fibroblast growth factors. They reported that supernatants from unstimulated macrophages contained no growth factor activity. In contrast, supernatants from macrophages stimulated with different particles and immune complexes contained a growth factor that caused a significant increase in the fibroblast replication rate. This mediator had an apparent molecular weight of 18,000 and appeared to be distinct from other previously characterized growth factors. In addition, in serum-free culture, this factor by itself did not promote fibroblast replication. The authors postulated that it acted as a progression factor.

Afterward, Bitterman et al.[47] analyzed the spontaneous release of the same mediator in ILD. For this purpose they studied the alveolar macrophages obtained from 65 patients with fibrotic lung diseases and 30 controls. Whereas none of the normal subjects had macrophages spontaneously releasing a growth-promoting activity for fibroblast, 82% of the patients showed alveolar macrophages that released this factor. This alveolar macrophage-derived growth factor (AMDGF) presented the same biochemical and functional characteristics as the molecule previously reported by Bitterman.[46] The authors suggested that the fibroproliferation within the alveolar structures in interstitial lung disorders may result, at least in part, from the release of AMDGF from cells stimulated *in vivo*.

Kovacs and Kelley[48] have demonstrated that a similar if not identical factor is found in the alveolar lining fluid obtained after acute lung injury induced by bleomycin.

Following the same line of thought, Bitterman et al.[49] postulated that since alveolar macrophages from most patients with interstitial fibrosis release two primary growth factors for fibroblasts (fibronectin which acts as a competence factor and AMDGF which acts as a progression signal later in G_1), both required for optimal cell proliferation, the combined action of them may explain the expansion of fibroblast number in the lung parenchyma. In this study they analyzed the influence of other alveolar macrophage mediators, which also play a role in modulating fibroblast replication, on the lung fibroblast response to fibronectin and AMDGF. They found that prostaglandin E_2 is able to inhibit fibroblast proliferation, while gamma interferon has no effect. In contrast, interleukin-1 slightly augmented fibroblast replication in response to fibronectin and AMDGF but was unable to provide a primary growth signal to nonreplicating fibroblasts.

It is important to remark that Duncan and Berman[50] suggested that gamma interferon is the lymphokine and beta interferon the monokine which inhibit collagen synthesis and late, but not early, fibroblast proliferation. On the other hand, concerning interleukin-1, Schmidt et al.[51] performing a series of biochemical analyses suggested that this molecule, under their assay conditions, stimulated fibroblast replication. Thus, the relative importance of interleukin-1 with respect to other macrophage-produced growth factors has not been clearly defined.

It has recently been suggested that the platelet-derived growth factor (PDGF), a potent mitogen and chemoattractant for fibroblasts and smooth muscle cells as well as a stimulator of collagen synthesis, can be produced by mononuclear phagocytes,[52] including alveolar macrophages.[53]

Mornex et al.[53] studied the spontaneous expression of mRNA complementary to the c-sis gene, which codified for one of the chains of PDGF from alveolar macrophages obtained from 14 normal subjects (smokers and nonsmokers) and 22 patients with fibrotic lung diseases. They found that all alveolar macrophage RNA samples evaluated expressed this factor. Concomitantly, based on several biochemical measurements they also demonstrated that this cell released a mediator with the properties of PDGF. The authors suggested that the c-sis proto-oncogene and its PDGF product are part of the armamentarium available to the alveolar macrophage for normal lung defense and for their participation in lung inflammation.

Nevertheless, in almost all of the studies discussed above, macrophages have been obtained from BAL, and it is unclear if these cells are really the same as those present in lung tissue. In this context, Lum et al.[54] demonstrated significant differences in a great number of parameters by detailed morphometric analyses on *in situ* macrophages and lavaged pulmonary macrophages from control and ozone-exposed rats. In contrast, studying some immune functions, Weissler et al.[55] found that macrophages obtained by BAL are largely similar to those obtained from whole lung minces, at least in their ability to stimulate T-lymphocyte proliferation and in their expression of HLA-DR antigens. Furthermore, great heterogeneity has been described in normal rat alveolar macrophages,[56] and up to now, we do not know if they are modified under pathological conditions.

Thus, with the purpose of gaining a better approach, we have recently examined the presence of PDGF *in situ* in patients with ILD.[57] Samples of 30 human ILD lung tissue from 15 patients with IPF, 13 with HP, 1 with DIP, and 1 with lymphoid interstitial pneumonia were studied by indirect immunofluorescence. Furthermore, 12 rats with experimental lung fibrosis induced by paraquat and O_2 were analyzed. Our results showed that 60% of IPF patients and 30% of HP cases had a positive immunofluorescent pattern with antisera against PDGF. In the experimental model a progressive interstitial inflammation was observed which coincided with positive immunofluorescence in all rat lungs examined. Normal lungs were negative for PDGF. These results partially support the finding by Mornex et al.[53] and provide the first direct evidence for the presence of PDGF in lung tissue of patients and animals with interstitial pulmonary disease. On the other hand, immunofluorescence was associated with mononuclear phagocyte cells present in interstitium and alveolar spaces (Figure 2).

However, interaction between macrophages and fibroblasts *in vivo* must be very complex since macrophages have the capacity to stimulate fibroblasts and to suppress them. Moreover, both cells produce collagenase and collagenase inhibitor,[58] "the other side of the coin", of which the actions on collagen turnover and on maintenance of proper connective tissue architecture are of noteworthy importance, as will be seen further on.

In this way, under certain specific stimuli, an imbalance must occur in lung parenchyma between the mediators which stimulate and inhibit fibroblast proliferation, collagen synthesis, and collagen degradation, resulting in the initiation of the fibrosing process.

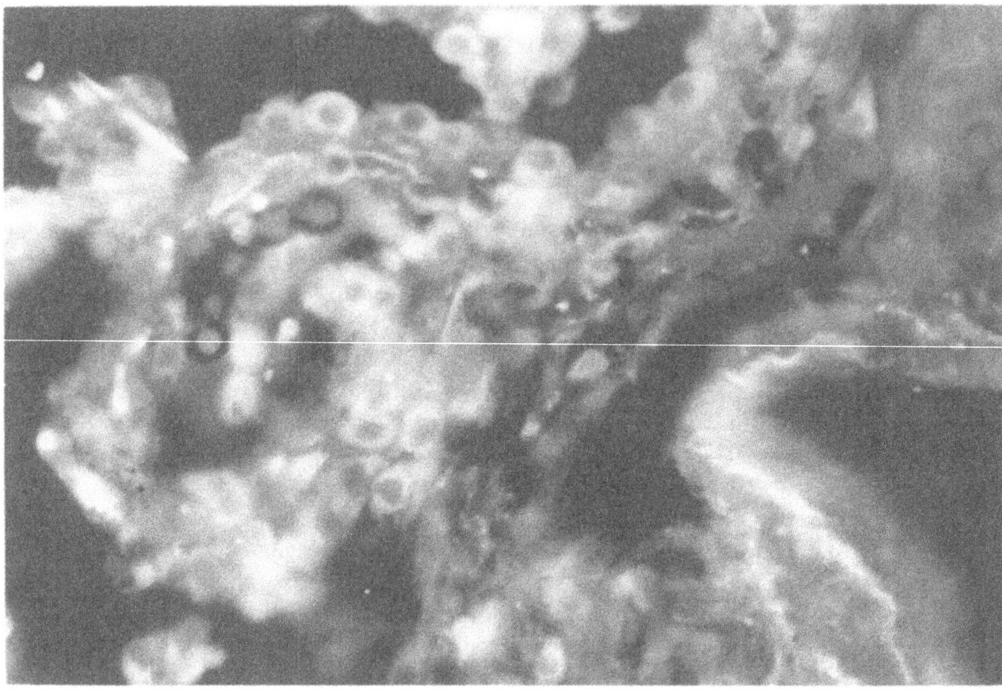

FIGURE 2. Indirect immunofluorescence staining pattern of a lung specimen from a patient with IPF. Numerous interstitial and alveolar mononuclear cells labeled with anti-PDGF antibody can be observed. Autofluorescence of elastin and collagen fibers is also seen.

In addition to macrophages, other cells may play a role in the early stages of fibrogenesis, although their possible participation in lung fibrosis has not been defined. For example, lymphocytes may activate fibroblast proliferation and collagen production. Thus, a 22,000 mol wt protein that acts as a fibroblast chemoattractant, a 60,000 mol wt factor which stimulates fibroblast division, and a 170,000 mol wt protein which increases collagen synthesis have been described.[40] Again, a 55,000-mol wt mediator of T-lymphocytes which inhibits collagen production has also been described. In fact, similar to macrophages, these cells are capable of releasing both a stimulator and an inhibitor of collagen synthesis, and the relative amounts of each of these factors appear to be influenced *in vitro* by several variables such as cell density, type of stimulant, and the culture period.[41]

Although these actions have not been studied *in vivo*, lymphocytes are believed to play an important role in the pathogenesis of several ILD with immunopathological abnormalities which can evolve into fibrosis. Interestingly, in experimental pulmonary damage the artificial decrease of lymphocytes through several procedures seems to inhibit the development of fibrosis markedly.[59,60] Thus, independent of their immunological and inflammatory amplifier cell role, the lymphocytes may be involved in the connective tissue alterations observed in lung fibrosis. Very recently Cathcart et al.[61] demonstrated that T-cell supernatants of IPF patients increase the proliferation of fibroblasts *in vitro*.

It has been recognized for some time that mast cells are increased in the interstitium of patients with pulmonary fibrosis,[22] although their real function is unclear.

Recently, Goto et al.[62] have studied the distribution, density, and histochemical subtype of mast cells in rats with pulmonary fibrosis induced by bleomycin. Pulmonary mast cell changes were present early in the fibrotic process, and by day 14 their density in the parenchyma was ten times higher than normal. In addition, lung histamine levels increased as much as 14-fold by day 50. Hawkins et al.[63] studied 12 patients with progressive systemic sclerosis (PSS), a multisystem connective tissue disease, and found that the number of mast cells in skin clinically involved with early stages was significantly greater than that in clinically uninvolved skin of the same patients. Moreover, the count of these cells in damaged skin from patients with late scleroderma was normal.

However, these are mainly morphological observations which do not clarify if and how mast cells are involved in the fibrogenic process.

Electron microscopic studies have demonstrated a close spatial relation between mast cells and fibroblasts, and improved techniques for the maintenance of living cells in culture have allowed the observation that mast cell granules appear to be phagocytosed by fibroblasts,[35] a phenomenon that has also been observed *in vivo*.[64] The ingestion and degradation of these granules by fibroblasts may represent an important mechanism in the regulation of these cells' activity. An indirect proof that this may occur in ILD has been reported by Bjermer et al.[65] who studied 69 patients with sarcoidosis and found increased levels of hyaluronate and type III procollagen propeptide in BAL, probably due to activated fibroblasts. This finding was strongly related to an elevation of the lavage mast cell counts. Additionally, at least histamine, one of the principal factors released by mast cells, increases collagen synthesis without stimulating fibroblast proliferation,[66] and augmented levels of histamine have been reported in BAL fluid from patients with IPF[67] as well as in bleomycin-induced lung fibrosis in rats.[62]

We can speculate, based on *in vitro* and *in vivo* observations, that the mediators of mast cells can influence fibroblast behavior, the extracellular matrix, and, as mentioned before, the integrity of BMs.

The dysfunction of endothelial cells may be involved, at least in some diseases, in the initial phases of fibrogenesis. In several pulmonary disorders there is damage to these cells, and it has been demonstrated that they can release PDGF-like mitogen.[68] Ward et al.[69] studied in irradiated rat lung the relationship between endothelial damage, monitored through lung angiotensin converting enzyme activity, and collagen accumulation, measured by hydroxyproline concentration. They found a linear dose-response curve for both parameters at 2 and 6 months (early and advanced fibrosis) after a single dose of irradiation. Nevertheless, the possible role of the endothelial cell deserves a more detailed study.

Finally, as to neutrophils and eosinophils, their great capacity to damage secretory products can severely affect lung structure and function. As we have seen, both types of cells have been implicated with worsened patients suffering from some ILD[9,25] and with the production of enzymes that affect the extracellular matrix and/or the BMs. However, their precise role, if any, in the linkage between inflammation and fibrosis remains unclear.

On the other hand, although much evidence points to the fibroblast as the cell directly responsible for synthesizing the increased collagen in pulmonary fibrosis, thus playing an essential role in the initial phases of fibrogenesis, it is possible that other cells from non-mesodermal origin may take part in this process as seems to occur in other tissues.[70] In this context, Crouch et al.[71] have reported that type II epithelial cells of lungs produce collagenous proteins. They used primary cultures of rat lung type II pneumocytes and found that three major bacterial collagenase-sensitive chains were synthesized and secreted into the medium by this cell. Biochemically, two chains seemed to be type IV procollagen. The third was a novel, low-molecular-weight collagenous protein which is immunologically and chemically distinct from previously described collagenous surfactant apoproteins. Although this finding may only reveal a production of BM collagens, we could speculate that under pathological conditions type II pneumocytes, which are usually observed in interstitial lung disorders in areas with varying degrees of fibrosis, play a role in interstitial collagen production. This possible mechanism deserves to be deeply studied.

Taken together, all these investigations suggest that there is a delicate cellular and molecular equilibrium which allows a correct collagen turnover in lung parenchyma. The loss of this balance during inflammation, both acute or chronic, may lead to the beginning of the fibrogenic process.

Clearly, it is impossible to believe that one or another cell is the culprit. This would be as much as accusing the second violin to be responsible for the turbulent music performed by a completely discordant orchestra. In this sense, all inflammatory and immunocompetent cells, as well as some parenchymal cells including the fibroblast itself, must play, to a greater or lesser extent and directly or indirectly, a role in the development of fibrosis in these lung disorders.

D. THE ABNORMAL ACCUMULATION OF COLLAGEN

The biochemical studies of collagens in pulmonary fibrosis are relatively recent and have produced some conflicting results, mainly due to the different methods used, the way in which the findings are expressed, and the limited number and clinical heterogeneity of the analyzed patients.

The behavior of collagen in pulmonary fibrosis will be discussed in relation to three questions.

1. Is There an Increase in the Amount of Collagen in Lung Fibrosis?

This question may appear somewhat irrelevant since the amount of this protein has been found to be augmented in all other fibrotic tissues studied.[72-75] In fact, the increase in collagen is considered the biochemical feature of fibrosis. Furthermore, pulmonary fibrosis is a general term used to describe diseases with histological evidence of diffuse thickening of alveolar walls, where the main protein stained is collagen.

Nevertheless, Fulmer et al.[76] reported that in IPF, a prototype of the fibrotic lung diseases, no increase in collagen content could be observed. The authors studied nine patients with IPF, measuring collagen through quantification of hydroxyproline. The results were expressed as collagen/dry weight and collagen/DNA. For both, there was a relatively broad dispersion of the data in patients with IPF, but the average was not different from that of controls ($n = 6$). This finding was unexpected, and Fulmer et al. speculated that, at least in IPF, there was not an increase in collagen concentration, but a shift in the ratio of interstitial collagens. This inference was based on results of a previous report by Seyer et al.[77] and not on the study of their own samples.

This statement, accepted for years with scarce critical thought, was nevertheless an astonishing proposition. Could it be that the fibrotic reaction of the lung differed from that in other tissues?

A previous study of acute lung fibrosis in man[78] and several studies performed in experimentally induced fibrosis[18,79-82] had shown an appreciable increase in total lung collagen. The human study was realized in 12 patients who died after severe acute respiratory failure. Collagen concentration was measured through quantification of hydroxyproline in several tissue samples taken from one to three lung lobes, and the amount was low or normal. When a correction was made substracting hemoglobin, collagen concentration increased but was not higher than in controls. Finally, when the results were expressed in grams per square meter of body surface area, the 10 patients who died after 10 d presented a significant increase of this protein. Nevertheless, this is a very acute disorder and does not necessarily reflect the events which occur in more chronic diseases.

More recently, three studies[83-85] performed in patients with chronic pulmonary fibrosis have demonstrated that collagen is augmented. In a first assay we[83] analyzed the lung tissues obtained from 15 patients with diffuse lung fibrosis of different etiologies and 6 controls. Afterward, Kirk et al.[84] studied 21 patients with IPF and 17 normal subjects, and finally, we studied 11 IPF cases and 6 controls.[85] The collagen content was measured by different hydroxyproline quantification methods and was expressed as concentration, that is, milligrams or micrograms of collagen by gram or milligram dry weight, respectively. In all these studies collagen concentration was significantly higher in patients than in controls. Furthermore, the concentration was greater in post-mortem lung samples, supporting the notion that collagen deposition is a progressive phenomenon.

The differences between the report by Fulmer et al. and the other studies are difficult to explain. In part, it might be due to the selection and amount of tissue studied, since the fibrotic lung disorders produce patchy, unevenly distributed lesions with varying degrees of fibrosis, and sampling errors may occur. Another problem relates to the manner in which the data are expressed, and that is not an irrelevant issue. For example, when lung collagen is calculated in relation to DNA, mistakes may be made because in these diseases there is always an influx of inflammatory cells with a concomitant increase of DNA. A similar misinterpretation occurs when collagen is expressed as concentration or as collagen mass because under pathological conditions lung weight, wet and dry, increases substantially. Unfortunately, we do not know any biochemical parameter which remains stable in the interstitial lung disorders. However, the last three human studies mentioned previously clearly demonstrate that there is an increase in collagen deposit, and this can well be evaluated as concentration.

Recently, Yamaguchi et al.[86] have suggested that collagen content may be calculated per lung volume, analyzing the degree to which a given lung is transformed from a standard state of expansion. They examined 14 lungs from patients who died of paraquat intoxication. Using a correction by morphometry, based on the sampling of the number of arteries, they found an increase in hydroxyproline content per unit volume, and their biochemical results correlated with the morphological study of the same sample. For example, in a patient who survived 102 d, the longest in their series, the histological examination of the left lung showed extensive fibrosis with correspondingly high hydroxyproline content, whereas the upper lobe of the right lung, where most of the alveolar structures were preserved, presented a normal quantity of this imino acid. If this study is confirmed by other authors for different types of diffuse pulmonary fibrosis, the method based on a morphometric correction could be very useful for calculating collagen content in these disorders.

2. Which Genetic Type or Types of Collagen Are Altered in Lung Fibrosis?

Our knowledge about collagen has changed remarkably in the last 15 years. One of the major discoveries is that there is a family of collagens, and each protein has distinctive structural and functional characteristics. Currently, at least 11 different types of collagen are thought to exist.[87]

In the pulmonary parenchyma, the target of interstitial fibrosis, there are four types of collagens which together constitute about 15% of the dry weight of the human lung.[88] Type I collagen is widely distributed in lung, including the alveolar interstitium, and type III seems to be part of the reticular fibers.[89] These are the most abundant under normal conditions, and they are apparently present in a ratio of about 2:1 in the alveolar septa.[77] Types IV and V collagens are codistributed in linear patterns in the epithelial and endothelial BMs and constitute important components of the integrity of these structures.[90]

The question cncerning the possible changes in relative distribution of the different types of collagens in human lung fibrosis has been investigated by several authors using two different approaches, each with advantages and disadvantages. Immunohistochemical analysis[32,90-92] has the great virtue of allowing the observation of the anatomic localization and distribution of the different types of collagen. However, it has three drawbacks:

1. It is not a quantitative method.
2. Some protein may not be stained because it can be masked by its complex interactions with other macromolecules.

3. There may be interbatch differences in specificity and affinity of anticollagen and fluorescein-labeled antibodies.

The biochemical techniques, generally based on collagen extraction by pepsin digestion and on the separation of the alpha-chain by chromatography or by sodium dodecyl sulfate polyacrylamide gel electrophoresis and/or lung homogenate cleavage with cyanogen bromide,[77,90,93] allow more quantitative measurements but present the following problems. First, the extracted collagen is not complete, and some types of collagens are more easily solubilized than others. In this context, we do not know if the collagen which is not extractable has the same characteristics as the extracted protein. This problem diminishes when several enzymatic digestions and additional extraction with a neutral buffer are performed or when a complete lung homogenate cleavage with cyanogen bromide is carried out. Second, the method does not allow a view of the topographic distribution of the collagens in the pulmonary parenchyma. Third, if the results are expressed as relative percentages of type I, to type III collagen, for example, the real absolute quantitities of each collagen may be wrongly estimated. In this way, an increase in the ratio of I/III may denote either that there is more amount of type I with low or normal content of type III or that there is an increase in both types of collagens, but more of type I.

In addition, the studies in human lung fibrosis are few, with a low number of patients in most of them, and sometimes there is great heterogeneity in the cases analyzed. Nevertheless, some light has emerged from these studies, and we currently have an approximate vision of what happens in these disorders.

An early report by Seyer et al.,[77] using digestion of whole lung tissue with cyanogen bromide and a chromatographic separation of the resultant alpha (I) and alpha (III) peptides, suggested an increase in the amount of type I collagen relative to type III in post-mortem samples of five patients with IPF. Unexpectedly, a later work headed by the same author[93] using the same method on lungs from three patients who died with PSS, showed that type I and type III collagens were present in these patients in identical proportions to normal lungs. This is surprising because pulmonary fibrosis, at least in advanced phases, is very similar in IPF and PSS. Unfortunately, none of these reports includes a detailed description of the morphological findings. The differences may be due to the small number of patients studied or, as Seyer postulates in the more recent paper, to the expression of a particular gene product in fibrosis, dependent on the specific stimuli operative at the fibrotic process.

Madri and Furthmayr[90] used indirect immunofluorescence to study the localization of types I, III, IV, and V collagens in lung tissues from six patients who died from respiratory failure due to pulmonary fibrosis of different etiologies. They found a marked increase of type I collagen in the thickened septae and a decrease of type III. Type V was remarkably augmented in the interstitium and located in areas of smooth muscle cell proliferation. These modifications were most prominent in advanced fibrosis ($n = 5$) than in the initial phase ($n = 1$), and their findings correlated well with biochemical analysis. It is important to note that the ratio of types I/III/V in normal lungs was 35:58:7, which is very different from the report by Seyer[77] of about 66:33 for types I/III, and serves to emphasize the methodological problems.

Bateman et al.[91] also studied by indirect immunofluorescence the tissue localization of collagens type I, II, III, and IV in normal fetal and adult lung, and in fibrotic pulmonary lesions. In addition, they analyzed the morphological alterations.On the basis of collagen type fluorescence, two distinctive patterns were identified. In sites with established, mature fibrous tissue there was an important increase of collagen type I. Type III was seldom detected in these locations. In contrast, in sites of early or active fibrosis the content of both types was increased, and type III was often similar to type I. The authors suggested that the immunohistochemical study of collagen is a useful and sensitive method for detecting small overall increases in lung collagen, and that the specific type may give a clue about the stage and activity of the disease. Afterward, Bateman et al.[92] in a very complete study of 25 patients with IPF, demonstrated a positive association between the presence of type III collagen and the stage of activity of the disease. Microscopy of the immunofluorescent sections was performed under transmitted ultraviolet light, which provides a contrast between the specific apple-green fluorescence of collagen and the nonspecific blue autofluorescence caused by the refractile property of the tissues. A strong correlation was observed between the presence of type III collagen in the lung tissue and the clinical course of the disease and the response to treatment. In this sense, 9 out of 11 patients with type III collagen improved in contrast to most of the cases with no type III collagen fluorescence, which had a stable course over the time of the study and did not improve with the therapy.

Recently, Raghu et al.[32] reported an investigation, also based on the indirect immunofluorescence technique, in which they analyzed 10 normal and 32 abnormal lung specimens from three different groups of patients. The first group consisted of 10 patients with ARDS, the second group was formed by 7 cases of IPF, and the third group consisted of 15 patients who had lung abnormalities but no interstitial fibrosis. Using polyclonal affinity-purified antibodies to human collagens type I, III, IV, and laminin, they found in normal lungs that type IV and laminin codistributed in a uniform, smooth linear pattern along the basement membranes. Type III collagen was present within the alveolar septa and interstitium in an interrupted ribbon-like pattern. Type I collagen was irregularly distributed within the alveolar wall and was less prominent than type III. In abnormal lungs without fibrosis, a similar distribution of these components was observed. On the other hand, in the groups with pulmonary fibrosis collagen types I and III accumulated in the interstitium. As well as in the Bateman reports, the accumulation of type III was predominant in the thickened alveolar septa during the early stages, whereas type I appeared to be the principal collagen at later stages of the course of the diseases. The abnormal distribution was identical in ARDS and IPF groups.

Taken together, all these studies, with their inherent limitations, suggest that a dynamic deposition of interstitial collagens exists in pulmonary fibrosis with a redistribution of collagen types which changes with the course and duration of the disease. The first to increase is type III and afterward type I and probably type V. The shift may be explained by a switch in the cells responsible for collagen synthesis or by the proliferation in the late stages of a distinct population of cells which mainly produce type I collagen. Nevertheless, it is necessary to look more deeply into the molecular factors and inflammatory mediators that modulate the relative amount of each type of collagen.

3. What Is the Reason for the Increase of Collagen in Pulmonary Fibrosis?

In the last instance, the common feature of fibrosis must be an imbalance in the normal homeostasis of collagen so that synthesis exceeds breakdown, resulting in an excessive accumulation of this protein. Therefore, it is very important to know the collagen turnover rates under both normal and pathological conditions. In this sense, our understanding of lung collagen metabolism has changed dramatically in the last few years. In the past, the prevailing view was that this protein turned over very slowly.[94] However, several recent studies, using improved methods, have suggested that synthesis and degradation rates are much more rapid. The analysis of the kinetics of this process have been studied *in vitro* and *in vivo* in animal lungs.

Bienkowski et al.[95] studied explant cultures of rabbit lung parenchyma and by the evaluation of the size distribution of [^{14}C] hydroxyproline- and [^{14}C] hydroxylisine-containing peptides, demonstrated that 20 to 40% of the alpha chains of newly synthesized collagen were degraded within minutes of being produced.

Laurent[96] and Laurent and McAnulty[97] have analyzed the rates of synthesis and degradation of collagen *in vivo* in normal and bleomycin-induced fibrosis in rabbit lungs. In both studies [^{3}H] proline was administered by intravenous injection together with a large amount of unlabelled proline, and the synthesis was determined by the specific radioactivity of hydroxyproline in the protein-pool. In the later work degradation rates were assessed by several independent methods, including measurements of hydroxy ^{3}H-proline in the tissue-free pool as an index of the digestion of newly synthesized collagen. Their results demonstrate that in normal lung, collagen was synthesized and destroyed at a rate of about 10% each day, and approximately one third of the newly produced collagen was degraded rapidly after its production, which agrees with the *in vitro* studies of Bienkowski. These findings support the hypothesis that the lung is an active tissue in terms of collagen metabolism and suggest that the cells responsible for collagen production are continually synthesizing and degrading this protein. In this way, modifications in anabolism or catabolism may play a major role in the accumulation of collagen in pulmonary fibrosis.

Unfortunately, studies in normal and fibrotic human lungs are scanty.

Along with collagen synthesis, Fulmer et al.[76] quantified the rates of collagen and noncollagen protein synthesis in nine open lung biopsy specimens of IPF patients and five controls, using the uptake of radiolabelled proline into hydroxyproline in explant cultures. They did not find differences between either group.

More recently, we[85] analyzed the rate of collagen synthesis in lung tissue of 11 patients with IPF and 6 controls with the same method, and similar findings were obtained.

Both studies may have several imperfections. Again, the major problem relates to the way in which the results are expressed. Fulmer reported collagen synthesis related to concentration of DNA, but the increase of inflammatory cells may mask the increase of collagen synthesis. The same problem seems to exist when results are expressed as the ratio of collagen/noncollagen synthesis, since an increase in the production of other proteins, as occurs in lung damage, would not alter this ratio even though absolute collagen production is increased.

Furthermore it is important to note that the *in vitro* studies may underestimate the rate of collagen synthesis and lead to an erroneous interpretation of the results, because the rates seem to be slower than those obtained *in vivo*.

In this sense, Kehrer et al.[98] have compared the *in vitro* and *in vivo* kinetics of collagen synthesis in normal and damaged pumonary tissue and have demonstrated that the *in vivo* rate of hydroxyproline production is significantly greater than the *in vitro* rate in lung samples from normal and lesioned mice. The difference ranged from fivefold in normal lung to eightfold in pulmonary tissue damaged by the administration of BHT.

Obviously, this type of assay may not be applied to human beings. However, in spite of these limitations, our findings and those of Fulmer can be truthful because the patients in these studies were in advanced stages of the disease. It may be that this alteration in collagen metabolism occurs early in the development of these disorders and returns to normal levels later on.

A different picture is found in experimental lung fibrosis, where independent of the injurious agent and the method used there is almost always an increase in collagen synthesis. However, carefully observed the results, compared with the two reports in humans, may be reconciled.

For example, Phan et al.[99] examined collagen synthesis in rat lungs after intratracheal bleomycin. They found an increase in the synthesis rate, but the values returned to normal levels within 10 d.

Lee and Kehrer,[100] using a model of lung tissue lesion induced by cyclophosphamide in mice, demonstrated an increase of collagen synthesis on day 6 which was restored to control levels after about 3 weeks. This is a very interesting finding because Morse et al.[101] have reported, for the same species and with a unique low dose of cyclophosphamide, that there is an increase in collagen content and in the morphological lesions which progress at least for up to 1 year. Integrating both studies, one acute and one chronic, it is possible to deduce that while synthesis begins to diminish toward normalcy, collagen content continues in constant increase.

During 15 d, Kehrer and Witschi[102] studied the changes in collagen synthesis in a pulmonary fibrosis model developed with BHT and O_2. They observed an important increase in collagen synthesis by the seventh day which decreased progressively with time, although it remained somewhat higher than controls to the 15th day, the last of the study. However, the accumulation of hydroxyproline in lung was constantly increasing and reached its maximum value on the 15th day. Again, while synthesis showed a peak which decreased with time, collagen content continued to increase.

Hesterberg et al.[103] analyzed the biochemical, physiological, and histological changes in bleomycin-induced pulmonary fibrosis in rats. They found a dose-dependent increase of collagen synthesis which decreased after 3 weeks. However, the collagen content began to increase at 2 weeks and remained higher until the final point of this measurement, 8 weeks.

Thus, collagen biosynthesis, at least in these models, seems to be a transient state during the development of fibrosis.

On the other hand, it is possible that in some cases the maintenance or cessation of the noxious agent may play a role in the behavior of biosynthesis.

Thus, Collins et al.[104] demonstrated in baboons treated with bleomycin that the synthetic rates of total protein and of collagen were markedly stimulated in the early biopsy specimens, but were similar to those of control animals in the later biopsy samples obtained 3 months after withdrawal of the drug.

Comparable results have been reported in other tissues with fibrosis. Ehrinpreis et al.[105] found in rats with cirrhosis induced by carbon tetrachloride (CCl_4) that collagen synthesis increased at least twice more than in controls during CCl_4 treatment, but the animals sacrificed after CCl_4 treatment showed normal values in spite of higher content of collagen in the cirrhotic liver from the second week until the seventh week.

In summary, I think that collagen biosynthesis increases in human and experimental lung fibrosis, but in the former it must be studied at earlier stages of the disease because it seems that in these disorders (at least in IPF) the values may return to normal in the later stages, as has been suggested by several animal models.

In comparison to biosynthesis, there are even fewer studies related with collagenolytic activity in either human or experimental pulmonary fibrosis. This is discouraging because this pathway seems to be very important for the maintenance of the collagen mass. In addition, recent observations suggesting that collagen degradation in the adult lung is more rapid than was generally believed and that collagenolysis can occur with intracellular procollagen or alpha chain molecules as well as in the extracellular space (before or after incorporation of collagen into a fibril)[95,106] open new avenues concerning the role of degradation in the regulation of collagen turnover and the different steps involved in its control, both under physiological and pathological conditions.

In experimental pulmonary fibrosis, a decreased rate of degradation of newly synthesized collagen has been proposed.

Kehrer and Witschi[102] found, in lung fibrosis developed in mice treated with BHT and 70% O_2, that the degradation of newly formed collagen measured by the percentage of total ^{3}H-hydroxyproline which was acid soluble was significantly decreased in fibrotic mice in relation to controls.

Laurent and McAnulty,[97] in bleomycin-induced pulmonary fibrosis in rabbits, demonstrated that 6 d after drug administration the degradation rate of newly synthesized collagen, estimated from the difference between synthesis and growth rates and from the relative amounts of hydroxyproline in the tissue-free and protein pools, had decreased by about 30% from normal values.

In human pulmonary fibrosis, we have found a remarkable decrease of collagenolysis.[85] Eleven patients with IPF and 6 controls were studied, and the degradation was evaluated in homogenized tissue samples by the solubilization of polypeptide fragments smaller than the alpha chain. With this technique, the values obtained for normal lungs were 0.23 ± 0.04 μg collagen degraded per milligram collagen incubated per hour. In constrast, the IPF patients showed a mean of 0.07 ± 0.04, that is, a threefold decrease. This study demonstrates an important diminution of collagenolytic activity, but it does not define the pathway where the main alterations occur.

A substantially different result was reported in a previous study by Gadek et al.[107] They used the fluid of BAL from 24 patients with IPF, 9 cases of sarcoidosis, and 18 controls. Collagenase activity was determined by measuring the amount of collagenase-specific peptides released from soluble [^{3}H] proline-labeled type I collagen substrate obtained from human lung fibroblasts in culture. Fifteen of the patients with IPF showed collagenase activity, whereas normal subjects and patients with sarcoidosis showed none. Furthermore, in two cases with IPF which were studied again after 8 to 24 months, this activity persisted.

However, the source of collagenase (BAL) and the use of exogenous soluble type I collagen as a substrate can produce some findings which do not reflect the real interstitial events occurring in the microenvironment of the pulmonary parenchyma.

In this context, a similar work performed by Christner et al.,[108] who measured the collagenolytic activity against type I and type III collagens in the BAL from patients with ARDS, found type I collagenase activity in 12 out of 17 patients, and type III collagenase was detectable in 12 out of 13 cases with the same syndrome. They used collagens extracted from placenta, labeled with ^{125}I. The procedure for the assay of collagenolysis was based on sodium dodecyl sulfate polyacrylamide gel electrophoresis, and after running the samples, they cut the slices and measured the radioactivity. Interestingly, there were no differences between the ARDS survivors and nonsurvivors. In addition, collagenase was detectable even when patients were ambulatory and improving clinically. Therefore,, the presence of collagenase in the lower respiratory tract does not necessarily play a role in the development of interstitial lung fibrosis and may only reflect an initial inflammatory response. In this sense, collagenolytic activity in the patients with ARDS was generally associated with a markedly increased percentage of neutrophils (a cell well known to produce collagenase mainly for type I collagen) in BAL fluid.

Moreover, collagen degradation has been studied in animal and human cirrhotic livers by immunofluorescence using a monospecific antibody against purified rat and human collagenase,[109] and the results agree with ours. In the late stages of hepatic fibrosis the enzyme decreased, and the same modification occurred in the skin of patients with PSS, a human disease characterized by an increase of collagen in the dermis and other organs.[110]

I believe that in early phases of fibrotic lung disorders an increase of collagenolysis, as well as of biosynthesis may take place as part of the inflammatory and fibroproliferative response. However, afterward, a progressive decrease in collagen degradation can occur, so that, although collagen synthesis has returned to normal, the production exceeds the breakdown.

However, the decrease in degradation may depend on a complex series of variables including the rate of degradation of newly synthesized collagen, the production of active collagenase and/or the activation of the enzyme precursor, the action of serum and/or tissue collagenase inhibitors, and the susceptibility of the substrate. Currently, we do not know their relative roles *in vivo* and which or in what combinations they contribute to the decrease of collagenolytic activity in chronic pulmonary fibrosis.

In summary, the overall conclusion of the examined reports supports that the reason for an accumulation of collagen in lung fibrosis is the alteration of the protein turnover, probably due to an increase in anabolism in the early stages and a decrease in catabolism in the advanced phases of these diseases.

III. CLINICAL MANIFESTATIONS

Although there are more than 130 disorders that may lead to pulmonary fibrosis, the spectrum of symptoms and signs are relatively limited.[111]

Uniformly, the main symptom is dyspnea. Usually it begins as an insidious and progressive exertional breathlessness which becomes more severe with time and evolves until there is dyspnea even at rest. The mechanisms underlying the development of dyspnea are yet unclear. However, they may be related (1) to alterations in inspiratory muscle contractility by impaired mechanical properties with the development of inspiratory muscle fatigue and/or (2) to increased respiratory muscle output to sustain a given ventilation due to defective gas exchange function.

Cough is frequently present and is mainly evident during exercise or deep breath. This cough may be slight or extremely troublesome and is usually nonproductive, but occasionally may be associated with mucus or even hemoptoic expectoration.

After several months, most patients develop hyporexia, weakness, and easy fatigability to some degree. In addition, secondary to hypoxia they show a variable but significant weight loss.[112]

On physical examination the patient is almost always tachypneic, and although the mechanism responsible for this sign is unknown, it has been proposed that it is mediated by a tonic intrathoracic reflex. In addition, the increase of respiratory rate that occurs during lung inflammation can be prevented if the *vagi* are blocked.[113]

We have recently observed that the apnea period, that is, the lapse of time which a patient is capable of holding the air after a forced inspiration, relates closely with the overall lung disease. Normally, although it may vary with sex, age, size, and physical condition, a control subject shows an apnea period greater than 45 s. In ILD patients, this period decreases according to the extension and severity of the pulmonary damage.

Digital clubbing is a common physical sign, mainly as fibrosis progresses.[112] Occasionally, hypertrophic pulmonary osteoarthropathy with proliferative periostitis of the long bones and polysynovitis has also been described,[114] but seems to be an uncommon feature. In our experience, digital clubbing appears early in IPF and late in other ILD.

Physical examination of the chest reveals late inspiratory crackles,[111,112] mainly but not exclusively over the lower portions of the lungs. Nevertheless, the auscultation of the chest can be normal in the early stages of these disorders. Some patients may have bronchovesicular breath sounds and/or rhonchi and occasionally sibilances.

In the middle and late phases of these diseases, there can be signs of arterial pulmonary hypertension; for instance, the second heart sound may be narrowly split with a loud pulmonary component. As the right ventricle responds to increased afterloads, signs of right ventricular hypertrophy, such as sternal lift or right-side gallops may occur. Finally, when pulmonary hypertension leads to cor pulmonale, further signs can appear, such as neck vein distention, hepatomegaly, and peripheral edema.[115]

Secondary polycythemia is less frequent than in chronic obstructive lung disease, but is present in approximately half of our patients and is more likely to occur if hypoxemia has been severe and progressive.

IV. RADIOGRAPHIC ABNORMALITIES

There are two roentgenologic manifestations common to almost all ILD: a shortening of the pulmonary fields due to a reduction of their volume and the engagement of both lungs in a diffuse or multifocal pattern.

On the other hand, several types of radiologic patterns can be seen (Figure 3).[111,112,116]

The ground-glass pattern is frequently associated with the initial phases of these diseases.

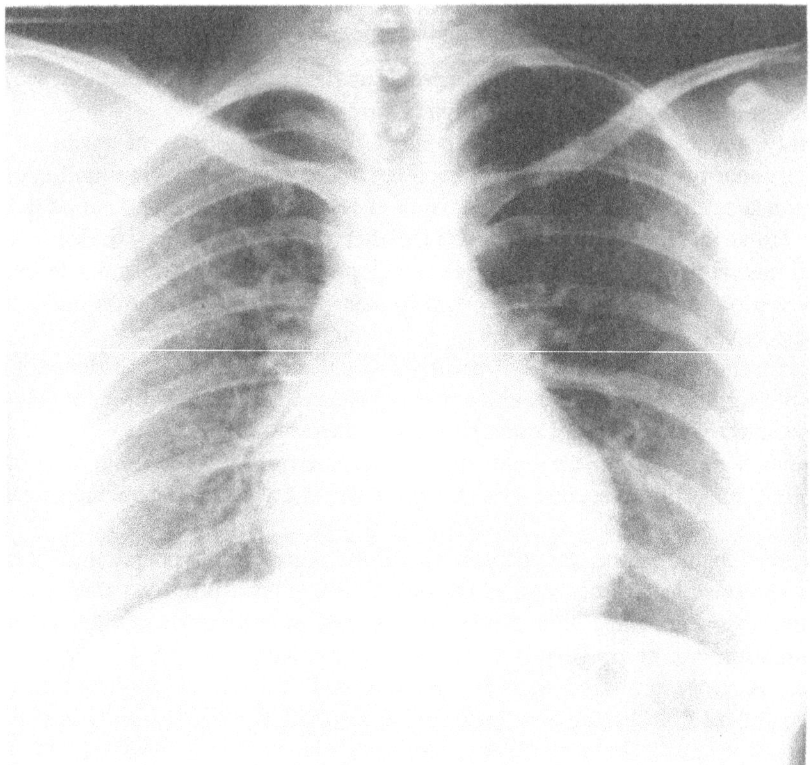

A

FIGURE 3. Different posteranterior roentgenogram abnormalities observed in ILD. (A) Basal ground-glass pattern in a hypersensitivity pneumonitis patient with 3 months evolution. (B) Diffuse and prominent reticulonodular pattern in a IPF patient with 18 months of symptoms. Note the signs of pulmonary arterial hypertension. (C) Honeycomb pattern of an ILD of unknown etiology in the end-stage lung. Note the shortening of the pulmonary fields.

Their roentgenographic appearance is a relatively homogeneous clouding of the lungs and may show a replacement of air by liquid, proteins, and cells.

The nodular pattern is characterized by numerous, tiny, punctate, and uniform densities. They may be of several sizes and can reach 2 to 4 mm in diameter. This pattern usually represents an interstitial lesion and is generally observed with the presence of granulomas.

The reticular pattern consists of a network of linear densities and shows the thickening of the interstitial space. These images are randomly oriented, which differentiates them from pulmonary vascular markings.

The reticulonodular pattern is a mixture of the two previously described abnormalities and may result from the superimposition of nodular opacities on a reticular pattern.[112]

Finally, the honeycomb pattern, a late image in these diseases, refers to a coarse reticular pattern and consists of an irregular series of cystic spaces measuring 5 to 10 mm in diameter and surrounded by thick walls.

Occasionally, it is possible to observe a certain predominance in the location of these radiologic images that may suggest a diagnosis. For example, asbestosis, IPF, and collagen-vascular diseases show basal predominance particularly in the early stages, whereas silicosis, sarcoidosis and histiocytosis X initially show an upper zonal distribution. However, in the majority of patients the pattern and distribution of lesions are so similar that it is impossible to suggest a diagnosis.

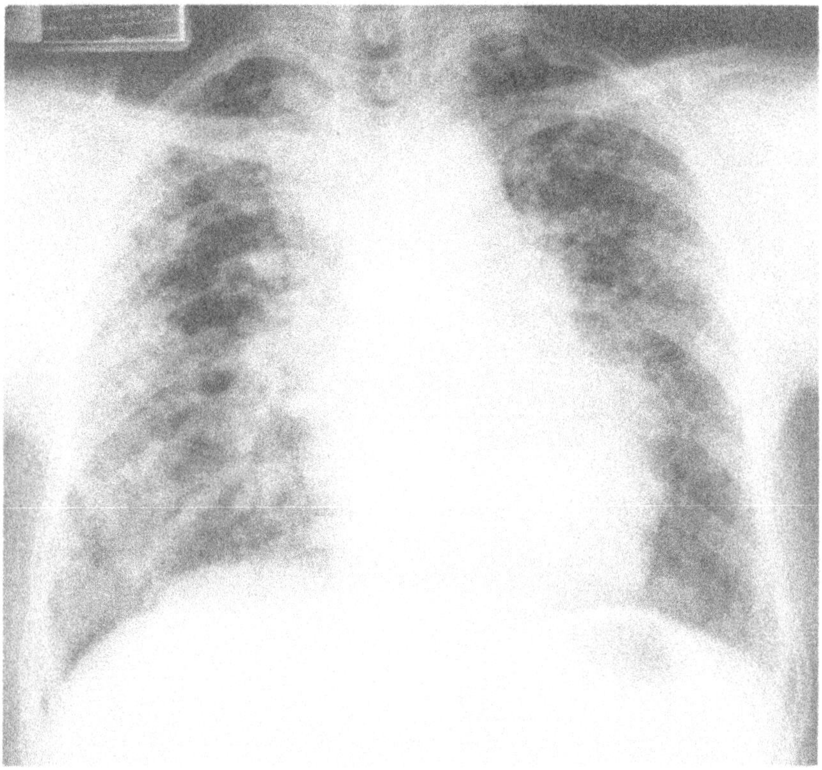

FIGURE 3B

On the other hand, it has been discussed whether or not the type and severity of roentgenographic abnormalities correlate with the course and morphological severity of the disease. In this sense, most authors agree that the chest film image correlates poorly with the stage of the lung disorder with the exception of honeycomb lung, which generally represents an end-stage lung.

Recently, several studies for evaluation of high-resolution computed tomography (HRCT) in ILD have been performed (Figure 4). Nakata et al.[117] analyzed this method in 15 patients with a variety of ILD. They found that the HRCT scan abnormalities correlate well with the histological findings in most of the cases, although mild alveolar wall thickening due to alveolitis in sarcoidosis and systemic lupus erythematosus was not detected. However, bullae, interstitial fibrosis, and small granulomas were clearly visible, and these findings were more specific than those with conventional chest radiographs. Thus, for example, persistent diffuse miliary granulomas were accurately shown by HRCT in the healing phases of HP, whereas only subtle abnormalities were seen on chest radiographs.

Wright et al.[118] studied the distribution of the disease in ten patients with IPF by several means including computed tomography (CT). Two patients had normal chest films after treatment, but one of them showed subpleural shadowing in both prone and supine views with CT. The other eight patients with abnormal radiographs showed changes in the central CT scans. However, the plain radiographs had no relationship to the CT findings, which seemed to be more sensitive and specific.

Additionally, Bellamy et al.[119] evaluated the sensitivity of this method in 100 patients treated with bleomycin, a drug known to be able to produce pulmonary fibrosis. The CT findings were compared with those of chest radiographs and lung-function tests. Lung damage due to bleomycin was detected by CT scans in 38 patients, whereas chest X-rays were

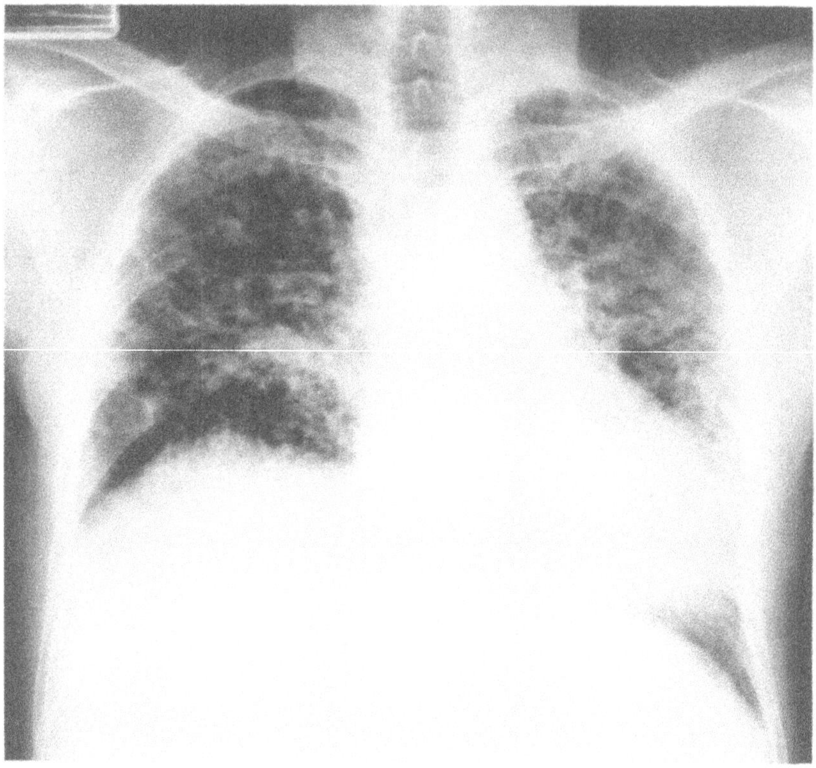

FIGURE 3C

abnormal in only 15 of them. Furthermore, there was a good correlation between the severity of damage shown on CT scans and changes in lung volumes. This is an interesting study because lung damage induced by cytotoxic drugs and metastases may coexist, and the differential diagnosis can be extremely difficult. Thus, sequential CT examination of the lungs should help to overcome this problem in individual patients.

The principal conclusion of these studies is that the information on the nature and the anatomic distribution of lung abnormalities is more accurate with these newer imaging techniques than that obtained from conventional chest radiographs.

On the other hand, McFadden et al.[120] employed magnetic resonance imaging (MRI), a novel method to obtain sectional images of the body, to analyze its usefulness in separating active alveolitis from predominant fibrosis. They studied 34 patients with various forms of ILD. They found that MRI was often a reliable means for differentiating both groups and correlated with patients in which specific treatment was or was not indicated. Median values for the profusion scores of individuals receiving a trial of therapy were significantly greater than those of patients which remained stable or spontaneously improved. The authors propose that MRI may provide useful information with respect to disease activity and therapeutic decision.

A previous study by Vinitski et al.[121] in bleomycin-induced lung damage in rats, showed that nuclear-magnetic-resonance signal intensities were significantly elevated in alveolitis and fibrosis stages. Moreover, MRI could be used to separate the alveolitis stage from pulmonary fibrosis because in the latter, both T_1 and T_2 relaxation times tended to drop.

Nevertheless, this technique has several disadvantages for the study of lung parenchyma, mainly the lung imaging time and its susceptibility to artifacts caused by respiratory movements and cardiac motion. Thus, a precise pathological correlation as well as more pro-

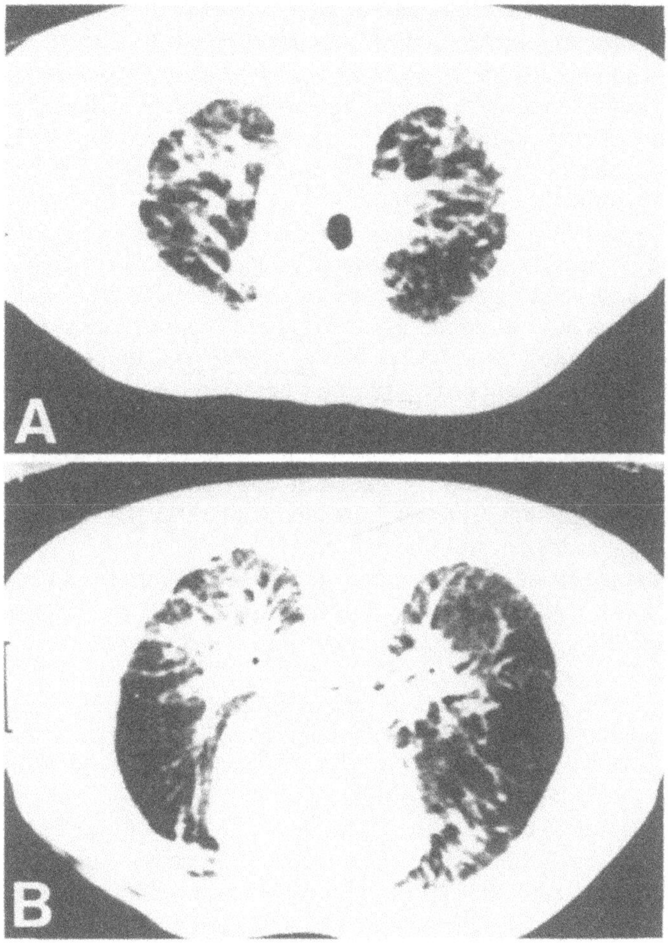

FIGURE 4. CT scans taken at the level of aortic arc (A) and pulmonary hilium (B) in a patient with IPF. Severe lung damage can be clearly observed with linear and irregular nodular opacities in the lung fields, and areas of destruction presenting peripheral honeycomb-like air spaces.

spective trials are required to determine the exact relevance of this method for the evaluation of the disease activity and its clinical course.

However, both CT scans and MRI imaging of pulmonary fields may, indeed, clarify the anatomic extent and the predominant type of pattern lesion present in the parenchyma during the development of the disease. The techniques also can probably be incorporated in the future as usual studies in ILD patients.

V. PULMONARY FUNCTION TESTS

Pulmonary function tests are widely employed by physicians mainly to assess the degree and extension of the respiratory damage and to evaluate the therapy during the follow-up. In addition, some of these tests may, at least partially, reflect some of the morphological changes.

A. LUNG CAPACITIES AND COMPLIANCE

A restrictive functional defect characterizes the main alterations in lung mechanics.

Normally, the patients present a decrease of total lung capacity (TLC), vital capacity (VC), and compliance, with an increase in the elastic recoil pressure (ERP).

Although the measurements of these parameters require expensive equipment, the use of a spirometer may suggest a restrictive abnormality with a high degree of certainty. In this context, Gilbert and Auchincloss,[122] in a study on 211 subjects, demonstrated that restriction can be diagnosed spirographically when a low forced vital capacity (FVC) and normal ratio of 1-s forced expiratory volume to FVC (FEV_1/FVC%) exists. With this criteria, and using 70% as the lower normal limit for FEV_1/FVC%, the spirographs showed 93% sensitivity and 82% specificity for the detection or exclusion of a restrictive impairment. Similar findings have been reported by other authors and agree with our results. However, this ratio only gives a general notion of the existence of a restrictive defect, but correlates poorly with the degree of alveolitis and/or fibrosis and with the clinical course of the disease.

Thus, all parameters mentioned before must be measured in a patient with ILD. On the other hand, although it seems reasonable to assume that VC and TLC, commonly used as objective monitors of the disease process, would be sensitive to assess the integrity and function of pulmonary parenchyma, an excellent study performed by Fulmer et al.[123] demonstrated that none of these parameters correlated with either the severity of fibrosis or the degree of cellularity in biopsy specimens.

Moreover, the alteration of FVC seems to correlate negatively with the prognosis and clinical course. Thus, Rudd et al.[9] analyzing 120 patients on whom data were prospectively collected, showed that patients responding to treatment had a more severe initial impairment of the FVC.

In contrast, almost all parameters of lung distensibility seem to correlate with the morphological assessment of fibrosis degree.[123] Lung compliance depends on the volume-pressure relationship, and its alteration is mainly related to lung volume changes. Classically, in ILD there is a shift downward and rightward. This means that whereas maximal transpulmonary pressure at TLC increases, the volume which the subject can introduce into the lungs decreases. In the Fulmer study, of all parameters derived from the static deflation volume-pressure curve, compliance showed the highest correlation coefficient in relation to the morphological degree of fibrosis. Nevertheless, the alveolitis did not correlate with compliance, comparative compliance nor static transpulmonary pressure at the TLC. In this way, lung compliance could be used as a reliable monitor of fibrosis progress but not of the inflammation in the follow-up of patients with ILD.

More recently, in our institute, Pérez-Neria et al.[124] have postulated that in relation to the time constants of the alveolar units, which depend on the transitional airway resistance and alveolar compliance, inflammation represents a slow unit whereas fibrosis can be considered a fast unit. Thus, the frequency dependence of compliance may be different in both states. With this hypothesis, they studied the relationship between lung inflammation or fibrosis, determined in open lung biopsy specimens, with the dynamic compliance/static compliance (Cdyn/Cst) index and found that 9 out of 9 patients with fibrosis predominance had a Cdyn/Cst index over 0.85, in contrast with 9 out of 10 patients with inflammation predominance who had indexes below 0.85. The specificity and sensitivity of the index was excellent with a $p < 0.001$. These results suggest that the Cdyn/Cst index may be a useful tool in predicting whether alveolitis or pulmonary fibrosis predominates and can be used as a prognosis index and in the follow-up of the patients with ILD.

Finally, the factors that contribute to modifying the volume-pressure relationship in ILD are not completely clear, but they may be two: a volume loss derived from the replacement of parenchyma with fibrous tissue (loss of the lung units)[125] and the increase of recoil pressure at all volumes by abnormal distensibility properties of the remaining normal alveoli.[126]

B. AIRWAYS

Although the ILD have classically been considered "pure" restrictive disorders, several physiological and morphological findings support that peripheral airways may be affected, at least in some of these diseases. Because most of these alterations occur in the "silent zone" of the lung, the narrowing of airways does not reduce $FEV_1/FVC\%$, a parameter normally recorded in these patients. However, the use of more sensitive tests has demonstrated that small airways are frequently involved.

Fulmer et al.[126] studied 18 patients with IPF and found that morphologically, 94% of them had peribronchiolar inflammation or fibrosis and/or bronchiolitis, and 67% had an overall narrowing of small airways diameter. From the functional point of view, 59% presented frequency-dependence dynamic compliance, and 50% had abnormal maximum expiratory flow-volume curves. Comparison of morphological and physiological data revealed a significant correlation between them, and the authors proposed that, at least in IPF, small airways are affected as well as alveoli.

In other interstitial disease, similar findings to those by Fulmer have been reported. Allen et al.[127] studied lung function in nine nonsmoke patients with HP. All of them had evidence, by frequency-dependence of compliance and forced expiratory flow during the middle half of the FVC, of small airways obstruction at the time of diagnosis. Moreover, three of them presented a progressive increase in the degree of airway obstruction during their follow-up, in spite of having ceased to be exposed to the antigen.

More recently, we have analyzed[128] the possible abnormalities of small airways, using the volume of isoflow, a sensitive method for detecting peripheral airway obstruction, in nine subjects with IPF and eight with HP. We found that approximately 40% of the patients showed changes suggestive of airway narrowing.

Comparable results have been reported in pulmonary fibrosis induced by bleomycin in nonhuman primates,[129] where 6 months after administration of the drug several morphological airway lesions were observed. In addition, physiologically, the animals showed findings of both lung fibrosis and airway obstruction.

Finally, other ILD may also involve airways. In summary, all these studies make evident that ILD are not "pure" restrictive functional disorders, but also include small airways obstruction.

C. GAS EXCHANGE

Although ILD patients may have different abnormalities in their alveolar walls, they present similar modifications in gas exchange.

One of the most remarkable alteration in patients with ILD is the hypoxemia that is generally worsened with exercise.[2] In initial phases arterial oxygen may be normal at rest, but hypoxemia becomes evident with exercise.

The mechanisms involved in respiratory failure in ILD have been largely discussed, and for a long time it was considered that an impairment of gas diffusion could be the principal cause through an "alveolar-capillary block". However, most physicians now agree that limitation of diffusion plays a small role in producing hypoxemia.[2,130]

Nevertheless, a reduction of the diffusing capacity for carbon monoxide appears to be one of the most sensitive tests to detect early alterations in lung parenchyma, although this abnormality must be taken cautiously because it is dependent on lung volume and hemoglobin concentration.[130] However, this alteration means that diffusion abnormalities do occur in the units of gas exchange, but they contribute minimally to the pathophysiology of hypoxemia, at least at rest. During exercise, when the transit of blood through the pulmonary capillaries increases, the alteration of the diffusion capacity may have a greater but moderate effect on the transfer of oxygen across the alveolar-capillary membrane.[112]

The major causes for hypoxemia in these disorders is a ventilation-perfusion imbalance, which is amplified with exercise.[112]

This disturbance may be due, from a theoretical point of view, to a shunt or venous admixture (lung units perfused but unventilated) or increased dead space (lung units ventilated but unperfused). Both modifications may occur to some degree in pulmonary fibrosis, that is, intrapulmonary shunting of blood and maldistribution of ventilation probably provoked by local alterations in compliance and increased resistance to airflow in small airways. However, it is important to note that Miller et al.[131] have recently studied the pulmonary shunting by injection of radiolabeled particles of macroaggregated albumin larger than 10 μm in diameter, and the mean shunt was not significantly different in IPF patients in relation to controls. This finding suggests that alterations in the distribution of ventilation contribute to a greater degree in the ventilation-perfusion inequality and the resulting hypoxemia.

Concomitantly, an increase of the alveolar-arterial oxygen gradient, which also worsens with exercise, is observed in almost all these disorders, and this analysis of gas exchange is a useful tool to verify with reasonable certainty the severity of the pulmonary lesions.

On the other hand, Fulmer et al.[123] have suggested that adjusting the change in PaO_2 during rest and exercise for the amount of work accomplished, measuring oxygen consumption, may give a more accurate estimate of the overall state of the disease.

With regard to the arterial pressure of CO_2 ($PaCO_2$), a low level of this gas is frequently found, mainly in the early stages of the disease. This modification is probably due to heightened ventilatory drive and tachypnea.[113] In general terms, during most of the course of these diseases, low or normal levels of $PaCO_2$ may be observed. However, in advanced phases resting hypercapnia may be found and usually implies a terminal stage (end-stage lung) and a very poor prognosis.

In summary, the overall pulmonary function tests, analyzed with caution, give a good idea of the severity and extension of pulmonary damage and allow clinicians to follow up the course of the disease. None of the individual parameters is better than the sum of them all, but the most important are pulmonary compliance and the different measurements of gas exchange. Finally, although some trials have been done to establish a relationship between inflammation and fibrosis with some of the functional tests, giving encouraging results, these studies need greater verification with more patients and in all different disorders which constitute the complex group of ILD.

In this context, with the notion that clinical, radiographic, and physiological parameters are usually studied in patients with ILD and the uncertain correlation of some individual data with the clinical status and prognosis of them, Watters et al.[132] have proposed a quantifiable form, composed of a clinical-radiographic-physiological (CRP) scoring system to study patients with IPF, which may be used for other ILD as well. Their preliminary results suggest that the initial determination of CRP scores correlates with the pathological findings, that is, inflammatory or fibrotic predominance, and concomitantly with the future response to treatment. Moreover, in none of the relationships compared by the authors did any individual component of the CRP score associate better with the histopathological index than the CRP score itself. These data suggest that this scoring system, perhaps amplified by adding gallium index and collagen markers, for example, may be a reliable means for the estimation of the severity of the lung derangement, predicting the prognosis and the longitudinal quantitative assessment of clinical impairment or improvement in ILD.

Pérez-Padilla et al.[133] studied breathing during sleep in 11 patients with ILD and 11 age-sex-matched control subjects. They found that sleep quality was worse in patients. They had more time spent in stage 1 and less time in REM sleep, together with more fragmentation of sleep. Patients with SaO_2 lower than 90% while awake had greater abnormalities in sleep structure. O_2 saturation fell during sleep, especially in REM and in periods of snoring. Sleep O_2 desaturations were present mainly in those hypoxemic while awake. These results suggest that sleep quality is poor in patients with ILD, similar to what occurs in other pulmonary patients.

VI. PULMONARY HEMODYNAMICS

It is well known that lung fibrosis sooner or later leads to pulmonary arterial hypertension. Afterward, a right ventricular dysfunction and failure are common complications of the ILD.

The increase of pulmonary vascular resistance is mainly due to anatomopathological lesions characterized by a diffuse medial hypertrophy of the muscular pulmonary arteries associated with intimal smooth muscle proliferation, which cause obliteration and compression of the pulmonary vascular bed. However, a hypoxic vasoconstriction, surtout in the early stages, also contributes. This latter notion is supported by the finding that mean pulmonary arterial pressure increases during exercise, and this modification significantly correlates with the fall in PaO_2.[134] Additionally, Weitzenblum et al.[135] have demonstrated that this phenomenon occurs mainly in IPF rather than in other ILD. They studied 31 patients with IPF and 34 cases with a variety of interstitial disorders and found that in IPF the PaO_2 regularly worsened during exercise, whereas it did not significantly vary in the other patients. Concomitantly, all indexes of pulmonary vascular resistance were higher in the IPF group and markedly increased with exercise, in contrast to a modest elevation in the cases with other interstitial diseases.

Nevertheless, simultaneous to the progress of ILD toward fibrosis, pulmonary arterial alterations will increase, and secondary to pulmonary arterial hypertension several anatomic changes will occur in the right ventricle, leading to the cor pulmonale which always accompanies chronic respiratory failure due to lung fibrosis. Moreover, decompensations of cor pulmonale are frequently associated with the cause of death in these patients.

VII. LUNG BIOPSY

The lung biopsy is presently the best method available to obtain a precise diagnosis in ILDs, and I would like to treat this subject in relation to three questions: why, how, and where.

A. WHY

When we have a patient with the CRP alterations mentioned before, we can be reasonably certain that we are in the presence of an ILD, but the definitive etiologic diagnosis is almost always only presumptive.

A detailed history of occupational and enviromental factors as well as other situations known to be associated with ILD (such as drugs), the exclusion of generalized disorders (CVD, sarcoidosis, etc.), and the use of other techniques (BAL, for example), only offer some clues which may orient the diagnosis but frequently do not determine it with precision. Mistakes may occur more often than one would like to believe.

In this context, we have observed several patients with many manifestations suggestive of HP, a frequent ILD in Mexico, including the exposure to avian antigens and the presence of circulating specific antibodies. However, the final diagnosis was lymphoid interstitial pneumonia or bronchioalveolar carcinoma, among others. Likewise, some patients with high risk for tuberculosis, fever, lung X ray suggestive of miliar tuberculosis, without apparent contact with antigens, turned out to suffer from HP and healed with corticosteroid therapy exclusively.

Finally, most patients have an ILD which does not involve other organs and have no antecedents suggestive of a cause. In these conditions the presumptive diagnosis is an ILD of unknown etiology.

Therefore, the best way to obtain an accurate diagnosis is the lung biopsy with the subsequent observation of the samples by expert pathologists. Of course, as in every "rule", there are some exceptions. For instance, if we have a young patient with recurrent hemoptysis, breathlessness, anemia, bilateral pulmonary infiltrates, hemosiderin-laden macrophages in lavage cellular analysis, and circulating anti-BM antibodies, this would strongly suggest the diagnosis of Goodpasture's syndrome.

In general terms, however, the lung biopsy should be done to a patient with ILD always or almost always.

On the other hand, the tissue obtained by this method is very useful for several reasons:

1. To establish a specific diagnosis
2. To plan the treatment, including drugs and time
3. To stage the disease and assess the prognosis
4. To correlate the morphological changes with the CRP, gallium scan, and BAL fluid abnormalities
5. To carry out some biochemical, immunological, and ultrastructural research analyses which allow us to know something more about the pathogenesis

B. HOW

A variety of methods has been used to obtain lung tissue, which range from thoracotomy and open lung biopsy to less invasive techniques such as transbronchial lung biopsy, and the comparison of their effectiveness has produced some controversial results.

Open lung biopsy has been the technique most frequently used, and I will only comment on some of the larger series which, in my opinion, summarize the general thought about this technique.

Ray et al.[136] reported their experience with 416 consecutive open pulmonary biopsies through limited thoracotomies performed between 1955 through 1973. In all patients adequate tissue for diagnosis was obtained. However, in 122 cases (29%) the histological report was of nonspecific pulmonary disease which included nonspecific interstitial pulmonary fibrosis, nonspecific granuloma, and pneumonitis. With respect to this, it is important to consider that many of these biopsies were performed at a time when the pathologists were less acquainted with these type of disorders, and many of the "nonspecific pulmonary diseases" may have presented the morphological findings compatible with IPF, a relatively well-defined entity. Unfortunately, there were no detailed histological analyses in this report. Nevertheless, specific diagnoses that included HP, silicosis, neoplastic disease, sarcoidosis, eosinophilic granuloma, bronchiolitis, vascular disorders, and several infectious diseases were made in 71% of the patients. Pneumothorax was the most common complication (23%), but only 6% required the placement of a chest tube. Pleural effusion occurred in 106 patients and was minor. There was an overall mortality of 4.5% (19 patients), but it was related to the procedure in only two cases.

Another important series of patients with ILD who underwent open lung biopsy was reported by Gaensler and Carrington.[137] They analyzed the results obtained from 502 cases biopsied between 1950 and 1980, and the diagnostic yield was 92.2%, which included a great variety of different ILD. Mortality was 0.3%, and the rate of complications was 2.5%, but only 7 patients presented major complications such as empyema, respiratory insufficiency, myocardial ischemia, and pulmonary edema.

Early et al.[138] obtained similar efficiency in 35 pediatric patients in which the diagnosis was obtained by the morphological study of the open lung biopsy specimen. Moreover, in 43% of the total group the results of the biopsy caused a change in therapy, although in this series most patients were immunocompromised.

Many other recent reports, whose aim was an investigation other than the usefulness of the type of biopsy, based the diagnosis of the studied population on an open lung biopsy with results similar to Gaensler and Carrington, that is, a high percentage of specific diagnoses close to 100%, with a very low percentage of mortality and morbidity.

Transbronchial lung biopsy (TBB) offers an attractive alternative to open lung biopsy and has been used in ILD patients for a long time. In addition, more recently it has attained greater popularity with the advent of the flexible fiberoptic bronchoscope. However, its sensitivity and specificity are controversial.

Andersen and Fontana[139] reported in 1972 the results obtained with this technique in 450 patients with diffuse pulmonary disease. Satisfactory lung tissue for diagnosis was obtained in 378 cases (84%). Complications included pneumothorax in 14% and significant bleeding in about 1% of patients, and there was only one death.

Afterward, Andersen[140] analyzed the findings obtained in 939 patients. Insufficient tissue was reported in 194 cases (20.6%) for several reasons, and normal lung was diagnosed in 35 patients, but this was not commented on further in this study. The complications of the technique were the same as reported previously by this author.

Zavala[141] reviewed this method and proposed some technical corrections to optimize the TBB, including the selection of the lung segment, the use of fluoroscopy during the procedure, and a wedge method to control the hemorrhage.

Two important aspects arise from this review. One of them is that the diagnostic accuracies are only of 62, 64, and 67%, respectively in infectious, interstitial, and malignant lung diseases. The other one is that the potentially serious complications are not only pneumothorax and hemorrhage, but also laryngospasm, bronchospasm, hypoventilation, cardiac arrhythmia, and myocardial infarction, among others. Nevertheless, most of them may be prevented with a proper preparation of the patient and skillful biopsy technique.

Poe et al.[142] examined 144 consecutive patients with localized or diffuse infiltrates who underwent diagnostic fiberoptic TBB, and the technique allowed the diagnosis in 48% of immunosuppressed patients and in 67% of nonimmunosuppressed cases.

Since one of the major problems is the relatively low specificity of TBB, Gilman and Wang[143] have studied the number of biopsies that may be useful to increase the diagnostic yield. From all of their study population, they selected individual specimens from six biopsies in patients with stage II sarcoidosis and found that the probability of obtaining a positive diagnosis per biopsy was 46%, but it increased to 90% when at least four biopsy specimens from the same patient were examined. Nevertheless, sarcoidosis is probably one of the diseases that may be diagnosed more reliably by TBB.

More recent studies support the notion that the specificity and sensitivity of TBB is relatively low, at least for some ILD. Wall et al.[14] demonstrated in 53 patients that TBB was diagnostic only in 20 (37.7%), and the histological findings in the remaining cases were normal lung, nonspecific abnormalities, interstitial pneumonia or fibrosis, and inadequate specimen. Open lung biopsy in these 33 patients resulted in specific diagnoses in 92%, and the histological changes have little relationship with the original TBB results. In addition, a review of 15 of the largest series of open pulmonary biopsy, including 2290 patients with ILDs studied from 1940 until 1978, indicates a diagnostic result in 94% of cases with a relatively lower rate of complications.

Similar results have been obtained by us[145] in 12 consecutive patients with ILDs in which a TBB and open biopsy were performed simultaneously. Only in two cases did the TBB allow the diagnosis, compared to 100% of the open lung biopsy samples.

The major problems of the specimens obtained by TBB are that they are taken blindly, that is, in an unselected region of the lung, with a strong possibility of sampling error. Also, the tissue specimen is usually crushed and comes from areas adjacent to bronchi. In this sense, interstitial inflammation and fibrosis, often accompanied by inflammatory cells within the alveolar spaces, are common nonspecific changes that can be found in many individuals without evidence of diffuse interstitial disease. More importantly, however, the size of tissue, usually 1 by 2 mm, is very small and often does not represent the extent or intensity of the alveolitis and/or fibrosis, nor does it fulfill all the purposes mentioned before for which a biopsy is necessary.

In contrast, in an open lung biopsy the surgeon may select the more representative areas and can take greater samples from two or three different lung segments, giving a broader overview of the pathological disorder and allowing the use of part of the specimen for several other studies.

On the other hand, some authors have proposed the use of needle or trephine lung biopsy,[146,147] but this procedure is seldom used and presents the same problems as TBB, mainly the small, and sometimes very small, sample obtained.

Finally, we have recently initiated the thoracoscopic transpleural pulmonary biopsy for the diagnosis in ILD[148] in patients with severe lung function impairment. We reported 31 cases which underwent thoracoscopy where two to five samples per patient, under visual observation, were obtained. In 30 patients (96%) specific diagnoses were made, including HP, lymphoid interstitial pneumonia, bronchiolitis obliterans, bronchioloalveolar carcinoma, histiocytosis X, histoplasmosis, and IPF. These results suggest that lung biopsy by the thoracoscopic technique may be an alternative procedure in patients with ILD and severe respiratory insufficiency. Again, the size of the sample is relatively small (5 by 3.6 mm in our hands) and may only be useful for diagnosis, and sometimes to evaluate the degree of inflammation and fibrosis.

Summarizing, currently the open lung biopsy is the best method available to obtain the most adequate specimens for diagnosis and to carry out a variety of other analyses which orient physicians about prognosis and treatment (degree of alveolitis and/or fibrosis; type of cells involved, etc.). It also allows several studies for further research as ultrastructural examination, immunohistochemical analyses, identification of cell-surface antigens to distinguish specific cell subtypes, isolation of different types of cells for functional activity studies, and the analysis of concentration, metabolism, and genetic types of collagens involved in the process, among other things.

C. WHERE

An important conclusion of the Gaensler and Carrington study[137] is that often the most abnormal regions are biopsied by the surgeon, and this selection is frequently the cause of meaningless histological findings because these areas usually show only end-stage disease of unrecognizable origin. The authors proposed that average lung, or in patients with advanced disease the least involved regions, present the pathological process at an active and recognizable stage, and these sites must be selected for the biopsy. This notion is universally accepted, and most surgeons take lung samples from different sites and with diverse types of macroscopic lesions.

Another interesting problem relates to which segment must be or must not be selected for the biopsy. In this context, it has been thought for a long time that the lingular or middle lobe segments are inappropriate for diagnosis purposes in patients with ILD. Early, Ray et al.[136] and Gaensler and Carrington[137] proposed that these portions of the lung are common sites of nonspecific inflammatory changes and fibrosis unrelated to the ILD, and that they are also excessively affected by passive congestion. Therefore, their morphological study may lead to erroneous interpretations.

More recently, Newman et al.[149] examined specimens of the lingula and left lower and right upper lobes from 50 consecutive unselected autopsies. Using a morphometric and a semiquantitative grading system, they found that both fibrosis and pulmonary vasculopathy were more notorious in the lingula than in either of the other two lung segments. The researchers concluded that the lingula shows nonspecific changes and for this reason may not be a valid biopsy site. It is quite unfortunate that this should be so, because biopsy of the lingular segment is a simple and relatively minor surgical procedure.

However, recently Weinstein[150] reported a completely different result. He analyzed 20 consecutive patients with ILD in which biopsy of the lingula and other lung segments were taken. The histopathological findings of the lingular biopsy samples were identical to those from the other segments of the lung, and in all cases the lingular specimen helped in establishing, confirming, or excluding the diagnosis. Thus, this segment presented a sensitivity and specificity of 100%. In addition, Weinstein supports that the lingular biopsy is a rather benign and quickly operative procedure that may be performed by the majority of thoracic surgeons. Obviously, this report clearly refutes previous studies and places in discussion once again a classic concept about the representativeness of the lingula in ILD biopsy, which is of great importance to clinicians when accepting the surgical decision. This result should be taken cautiously, and further studies need to be performed to clarify this dilemma.

VIII. METHODS FOR STAGING CLINICAL ACTIVITY OF ALVEOLITIS AND FIBROSIS, AND FOLLOW-UP

A. BRONCHOALVEOLAR LAVAGE

BAL is a widely used procedure which serves different purposes such as the study of the lung disease activity for therapeutic decisions and follow-up, research into pathogenic mechanisms, and occasionally for diagnosis. (For review see References 151 and 152.)

The aim of BAL is to sample cells and extracellular molecules present in the lower respiratory tract (Figure 5), assuming that these analyses reflect the events which occur in lung parenchyma more accurately than the study of other fluids such as blood, for example.

The general method employed is relatively standard. Physiological saline solution is introduced through a fiberoptic bronchoscope wedged into a segmental or subsegmental bronchus, usually in the lingula or right middle lobe, and recovered by gentle aspiration. The sample obtained is filtered through a layer of cotton gauze and centrifuged, normally at low speed, for 10 to 15 min at room temperature. Supernatant fluid is decanted, frequently concentrated, and used for the study of different molecules. Cell pellets are resuspended in a solution which depends on their subsequent use (modified Hank's balanced salt solution, culture medium, etc.). Afterward the cells are counted using a hemacytometer, and the cell-types (differential count) are determined on preparations stained with Wright-Giemsa or an analogous stain. Later the cells may be used for different purposes, such as the study of subpopulations through monoclonal antibodies, functional activities, etc.

At first sight this seemed a very useful technique; however, a progressive number of controversies have appeared in the literature subsequent to the adoption of this method in the study of ILD patients. The controversies mainly derive from disagreements as to standardization of the methodology, the reproducibility of results, their real sensitivity and specificity, the form in which data should be expressed, and the number of lung segments to be lavaged. This has polarized the feelings of our colleagues as to the relevance of this technique to the point where opinions oscillate like a pendulum: from those who assert that analysis of BAL fluid is an excellent method which must always be used in ILD patients to those who show frank skepticism about its rightful place in their study.

1. Some Methodological Problems

a. Analysis of Extracellular Molecules

Since the epithelial lining fluid (ELF) is diluted by a certain amount of solution and the liquid recovered is variable (sometimes very variable from patient to patient), the first problem is how to express the results so as to allow an estimate of the real concentration of the molecule in study.

For a long time, albumin has been used as a denominator for protein ratios, but it seems not to be reliable if alveolitis or capillary leaking exist,[151] as occurs in ILD.

Thus, in search of other markers, Rennard et al.[153] have recently proposed that urea, a

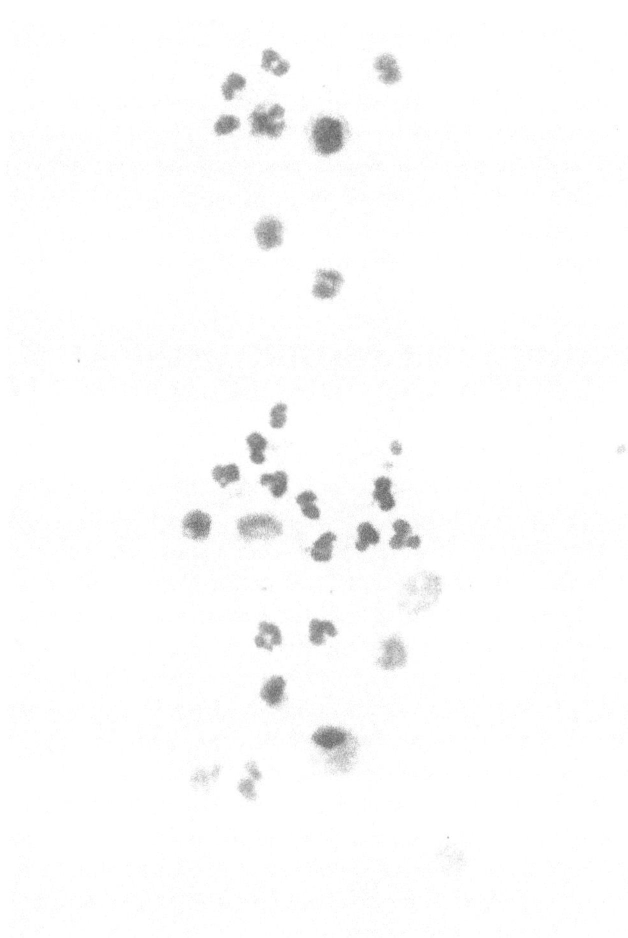

A

FIGURE 5. Hematoxylin-eosin stained smear of cells recovered by bronchoalveolar lavage in two cases with ILD. (A) Patient with IPF; a great number of neutrophils can be observed. (B) Patient with hyprsensitivity pneumonitis; an increased number of lymphocytes are present.

molecule freely diffusable through most body compartments including the lung, can be used as a reliable endogenous marker of ELF dilution. They studied 13 normal nonsmoking volunteers and found that the use of urea allows an estimate of the actual volume of ELF recovered. In this way, they proposed that it is possible to know the absolute concentrations of different molecules and the density of the cells present in the lower respiratory tract.

However, a few months later, Marcy et al.[154] reported a completely different result. They studied six healthy nonsmoking subjects with two distinct BAL protocols, using 100 and 300 ml of saline solution and analyzing the individually aspirated aliquots (six per subject) separately. They found that urea concentration increased progressively from the first to the sixth aliquot when 300 ml were used, and from the first to the fifth aliquot when 100 ml were used. The authors suggest that a significant amount of urea diffuses into the instilled BAL fluid as it dwells in the alveolar space and that this phenomenon may cause mistakes in the calculated ELF recovery.

It is important to note that both studies, in addition to their noteworthy differences, were performed on normal individuals, and the problems can be more serious with damaged lungs.

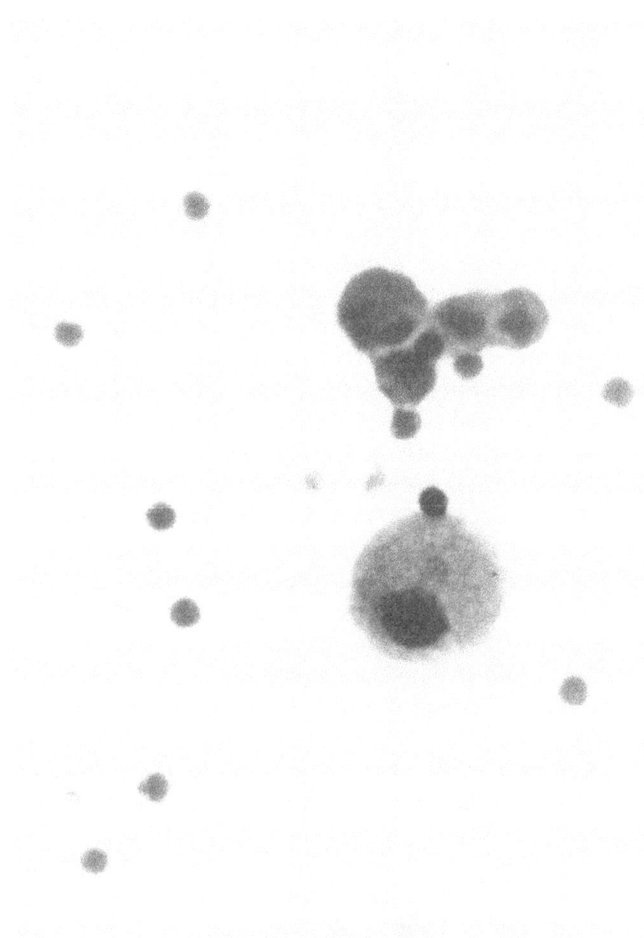

FIGURE 5B

Baughman et al.[155] have proposed to use methylene blue in the introduced fluid as an external marker. They added a known amount of this substance to the saline solution and determined its concentration in the aspirated material, which allowed them to calculate the dilution of the introduced liquid by the lung fluid. They studied 9 controls, 11 patients with sarcoidosis, and 2 cases with fibrosing alveolitis. After correction by the dilution factor (based on the concentration of methylene blue in the introduced and aspirated liquids), they obtained a volume of alveolar lining fluid of about 10 ml in average, both from controls and from sarcoidosis patients. In my opinion and that of others[151] this value seems too high, considering the total volume aspirated. However, the use of an exogenous marker could be a better solution than the use of an endogenous one, whose variability during lung disease is uncertain.

Finally, standardization on the basis of potassium concentration is seldom used because of the existing doubts regarding active transport by lung cells.[152]

Thus, in general terms, there is not one reliable form to quantify the relative or absolute amounts of extracellular molecules or for measuring the volume of lung surface fluid recovered after a BAL. Of course, this is a great limitation to this procedure.

b. Effect of the Amount of Instilled Volume

Volumes ranging from 50 to 300 ml are normally used, but it is unclear whether the results of the analyses are affected by the amount of instilled volume or, if so, how they affect them.

Merrill et al.[156] studied 14 normal volunteers and found that low volumes (less than 100 ml) preferentially recover airway proteins, but with 100 ml of lavage instillate, lung proteins are efficiently and homogenously sampled. Larger volumes add more proteins but do not alter the protein ratios.

Nevertheless, a different picture arises from the report by Davis et al.[157] who studied 14 healthy normal subjects and found that proteins, carbohydrates, lipids, and potassium concentrations decrease in sequential lavages of 60 ml, but each molecule follows a different kinetic pattern. Thus, measured values for a component in a 120-ml lavage may be substantially different from the values in a 240-ml lavage, and as introduced volume increases, the amount of soluble factors decreases in a way that seems unpredictable by simple dilution.

In relation to the cells recovered, Dohn and Baughman[158] have reported that the volume used may also influence the results. Since the standard volume of instilled fluid is 100 ml, they compared the values obtained after a first and after a second instillation of this amount. In the second 100 ml of the ILD group (n = 8) a significant increase of lymphocytes was observed in five patients, a decrease in two of them, and no change in one. On the other hand, two patients presented a significant increase of neutrophils in the second 100 ml, and the others showed no significant changes. Furthermore, the ILD patients as a group had more cells aspirated after the second lavage. The authors concluded that differences in the cell population of the BAL fluid are more striking when a larger volume is used.

These and subsequent studies demonstrate that further research into the kinetics of cells and molecules recovered with different instilled volumes is necessary.

c. Variability Provoked by the Processing and Staining of the Recovered Cells

From the several methods that have been described to concentrate cells before counting and differentiating them, cytocentrifuge preparations stained with Wright-Giemsa are very commonly used.

However, Saltini et al.,[159]; studying 10 normal vounteers and 39 ILD patients, demonstrated that the percentage of lymphocytes identified in cytocentrifuge preparations is significantly less than that identified on filter preparations. In this analysis, the magnitude of the loss was variable and affected these cells selectively. Concomitantly, a greater percentage of macrophages was seen. In addition, they observed that concentration and washing of lavage cells, a frequently used technique, resulted in a cell loss of about 22 ± 20%. In this case, the cell loss was variable in magnitude and was not restricted to a single cell type. Thus, with this procedure there is a loss of all cells types, whereas with cytocentrifugation it is particularly a loss of lymphocytes. The authors proposed that the quantification of total cells is better if an aliquot of the original cell suspension is used and that more accurate differential cell counts are obtained with millipore filter preparations.

Mordelet-Dambrine et al.[160] also analyzed the effects on cell counts of different forms processing BAL. For this purpose they studied 19 controls and 58 patients with sarcoidosis and found that total cell number decreased by 34% after 2 washings of the BAL pellet cell. With regard to cytocentrifugation they used two different speeds and noted that fewer lymphocytes are counted after cytocentrifugation at 23 *g* than at 90 *g*. The changes in lymphocyte percentage were accompanied by an increase or a decrease in macrophages and no variations in polymorphonuclear cells.

On the other hand, they observed that the percentage of macrophages stained with neutral red was lower than that determined after May Grunwald Giemsa staining. Again, there is another problem here which may produce artificial results and must be standardized.

d. Interlobar Variations

Normally the site selected by most authors to perform the BAL is the lingula or right middle lobe,[151] probably, among other advantages, because in these places it is possible to recover more lung fluid and cells. In my opinion, this is very astonishing if we remember that the lingular segment is a zone usually refused for biopsy because this site is "supposed" to present nonspecific inflammatory and fibrotic changes unrelated to ILD. Then, I do not understand why the same argument is not applied to the BAL selection. Nevertheless, lower lobes are scarcely used, and except for specific purposes, the upper lobes are never used.

However, a report by Pingleton et al.[161] supports the notion that other lobes can be useful for BAL analysis. They studied, among other things, the amount of cells and proteins recovered from the right middle and lower lobes, lingula, and left lower lobe in seven normal volunteers. Whenever they used the same solution and volume, although the percentage of fluid recovered was greater in the lingula and right middle lobe, the cell count per milliliter and total proteins (mg/ml) were not different in any lobe lavaged.

On the other hand, a few studies have been performed in patients with ILD, with the aim of assessing the possible interlobar variations, mainly in relation to the differential cell count.

García et al.[162] analyzed 53 bilateral lobar lavages (right middle lobe and lingula) in patients with different types of ILD. Patients with sarcoidosis, who predominantly have T-lymphocytes, showed an excellent interlobar correlation. The group with "IPF-CVD" showed good interlobar correlation, but 35% had a greater than 10% discrepancy in the percentage of neutrophils. Finally, the patients with other ILD, such as silicosis, HP, and drug-induced disease, presented a poor interlobar correlation.

Nugent et al.[163] examined this problem performing bilateral lavages in 20 patients with sarcoidosis, 11 with IPF, and 14 with fibrosis associated with CVD. In sarcoidosis the overall correlation between both sides was good. Only in 25% of these patients one site was considered active and the other inactive; however, in three of them the findings correlated with differences in chest radiographs from the region of lavage. In the other two diseases, the overall correlations were poor.

Taken together, the reports by the García and Nugent groups suggest that the interlobar cellular findings are similar and reliable only in sarcoidosis. In this way, for a better approach and more accurate results, BAL must be carried out in at least two different areas, since one zone might not reflect the average disease process and the findings may lead to an erroneous interpretation about the activity and stage of the lung disease.

Obviously, all these possible variations, which depend on the handling conditions of this procedure, are responsible for many disagreements reported in the literature. For instance, while Keogh et al.[164] found that the precentage of lymphocytes may have a predictive value in patients with sarcoidosis, in the sense that those with more than 28% of T-cells (high-intensity alveolitis) will have impairment in lung function over the next 6 months, Lawrence et al.[165] examined 12 patients with clinically active sarcoidosis, and 11 of them improved showing minimal changes in the BAL lymphocytes.

Moreover, Ceuppens et al.[166] studied 35 patients with the same disease, from which 28 were lavaged at the time of diagnosis and remained untreated and the other 7 patients received therapy. The alveolar lymphocytes were similar in both groups,independent of the stage of the disease, and the percentage of these cells did not decrease in treated patients. In addition, steroid therapy induced normalization or even reversal of the T-helper/T-suppressor ratio, whereas the percentage of lymphocytes remained high.

Thus, to compare results from one or another laboratory is difficult and must be taken with great caution.

Finally, smoking and perhaps air pollution produce considerable changes in cell counts and proteins,[151] and these factors must always be taken into account in the analysis of results.

2. Do Cells Recovered from BAL Reflect Those Present in the Interstitial Tissue?

This is a very important question, since most of the activated cells with abnormal functional behavior and the injurious soluble factors are located in the interstitial microenvironment. In fact, the main pathological alterations in ILDs involve both cells and connective tissue components of this site. In addition, in almost all ILDs the number of inflammatory alveolar cells (some of which are obtained with BAL) are very few compared with those in the interstitium. Currently we do not know if the former are participating *in vivo* in the derangement of the alveolar structures or if they are "terminal cells" which, having finished whatever they were doing, are now simply on their way out.

Thus, the morphological and functional comparison of interstitial and alveolar cells is very important, and the few investigations done concerning this question have also produced some conflicting results.

Lum et al.[54] using electron microscopy and morphometric procedures, analyzed in normal and ozone-exposed rats several morphological aspects of macrophages obtained by BAL and of those lying *in situ*. The comparison of both showed significant differences in a series of parameters including volume fractions of nucleus, cytoplasm, ectoplasm, mitochondria, lysosome-like structures, lipid droplets, vacuoles, and phagosome/autophagosomes. *In situ* macrophages of ozone-exposed rats had a greater number of components which were significantly modified, and these alterations were clearly more notorious here than in macrophages from the lavage. The authors conclude that these differences indicate distinct cell populations sampled by these two techniques.

Thrall and Barton[167] compared lymphocyte populations in BAL and in lung tissue during the development of bleomycin-induced pulmonary fibrosis. In both samples these cells were identified by immunofluorescence using specific antibodies to cell surface antigens. After bleomycin administration a significant increase in the percentage of B-cells in lung tissue was observed, with a peak at 7 d. In addition, changes in the T-cell subpopulations were also found in lung tissue. Fourteen days after bleomycin the helper/suppressor T-cell ratio was 2:1 (normal 1:1), and after 120 d their ratio was reversed to 1:2. In contrast, although significant populations of lymphocytes were found in the lavage fluid at 3, 7, and 14 d after drug treatment, these cells showed the same percentage of B and T-cells as normal controls, and no modifications in T-cell subpopulations were observed. Moreover, no lymphocytes were obtained in lavage fluid 30 and 120 d after bleomycin administration.

These results demonstrate that, at least in this experimental model, specific lymphocyte populations are actively changing during the lung damage, whereas this shift does not occur in the lymphocyte populations in BAL.

A somewhat different result has been reported by Weissler et al.[55] They analyzed some functional activities from human pulmonary macrophages obtained by BAL and from whole lung minces. In this study, both populations of macrophages were similar in their ability to stimulate T-lymphocyte proliferation, in their expression of HLA-DR antigen, and in the production of interleukin-1. However, macrophages from lung tissue stimulated a significantly higher mixed leukocyte reaction than those from BAL. The authors suggest that macrophages from BAL and lung tissue are similar in most immunological functions, but some differences may also exist.

Another early, but interesting result was reported by Davis et al.[168] They analyzed the relationship between the number and types of mononuclear cells present in the airspace and interstitium, using a semiquantitative method based on light and electron microscopic studies of lung biopsy samples. Their attempt to find a direct correlation between the intensity of interstitial inflammation and lymphocytes present in the alveolar spaces was not successful. Although patients with greater numbers of lymphocytes tended to have higher scores of interstitial mononuclear cells, there were many exceptions. The authors proposed that the lack of association between the degree of interstitial mononuclear cell infiltration and the percentage of airspace cells suggests that this relationship is not determined by a simple gradient of cell concentrations, but certain stimuli may attract cells and retain them in determined anatomic compartments. Even though a more precise means of quantification is necessary, these results suggest that the proportions and type of cells present in interstitium are different from those in alveolar spaces.

Finally, other studies, for example, those performed in HP, showed that, at least for the percentage of T-lymphocytes and T-subset populations, there is a strong association between BAL and lung tissue findings,[169,170] although in HP with very high intensity alveolitis there are some discrepancies probably due to an increased recirculation of lymphocytes from the interstitium to the alveolar spaces, which leads to higher concentration of these cells in the lavage than in the corresponding biopsy sample.[171]

Summarizing, this question has not been resolved yet, and it is possible that, although the controversies are doubtlessly of methodological origin, they may also be due to the fact that only in some ILD are the BAL cells representative of interstitial inflammation, whereas in others this does not occur. Moreover, the cells obtained in BAL and lung tissue may be similar in only some morphological or functional activities and different in others.

Evidently, much more research is needed to provide a better understanding of this problem.

3. Changes in the BAL Inflammatory Cell Profiles and the Prognosis of Patients with ILD

In spite of its numerous limitations, this procedure has been used to find out its possible value for the clinical follow-up, and several recent reports have been published.

Watters et al.[172] examined the correlation between BAL cellular constituents and the subsequent clinical response to steroids in 26 untreated IPF patients. Their results show that lymphocytosis was present in 5 out 7 patients who improved and in none of the 12 patients who deteriorated or remained stable. In contrast, 3 out of 9 cases with BAL neutrophilia in the initial evaluation ameliorated, and 4 out of 10 patients without increase of this cell type had a similar response. Finally, 4 out of 9 cases with BAL eosinophilia and 3 out of 10 patients without it also improved. On the other hand, BAL lymphocytosis was associated with variable degree of septal inflammation and a relative lack of histological honeycombing. The authors proposed that BAL lymphocytosis identifies a subtype of IPF patients who have a better prognosis and are likely to improve in response to therapy, and neither neutrophil nor eosinophil content relate to the subsequent clinical evolution.

With regard to the hypothesis that BAL lymphocytosis in the initial assessment may be an indicator of a better prognosis, similar findings have been reported by Turner-Warwick and Haslam.[173] They studied 32 patients with IPF and fibrosis associated with CVD and found that subjective and objective improvement with prednisolone was observed in 6 out of 7 patients who had an initial increase of BAL lymphocytes. However, unlike the report by Watters et al., in most patients who failed to recover, neutrophil and eosinophil counts tended to remain elevated, although some of these patients had clinically stable disease, suggesting that the persistence of granulocytes in BAL not always correlated with clinical deterioration.

Thus, it seems that, at least in IPF, the increase of BAL lymphocytes predicts a better response to treatment, and the increase of neutrophils or eisonophils is accompanied by a poor prognosis. For this reason elevated amounts of granulocytes are considered to imply a high-intensity alveolitis in this disease and probably in others similar to it as well.

A different situation, also with conflicting findings, is observed in other ILD in which lymphocytes seem to play a major pathological role, such as HP and sarcoidosis. As mentioned previously[164-166] in sarcoidosis the results relating high or low levels of BAL lymphocytes with response to treatment are contradictory. However, the prevalent notion is that in this disease a high percentage of these cells, mainly T-helper, represents an active form of disease and may predict an impairment if patients are not treated. Regarding HP, Cormier et al.[174] have followed up 14 patients with farmer's lung and have demonstrated that, independent of their clinical response, increased levels of lymphocytes persist for at least 2 years, and, therefore, high-intensity lymphocytic alveolitis does not predict the outcome of the disease. This is an interesting finding because the same authors[175] have demonstrated that an increase of lymphocytes in BAL may be found in asymptomatic subjects exposed to organic particles for long periods without development of the disease. Here are the questions:

1. When lymphocytes increase do they represent a controlled defensive mechanism?
2. When and why do they become involved in an abnormal response?
3. When may they be considered as part of an alveolitis?

4. The Role of BAL in Diagnosis

With respect to this point the accepted rule is that BAL has no place in ILD diagnosis. Probably, as in any rule, there are some rare exceptions, and one of them seems to be histiocytosis X.

In this sense, Basset et al.[176] studied four patients with this disease and found in the BAL fluid of all of them characteristic cells whose ultrastructural examination revealed the typical cytoplasmic marker (X body). In 3 of them the diagnosis was corroborated by lung biopsy.

However, BAL must not be considered a diagnostic tool because its results, in addition to the difficulties in interpretation, on many occasions give overlapping findings between different ILD and are not pathognomonic of a determined interstitial disorder.

I would like to finish the discussion about the relevance of BAL, emphasizing that, even though it has thrown some light into the knowledge of ILD, it has also created too many expectations, and its inherent limitations have not been considered carefully. I believe that, as in any oscillating pendulum this method will arrive at a middle point, and we will then be able to value its real significance in the study of ILD patients. Nevertheless, for this it will be indispensable to reach one common criterion related to the standardization of all procedures performed during the development of the technique and the processing of the samples.

B. GALLIUM LUNG SCANNING

^{67}Ga is a cyclotron-produced radionuclide with a half-life of 3 d, which is not normally localized in healthy lung tissue. Gallium lung scanning has been used to detect a variety of pulmonary diseases including lung cancer, infectious diseases, and more recently ILD (Figure 6).[164,177-182]

Although the mechanism by which ^{67}Ga identifies neoplastic and inflammatory cells is not known, isotopic activity has been found in macrophages, lymphocytes, and leukocytes, all of them present to a greater or lesser extent in ILD. Thus, this scan is useful because it detects inflammatory processes with simple and noninvasive techniques. However, an abnormal uptake of this radionuclide is not specific for a determined disease and only reflects the degree of alveolitis common to all of them. In this sense, it is a useful tool for the staging and follow-up of the inflammatory activity, but does not have a place in the differential diagnosis and can be misleading if it is not associated with other clinical, laboratory, and morphological data.

Some years ago, Niden et al.[177] studied 12 patients with diffuse pulmonary images on chest films. They found that in sarcoidosis there was a close correlation between pathological activity in biopsy specimens and gallium lung uptake. Moreover, the active diffuse inflammation did not correlate with conventional means of clinical assessment such as symptomatology, chest roentgenogram, and pulmonary function test. Similar results were demonstrated

FIGURE 6. Posterior ^{67}Ga scan of the thorax in a 31-year-old woman with IPF. Diffuse moderate radionuclide uptake, greater than the background area can be observed.

in the other cases of the interstitial disorders with active inflammation studied, and the one patient with a normal gallium scan had little if any evidence of alveolitis from open lung biopsy.

Afterward, Line et al.[178] studied 30 patients with IPF and 19 control subjects. Since gallium uptake is usually interpreted in a qualitative fashion, the authors attempted to develop a quantitative index, and for this purpose each region of increased gallium-accumulation was characterized and computed by recording the area of gallium uptake, its image area intensity, and its textural features on a diagram of the posterior pulmonary segments. With this method, 100% of controls showed a ^{67}Ga index of less than 50. In comparison, 70% of the patients with IPF showed significant accumulation of lung gallium. When the results were compared with lung morphological studies, this index correlated significantly with the degree of interstitial and alveolar cellularity. Interestingly, when compared with cellular analysis of BAL fluid, in 17 patients the gallium index correlated with the differential percentage of neutrophils, but not with lymphocytes, eosinophils, or macrophages. Similar to the findings by Niden, there were no significant associations with pulmonary function tests nor with patient age, sex, duration of symptoms, smoking history, or drug therapy.

Subsequently, Line et al.[179] analyzed the ^{67}Ga scanning to stage the alveolitis of sarcoidosis. For this purpose, they studied 41 patients with this disease and found a strong correlation between ^{67}Ga uptake and the number of lymphoctes and T-lymphocytes recovered by BAL.

In regard to this, Keogh and co-workers[164] have suggested that the gallium scans and BAL can be used in sarcoidosis to separate high- and low-intensity alveolitis and to predict the response to treatment. Thus, they have shown that patients with high-intensity alveolitis, that is, a positive gallium scan and an increased number of lymphocytes in the BAL fluid, will have impairment of the pulmonary function if not treated, whereas those with a low-intensity alveolitis will show no change in respiratory tests.

In a similar way, Baughman et al.[180] studied 16 patients with sarcoidosis and, using a different technique to quantify the gallium scan, demonstrated a strong linear correlation between gallium uptake and the response to treatment, measured by the degree of recovery in vital capacity.

In relation with gallium uptake and the response to treatment, a somewhat different picture has been reported by Gelb et al.,[181] who studied 16 patients with predominantly intraalveolar or interstitial cellular disease and 4 patients with a clear fibrotic predominance. The gallium scan was useful for separating both groups and was clearly associated with the cellular phase of the disease, but was not reliable in the identification of a therapy-responsive group.

Nevertheless, IPF and sarcoidosis cannot be compared with respect to prognosis and therapy response, since the latter may dramatically improve in the active stage with corticosteroids whereas IPF usually seems to lead to progressive lung damage in spite of therapy. In this sense the differences reported are only reflecting the behavior of two very different diseases.

^{67}Ga lung scanning has been also used in asbestosis. Begin et al.[182] studied 58 asbestos workers and a sheep model of this disease. The gallium lung index was significantly higher in patients with asbestosis, and, more importantly, it was clearly positive in 43% of long-term asbestos workers before usual diagnostic criteria for the disease were met. In the animal model, gallium uptake was increased in the high-exposure group, and this index was clearly related to the intensity of the alveolitis. Moreover, ^{67}Ga was localized in the lung through enhanced protein-bound leakage into the bronchoalveolar milieu and through its accumulation in the macrophages at the disease site. This is an interesting finding because it suggests that the cell responsible for gallium uptake is different for IPF, sarcoidosis, and asbestosis.

These studies indicate that gallium scanning gives a very good notion of whole lung alveolitis, and thus represents another useful test for the assessment of inflammatory activity of these disorders and their response to therapy. Still, additional studies defining the value of ^{67}Ga lung scanning in other ILD are needed.

C. MARKERS OF COLLAGEN BIOSYNTHESIS

Several authors have recently studied the usefulness of some markers of collagen synthesis as indicators of pulmonary fibrosis and their possible clinical relevance. Among them, type III procollagen peptide (PRO [III]-N-P) has been analyzed both in sera and BAL fluid.

Low et al.[183] studied 9 patients with IPF, 11 with sarcoidosis, and 14 controls, and the levels of PRO(III)-N-P were immunoassayed in sera and BAL fluid. They found a significant increase in both patient groups, and the values were higher in BAL than in serum, suggesting active lung production. In this study the PRO(III)-N-P was higher in IPF samples than in sarcoidosis, and the correlation with clinical severity of the disease was poor, but the patients were not followed up.

In this sense, a relatively different result was reported by Kirk et al.,[184] who measured the serum concentration of this peptide and also found an increase in IPF patients, but an excellent correlation with the clinical course and the response to therapy. They quantified procollagen peptide by radioimmunoassay in 23 IPF patients without treatment, 10 patients with nonfibrosing lung disease, and 21 control subjects. In addition, in 12 cases the percentage of type III collagen was analyzed biochemically in open lung biopsy or post-mortem samples. In this context, they observed a positive correlation between the percentage of type III of the total type I + III collagen in lung tissue and the serum procollagen peptide concentration. Furthermore, patients with the highest levels of PRO(III)-N-P tended to show a better response to treatment, and this response was associated with a decrease in serum peptide levels.

Watanabe et al.[185] studied this peptide in sera and BAL fluid from rabbits with bleomycin-induced pulmonary fibrosis and in sera from patients with different respiratory diseases. They found a significant increase of PRO(III)-N-P in rabbit lavage fluid at 7 d, which nearly returned to control levels after 21 d, again suggesting as Kirk et al. did that the elevation of this peptide may be an early phenomenon in lung fibrosis. In contrast, the average mean of serum levels in human pulmonary fibrosis was no different from controls. Moreover, patients with some nonfibrotic disorders showed significant increase of this peptide. This latter finding suggests that serum measurements may not be reliable for the study of the local lung process and agrees with the work of Low et al.,[183] who demonstrated that BAL fluid concentrations are higher than in serum. On the other hand, the finding that patients with lung cancer, chronic bronchitis, and tuberculosis showed higher values than controls is difficult to explain and needs further studies.

Begin et al.[186] studied the PRO(III)-N-P in 7 asbestos workers without disease, 8 asbestos workers with evidence of alveolitis, and 19 asbestos workers with fibrosis. In addition, they analyzed three groups of 24 sheep receiving PBS (controls), latex beads, and chrysotile fibers, respectively. In the asbestos workers without disease peptide concentrations in BAL fluid were comparable to controls, but these levels were significantly elevated in workers with asbestos-associated alveolitis or asbestosis. In the animal model of asbestosis a significant increase of the peptide was observed at 2 months, and these values returned to normal at 4 months and remained so until the end of the experiment (12 months). The increase correlated with lung inflammation which progressively regressed, whereas peribronchiolar and endobronchiolar fibrosis progressed, severely distorting the pulmonary architecture. Interestingly, the animals injured with latex also showed an inflammatory response which returned to normal at 4 months, without developing fibrosis, and no significant changes in peptide concentration were found.

Taken together, all these investigations suggest that increased levels of PRO(III)-N-P may be a useful tool for studying the fibrogenic activity of alveolitis, and support the notion that this type of collagen is the first to increase in pulmonary fibrosis and may indicate a reversible stage of the disease.

The galactosylhydroxylysyl glucosyltransferase (GGT), an enzyme catalyzing collagen biosynthesis, has also been explored in ILD.

Anttinen et al.[187] studied the GGT in sera of 101 patients with various pulmonary diseases including 45 cases of HP, 5 with pulmonary fibrosis, and 6 with sarcoidosis. Elevated values were recorded most frequently in progressive pulmonary fibrosis and in the acute stages of HP, but increased levels were also observed in nonfibrotic diseases such as infectious pneumonia. However, the authors proposed that elevated serum enzyme activity is related to those respiratory diseases in which pulmonary fibrosis is present or develops later, relatively often.

Afterward, Anttinen et al.[188] monitored the values of both collagen markers, GGT and PRO(III)-N-P, in sera of 40 patients with HP separated in two groups: with and without treatment. The mean GGT of all the patients at acute stages before any treatment was significantly increased. The levels decreased after 6 months in both treated and untreated groups. On the other hand, the initial average of PRO(III)-N-P was at control levels, but the individual values showed a slight and significant increase during the follow-up period in both groups. There was no correlation between either marker at the initial phase of the disease nor after 6 months in the whole group of patients. Interestingly, the patients who continued with abnormal pulmonary function tests 1 year after the acute stage had significantly higher initial GGT values than the rest of the patients. Thus, elevated GGT at the acute stage was observed in 90% of the patients who did not improve after 1 year, but in only 30% of the other patients. Therefore, these results suggest that high initial values of this enzyme are associated with long-term lung impairment and may be useful for early detection of interstitial pulmonary fibrosis. With regard to PRO(III)-N-P, the results are confusing and differ from the studies mentioned previously.

Therefore, although the use of serial measurements of collagen synthesis markers in the same patient during the development of the disease to assess the prognosis is very attractive and has given some encouraging results, much research is necessary to answer several questions that remain unclear.

IX. ANIMAL MODELS OF PULMONARY FIBROSIS

Experimental models of human disease have constituted a powerful tool for the progress of medicine. However, a model is only an approach to reality and in this sense presents two problems which should always be kept in mind. One of them is paradoxically also its principal virtue, the selective reduction inherent to it. Since this necessarily implies a simplification, it may be erroneously believed that the precise results obtained from the simplified model are an exact reflection of the complex reality (human disease) from which it was extracted. The other problem derives from our knowledge of reality which is frequently very intricate and our perception of it often subjective. Thus, the success of an animal model, in the end, lies in the precision of our knowledge of a given disease and in our ability to extract from it some real but sometimes insignificant parts, to build a model containing only its more relevant features.

The general criteria for a good animal model should be the greatest possible resemblance to the human disease, the reproducibility in any investigation laboratory, and, if possible, in several animal species.

Obviously, animal models have several limitations, but probably the main two are (1) that the obtained results may not be extrapolatable to humans and (2) that animal models do not provide a "final answer" to a determined problem, but offer only an approximation, since they commonly do not constitute an exact duplicate of the human disease under investigation.

However, a good animal model has many advantages, among which are

1. Genetic and environmental variables may be modified and controlled at will, and this allows the determination of such data as the cause-effect relationship for a given aggressor.
2. Complete tissues may be analyzed.
3. Animals may be dynamically studied, that is, at different times starting from a time zero.
4. Many animals can be studied in short periods, which modifies the time scale favorably.

The aim of this discussion is not the analysis of experimental models of the large number of ILD, but of diffuse interstitial pulmonary fibrosis (DIPF).

A. NATURAL MODELS

Undoubtedly, the best model should be one in which the disease occurs spontaneously, that is, in a natural form, as in human beings. Unfortunately, in pulmonary fibrosis such models are scanty.

In 1972, Pirie and Selman[189] reported 12 cases of a bovine respiratory disease which appeared to have most of the clinical and morphological features of human diffuse fibrosing alveolitis, but this finding has not been reported by other authors. More recently, an anecdotal case of progressive fibrosis in an African elephant was published.[190]

However, nobody would like to have a colony of elephants in his laboratory.

Thus, in general terms, to my knowledge there does not exist a natural model in a small species which develops interstitial fibrosis spontaneously, with a relatively high frequency to be used for research, as occurs, for example, with a colony of hemophilic dogs. An exception could be the motheaten mice,[192] which suffer from a single recessive mutation and phenotypically present several systemic immunological alterations. However, the homozygous (me/me) animals die in approximately 8 weeks from diffuse interstitial lung disease. Thus, the motheaten mouse can be regarded as a genetically determined model of ILD which shows natural progression to lung fibrosis.

In the same way, other ''genetic models'', not necessarily of DIPF, may be useful in the understanding of some pathogenetic mechanisms of this disorder. This has been so, for example, with the nude (nu/nu) mice,[59] and the genetically mast-cell-deficient W/W^v and Sl/Sl^d mice[192] could possibly serve to evaluate the physiopathological role of these cells *in vivo* in the development of fibrosis.

B. EXPERIMENTALLY INDUCED MODELS

In 1968, Carrington[193] proposed the main distinctive marks that must exist in a lung fibrosis model. The essential features were morphological. He included a very detailed list of them and also added some physiological characteristics. However, our current knowledge of these diseases, mainly of the aspects related with the modifications of collagens, renders this description somewhat incomplete.

We have recently[18] proposed that an adequate experimental model of human DIPF should fulfill the following requirements:

1. Be a chronic, progressive, and potentially lethal disease
2. Resemble the histological and ultrastructural features of the human disease, such as alveolar septal fibrosis, interstitial inflammation, cuboidalization of alveolar lining cells, etc. (fundamentally these lesions must be at least multifocal, and preferably diffusely distributed throughout the lung parenchyma)
3. Show an absolute increase in the total amount of lung collagen
4. Develop clinical and functional features of diffuse pulmonary fibrosis

The most important aspect is the development of inflammatory and fibrosating multifocal or diffuse lesions affecting both lungs, which must also be progressive. If this criterion is met, the rest of the requirements will develop of their own accord.

In the last two decades a great number of experimental models have been developed in several animal species, and I believe that they should be examined against the criteria proposed above in order to establish their adequacy as DIPF models for a better understanding of the human disease. The injurious agents have included irradiation, high concentration of O_2, ozone, paraquat, bleomycin, cyclophosphamide, methotrexate, busulphan, nitrofurantoin, melphalan, *N*-nitroso-*N*-methylurethane, monocrotaline, BHT, and several inorganic and organic particles (for review see Reference 194). More recently, the combination of two agents has also been attempted, mainly BHT plus O_2,[37] paraquat plus O_2,[18] and bleomycin plus O_2.[195] The animals used have mainly included mice, rats, hamsters, rabbits, dogs, and monkeys.

Clearly, this large number of attempts at finding a model demonstrates the difficulty in attaining a satisfactory one, and until now none of them seems to fulfill all the requirements of the scientists interested in the problem. If only one or two of them were adequate and reproducible, they could then be used by everyone to answer the endless succession of unknown facts related to the pathogenesis, development, and treatment of pulmonary fibrosis.

This is precisely what has happened with hepatic cirrhosis. Most scientists agree that the CCl_4-induced cirrhosis, albeit with some limitations, is a good model of the human disease, mainly from a morphological and biochemical point of view. Most investigators use it, and it seems to be easily reproducible in any laboratory.

An analysis of the different studies on DIPF reported in the last few years, in which experimental models have been used, demonstrated that the main agent utilized is bleomycin, which suggests that for many colleagues it is perhaps the best. This drug induces fragmentation of DNA, and its action has been attributed to the generation of free radicals.[196]

In a random revision of 60 papers in which bleomycin was used, only 18 of them (30%) included morphological pictures. Moreover, in several reports showing some histological features, these features appear with great magnification, which ignores a panoramic vision of the lung parenchyma. In addition, in some papers with pictures at lower magnification, the lesions are focal, and most of the parenchyma seems to be respected.[104,197,198] However, at least in three reports the lesions seem to be multifocal and involve most of the lungs.[199-201]

In our laboratory we have attempted to develop this model with a single intratracheal administration of bleomycin in Wistar rats,[18] and only focal perivascular, peribronchial, and subpleural lesions were obtained (Figure 7). However, there was a significant increase in collagen content, and in this context it is very important to note that this parameter on its own it not sufficient to talk of it as a good model. As air spaces are part of the lung, the occupation of some areas by fibrotic lesions can be enough to increase the amount of collagen, and this does not necessarily imply a multifocal or diffuse fibrosis.

For this reason, I maintain that any report which uses an animal model of DIPF must be accompanied by morphological features which allow the reader to recognize with greater precision the quality of the obtained model, since the increase in collagen content on its own may lead to an erroneous understanding and gives no security to the identity of the disease as DIPF.

On the other hand, differences in the magnitude of lesions obtained by different authors may be due to the species used or to strain variations, as has been suggested by Schrier et al.[202] for mice.

Finally, I think that in some animals,mainly hamsters and some strains of mice and rats, the intratracheal administration of bleomycin can produce patchy multifocal lesions which constitute a certain approach to the human disease.

Another experimental model widely used is that induced by paraquat, an herbicide which produces severe lesions in human beings, probably through an oxidant-initiated toxicity.

In 1974, Smith et al.[203] reported an excellent description of the morphological changes, both acute and chronic, produced in rats with intraperitoneal injections of paraquat, and this model has been used in several species for different purposes.

Similar to bleomycin-induced damage reports, about 65% of the revised papers do not show histological pictures, and in those where they do the use of high magnification and electron microscopy avoids a clear view of the magnitude of the lesions. Nevertheless, there are several reports which demonstrate that paraquat induces severe alterations in the lungs,[203-206] resembling those of human fibrosis.

In our experience, the main problem with this model, at least in rats, is that the doses usually recommended produce a high rate of mortality in the first days by acute damage, and with lower paraquat concentrations the pulmonary lesions are minimal or do not produce lung damage at all.[38,207] Thus, the dose and the frequency of administration (daily, twice weekly, etc.) are very important to obtain a good model and probably depend on the species used in each laboratory.

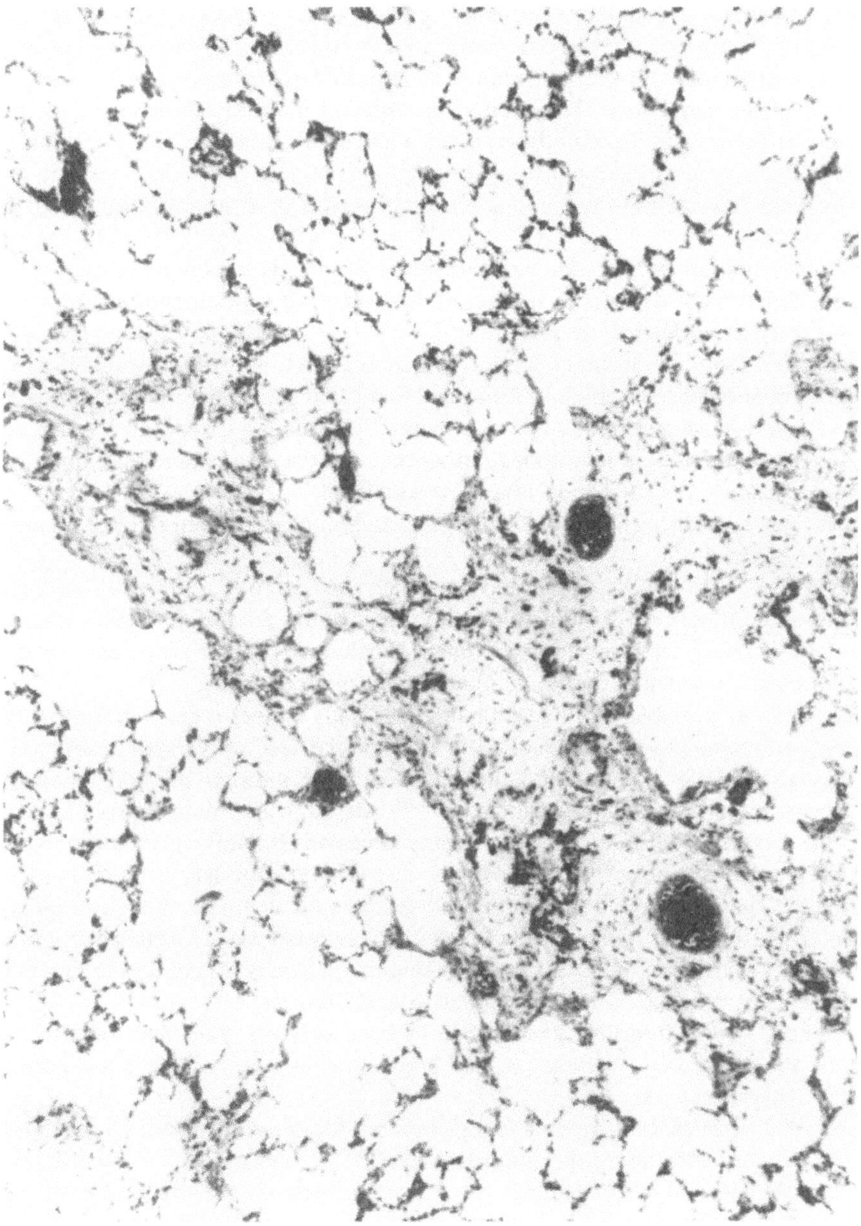

FIGURE 7. Histopathological features of a bleomycin-treated Wistar rat. There is a focal fibrosis, mainly perivascular and peribronchial. Approximately 50% of the parenchyma is undamaged. (H & E stain; magnification × 120.)

High concentrations of oxygen, which damage alveolar epithelial and endothelial cells, have been used by some authors,[80,208] but the lung alterations seem to be reversible, and after several weeks the lungs appear almost normal.[12] We used it[38] at 74% for 6 to 10 d, and no microscopic changes were observed. I believe that high concentrations of O_2 produce an initial alveolar-capillary injury of a severity that will depend on the concentration and time of exposure, but these modifications are rapidly reversible by the action of the antioxidant mechanisms and the subsequent regeneration of damaged cells. This notion is supported by a recent report of Coursin et al.,[209] who demonstrated that sublethal hyperoxia is capable of eliciting a significant antioxidant enzyme reponse. The results by Adamson and Bowden[210] may be interpreted in the same way. They found in radiation-induced pulmonary damage that acute endothelial injury may be rapidly repaired with few fibroblastic modifications.

In this context, the oxygen model is useful for an understanding of the initial events of some lung injuries and repairs, but does not seem to be an adequate model of progressive fibrosis.

With regard to irradiation injury, although interstitial fibrosis is a recognized sequel of thoracic radiation, this type of damage is seldom used to provoke experimental fibrosis. In addition there are some contradictory reports.

Shrivastava et al.[211] found in rats that 6 weeks after irradiation the morphological changes were atrophy, inflammation, edema, pleural fluid, and focal atelectasis, but without fibrosis. It is interesting to mark that lung compliance was markedly reduced (as would be expected for DIPF). This supports the notion mentioned before that one positive parameter cannot be used to determine the adequacy of a good lung fibrosis model.

Other authors, however, have found morphological and biochemical alterations suggestive of pulmonary fibrosis.[210,212]

On the other hand, whereas Adamson and Bowden[210] using autoradiography demonstrated an increased proliferation of interstitial cells (presumably fibroblasts) in irradiated mice, Vergara et al.[213] using stereologic-morphometric techniques found a decrease in the percentage of fibroblasts in irradiated rats, compared with mastcells.

Thus, to determine its possible usefulness this model needs a more precise definition of several aspects such as the most sensible species or strain and the dose of thoracic irradiation.

A good and reproducible experimental model is obtained through the inhalation or instillation of inorganic particles, mainly silica.[82,214-216] Although the main morphological characteristic is the formation of granulomas, on many occasions inflammation and fibrosis of adjacent alveolar walls are also seen. Of course, this is not rigorously a DIPF model, but silicosis probably shares many of the pathogenic features of other ILD which progress to fibrosis, and considering the simplicity with which it can be performed, it is in my opinion an experimental model which can be used to answer several questions regarding the kinetics of abnormal mechanisms related to lung inflammation and fibrosis.

On the other hand, other injurious agents[194] have been scarcely used and need to be validated to know their potential adequacy as DIPF experimental models. We have tried several of them[217] without success.

More recently, the proposition of a double aggression with appropriate timing has been explored with the aim of potentiating the lung damage.[18,37,195] Tryka et al.[195] and Rinaldo et al.[218] have suggested that high concentrations of oxygen increase the magnitude of the pulmonary lesions and the functional respiratory impairment induced by bleomycin.

We have assayed this method[18] and have observed that there are more lesions, but they are mainly characterized by hemorrhages and airway dilatation and are therefore unrelated to DIPF. Moreover, lung collagen content was significantly less than that obtained with bleomycin only. In addition, as Tryka and Rinaldo mentioned in their papers, there was higher mortality with this combination of agents. However, we only studied 12 rats and did not repeat the experiment.

On the other hand, Haschek and Witschi[37] proposed that the use of BHT plus O_2 in male BALB/c mice induces extensive intestitial fibrosis determined by hydroxyproline measurement and histological observations.In this report, animals were followed only until day 14.

Afterward, Haschek et al.[79] performed a long-term (1 year) study with the same protocol. In this work they demonstrated that elevated levels of collagen content persisted throughout the study, although the alterations of ratios of type III to I collagens observed in the course of the experiment returned to normal. However, the single morphological figure at low magnification after a year of follow-up only shows a dilatation of distal lung parenchyma structures and very slight interstitial modifications. Furthermore, this model has only been used by the same authors and does not seem to be easily reproducible in other laboratories. We have attempted several protocols with this combination in several species including BALB/c mice, and only modest and focal lesions have been obtained (Figure 8). Therefore, I presume that special conditions are required for this model to function, and these factors are not clearly explicit in their reports, although some of them have been analyzed in one of their papers.[219] In addition, the revision of several papers[37,79,220,221] using this model suggests to me that lesions occur in the first weeks, and afterward they seem to be reversible, at least concerning the morphological progressive interstitial fibrosis.

Finally, following the same hypothesis of a double aggression to induce DIPF, we have used paraquat plus O_2 in Wistar rats.[18,38] After several attempts using different doses of paraquat and O_2, we found that low paraquat doses (2 or 2.5 mg/kg) plus 74% oxygen produce extensive lung damage (Figures 9 and 10) with interstitial inflammation, predominantly with mononuclear cells, fibroblast proliferation, interstitial fibrosis, and a considerable distortion of the normal lung architecture. However, regardless of the scheme utilized, there were always some animals which failed to develop the disease, and many others died in the first days with severe respiratory insufficiency due to pulmonary lesions very similar to ARDS. Therefore, the main problem of this model is that only a few rats reach the end of the experiment (2 or 3 months), which implies that very many animals must be employed to obtain an appropriate and significant number of them with DIPF.

Summarizing, I believe that many of the proposed experimental models of DIPF do not fulfill all the criteria mentioned for a good model, and several of them are yet to be critically examined and validated from this point of view. Thus, the conclusions derived from experimental models must be considered with care and must not be given a definitive character.

Of course, it should not be forgotten that one of the main difficulties in obtaining a good model is that human DIPF is the consequence of many diseases from known and unknown etiologies and is far from being a neatly defined clinicopathological entity. Nevertheless, although many of the animal models currently used are not absolutely applicable to human DIPF, they may contribute, and, in fact, have contributed, to the knowledge of some of the pathogenic mechanisms involved in these disorders, in the normal collagen metabolism and the behavior of some cells. They have also allowed investigations on many possibly useful drugs which may at present not yet be applied to humans.

X. CURRENT THERAPEUTIC APPROACH AND SOME INSIGHTS FOR THE FUTURE

Two critical pathways in the development of the ILD should be treated: the alveolitis, which plays an important role in its pathogenesis, and, more importantly, the fibrogenesis with the subsequent overaccumulation of collagens in the lung parenchyma. However, until now most efforts have been directed to control the inflammation, and far fewer to stop the fibrotic process, which is normally considered an irreversible stage of the ILD.

The most commonly used drugs are corticosteroids,[1,3] but the results obtained are very

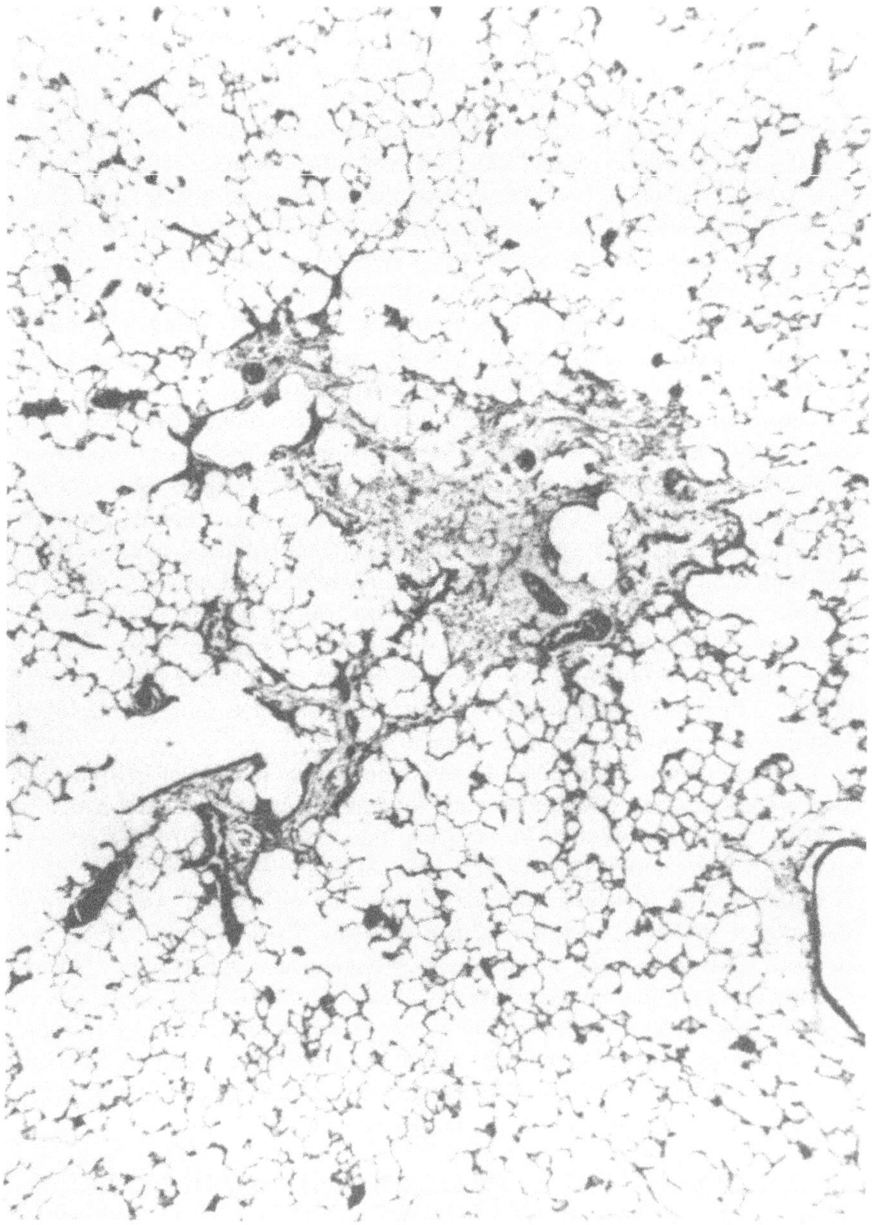

FIGURE 8. Lung section from a BALB/c mouse which received BHT plus oxygen. There is a small and irregular area of fibrosis in the center of the picture. The remaining parenchyma is unremarkable. (H & E stain; magnification × 60.)

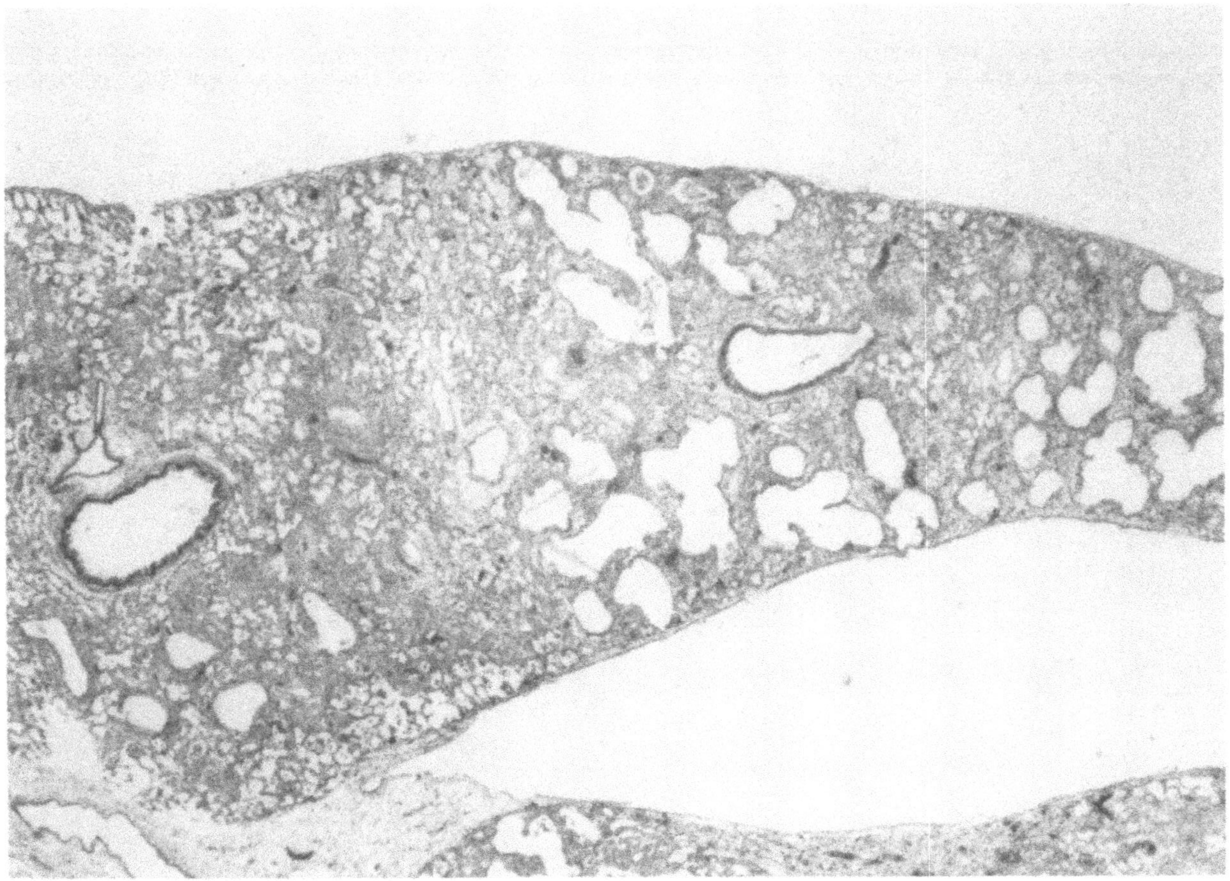

FIGURE 9. Lung specimen of a paraquat plus oxygen treated Wistar rat. At low magnificantion the diffuse nature of the lesion can be appreciated. There is interstitial inflammation and fibrosis with severe distortion of the lung architecture. (H & E stain; magnification × 10.)

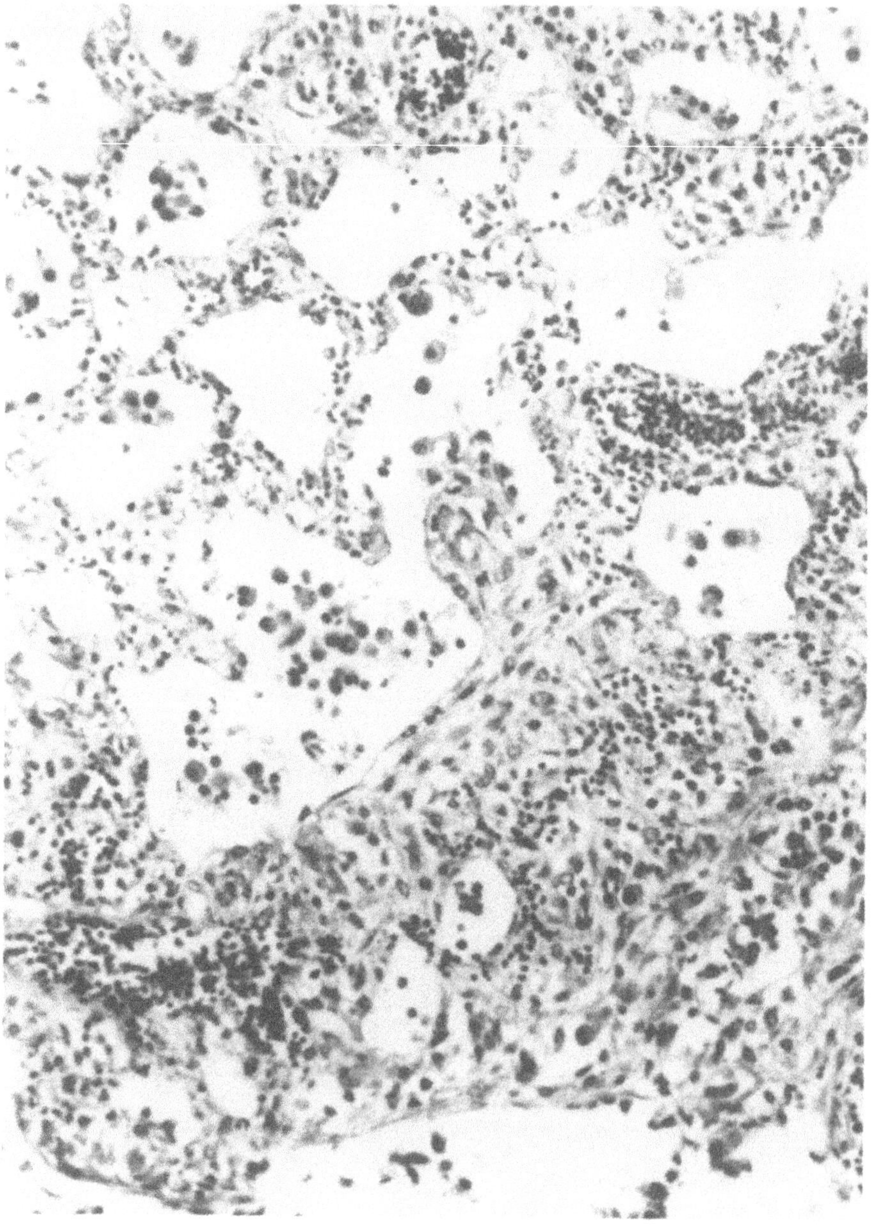

FIGURE 10. High-power view of a lung section from a Wistar rat injured with paraquat plus oxygen. There is an important interstitial thickening by mononuclear cells and fibroblasts. Note the increased number of macrophages in the alveolar spaces. (H & E stain; magnification × 360.)

variable, depending on the disease in question, on the stage of the natural history in which the disorder is diagnosed and treated, and probably on the type and intensity of the alveolitis. Thus, for example, in sarcoidosis and HP[29,31] the steroids may arrest the progression and even reverse the disease, although it is important to keep in mind that these disorders may present spontaneous remissions.

In IPF there are contradictory results,but, in general terms, in early stages with a clear predominance of the inflammation[3,30] longer survival with steroids has been reported. On the other hand, Turner-Warwick et al.[222] observed in a retrospective analysis of 220 cases that in addition to a more cellular histology, a better response is obtained in younger patients, in women, and in those with lesser degrees of clinical, radiologic, and gasometric abnormalities.

More recently, it has been reported that patients with an increase of lymphocytes in BAL have a better response to treatment,[172,173] whereas an increase of neutrophils and mainly eosinophils without elevation of lymphocytes is associated with failure to respond.[9,25] Likewise, elevated levels of procollagen type III seem to indicate a better prognosis.[184] In this context, it is important to consider that glucocorticoids, in addition to their potent antiinflammatory effects, seem to play a role in collagen synthesis. Thus, several investigations[223,224] have suggested that these drugs decrease collagen synthesis *in vivo* and in cell cultures probably by a selective reduction of polysome-associated messenger RNA.[225]

Nevertheless, in spite of their ability to modulate the inflammatory process and perhaps collagen synthesis, in advanced stages of IPF,[30] as well as in other ILD which progress to fibrosis, the steroids have little, if any, influence on the natural course of the disease.[226]

Therefore, and since some patients have a clinical response to steroids whereas many others do not, several studies on glucocorticoid receptors in ILD patient lung cells have been performed.

Ozaki et al.[227] reported that glucocorticoid receptors in BAL cells were significantly fewer in patients with IPF and other ILD than in normal volunteers and patients with localized lung cancer. Moreover, the patients who responded well to this drug had a normal number of receptors. In this sense, two patients with more than 10,000 sites per cell responded well to prednisolone, whereas four patients with less than 3000 binding sites per cell failed to respond. However, responsiveness was evaluated only 2 or 4 months after the onset of therapy.

A somewhat different result was reported by Lacronique et al.,[228] who studied the glucocorticoid receptors of alveolar macrophages in 36 IPF patients and the effect of this drug on the release of some fibrogenic factors by these cells. They found that the characteristics of dexamethasone binding by alveolar macrophages of IPF patients were similar to those of normal subjects, but the addition of this drug *in vitro* did not influence the spontaneous release of fibronectin nor AMDGF by macrophages. Furthermore, the production of these two factors was not different in treated or untreated patients.

The differences related to the number of receptors in both studies may be due to the fact that Ozaki's group analyzed all bronchoalveolar cells together, whereas in the Lacronique report only the macrophages were studied.

However, there are two reasons here which may explain the failure of IPF, and perhaps other ILD patients, to respond to steroids: fewer glucocorticoid receptors and/or the inability of this drug to avoid fibrogenic factor release.

On the other hand, there is no agreement on the dose or duration of steroid therapy, mainly because their efficacy has not been sufficiently evaluated in a systematic form. Again, the decision will depend on the particular disease and on its evolution during treatment.

In general terms, most authors use an initially high dose of prednisone (or an analogue), 1 to 1.5 mg/kg/d for 3 or 4 weeks, followed by a gradual diminution to reach a maintenance dose (10 to 20 mg/kg/d). If the patient shows an objective improvement after several months, this dose may be sustained, but if during the follow-up there is deterioration, the dose must be newly increased. If the disease remains without significant changes, that is, no progress and no regression, the maintenance dose should be continued indefinitely, as long as some of the severe side-effects of this drug do not appear. If the patient continues with lung impairment despite receiving high doses of corticosteroids, this therapy must be stopped. Finally, if it is reasonably certain that the patient is healed, corticoid therapy may be suspended and the patient must be followed up for several months.

Another form proposed for the use of steroids is the administration of higher doses of weekly intravenous methylprednisolone plus low doses of daily oral prednisone,[229] but this scheme has not been sufficiently assayed.

An alternative therapy to steroids are the immunosuppressive drugs, but their use has been more limited.

In 1971, Brown and Turner-Warwick[230] reported the results obtained from five IPF patients to whom some immunosuppressive drug was administered alone or in addition to low doses of steroids. Three patients improved and the other two, with advanced fibrosis, who had previously failed to respond to steroids, also failed to recover on immunosuppressant therapy.

Winterbauer et al.[231] treated 20 patients with diffuse interstitial pneumonitis with prednisone for 3 months, and afterward added azathioprine until 1 year of treatment was completed. With this therapy 12 patients showed improvement, and the single variable which allowed the researchers to distinguish those cases which responded from those which did not was the degree of interstitial fibrosis in the lung biopsy specimen. In addition, azathioprine per se showed a beneficial effect in 8 out of 13 patients who completed the combined therapy.

More recently, Weiland et al.[232] reported the results obtained with cyclophosphamide in five patients with IPF. Three of them had failed to respond, and the other two had presented severe side effects to steroids. Four of the five patients demonstrated a significant improvement in TLC, and the last one remained stable. During the course of the cyclophosphamide trial, two patients died, one of them of an oat cell carcinoma; the other three remained in clinical remission 9, 15, and 28 months after initiation of this therapy.

However, the three studies mentioned present the drawbacks of having lacked a control group and of a relatively short follow-up.

Similar results have been found in patients with lung damage secondary to CVD;[233] that is, some patients improved with immunosuppressive drugs whereas others did not respond.

More rigorous, prospective, and double-blind studies are definitely necessary to determine if immunosuppressive drugs, alone or associated with steroids, have a role in the treatment of ILD and exactly in which of them.

Other nonsteroidal antiinflammatory agents such as indomethacin have been anecdotally reported as beneficial and in an experimental model[234] seemed to decrease the mortality and diminish the accumulation of collagen in lung tissue. However, this medicament has only been used in a few cases in which other antiinflammatory drugs are contraindicated. Therefore, its real effect on alveolitis and/or fibrogenesis remains to be clarified, although I believe from the observation of four patients with this treatment that it has no important place in the therapy of ILD patients.

Nevertheless, corticosteroids, immunosuppressants, and inhibitors of arachidonate metabolism may help to control and reverse the alveolitis mainly, but the process which destroys the alveolar-capillary structures and severely distorts the parenchymal architecture with a definite loss of function is the abnormal accumulation of collagens. Thus, although a common assumption is that there is no solution to this problem, a few investigations, both in humans and in experimental models, have been made with the aim of avoiding or modulating or even reversing lung fibrosis.

Over the past 8 years we have used D-penicillamine,[235] which prevents formation of covalent intra- and intermolecular cross-links and, hence, conversion of soluble collagen to an insoluble and fibrillar structure.In a first attempt we studied 12 patients with advanced fibrosis, seven of whom had received steroids previously, despite which they showed progressive deterioration. All patients improved clinically and functionally during the first year of treatment. However, after 2 to 5 years most patients began to worsen, and eight patients died from different causes, but all associated with lung fibrosis. The remaining four patients were relatively stable during at least 5 years, and afterward a progressive deterioration was also observed. This work had two mistakes: there was no control group and the patients were selected in an advanced phase of the disease with severe fibrosis and honeycombing. Nevertheless, our initial impression was that D-penicillamine may have been beneficial and prolonged the survival time.

Steen et al.[236] reported the effect of D-penicillamine on lung fibrosis associated with PSS. They found that treated patients showed a significant increase in the diffusing capacity for carbon monoxide compared with those untreated. Moreover, this improvement was accompanied by arrest of dyspnea and chest radiograph abnormalities.

In addition, some studies performed on experimental lung fibrosis have also shown an antifibrosant effect of this drug, both on histopathological lesions and on collagen content.[237,238]

However, some severe side effects have stopped many colleagues from using D-penicillamine; the most severe of them, related to lung, is probably the development of obliterative bronchiolitis. However, this complication has been reported in a few cases with rheumatoid arthritis[239] and has not been observed in our patients, nor in the 44 scleroderma cases studied by Steen et al.

Another very interesting drug which can be used in humans and has given good results in other fibrotic processes is colchicine. This drug seems to present several theoretical advantages which should make it very useful for lung fibrosis therapy. In contrast to D-penicillamine, which interferes with some biosynthetic steps of collagen, colchicine acts in the two pathways of collagen metabolism, that is, in addition to preventing the extracellular deposit of this protein through the inhibition of microtubule assembly and the transcellular movement of collagen, it is able to increase the rate of collagenase synthesis at least *in vitro*.[240]

Kershenobich et al.[241] studied the effect of colchicine on 43 patients with liver cirrhosis for which 20 received placebo and 23 colchicine. They found that patients with colchicine treatment presented a lower mortality than those receiving placebo, although the difference was not significant. On the other hand, several of the colchicine-cases improved clinically with disappearance of ascites and edema and showed an increase in serum albumin and alkaline phosphatase levels. In contrast, all the survivors on placebo showed clinical and laboratory data of deterioration. In addition, in three patients using colchicine a noteworthy decrease in liver fibrosis was observed in serial biopsies.

Moreover, this drug has also been used in experimental lung fibrosis.

Dubrawsky et al.[242] demonstrated that colchicine modifies the collagen deposition in irradiation-induced lung fibrosis in rats. Furthermore, fibrosis was only reduced if colchicine was injected before the 90th day of irradiation, but not if the treatment was given 2 weeks before 120 or 150 days after irradiation. The authors interpreted this finding in the sense that colchicine is useful only when collagen is actively being synthesized and not after this protein has been deposited in the extracellular space.

Encouraged by these results, and by several informal conversations with Rojkind, we initiated 3 years ago a double-blind randomized trial with prednisone, prednisone plus D-penicillamine, prednisone plus colchicine, and prednisone plus both antifibrotic drugs, in patients with IPF and chronic HP.

A preliminary picture suggests that either of them, D-penicillamine or colchicine, added to corticoids produce a better clinical and physiological response than prednisone alone in HP. In contrast, in IPF, independent of the stage, only the use of both antifibrotic drugs together plus steroids seems to give a better result than any other of the combinations mentioned. However, it will be necessary to complete a rigorous follow-up of at least 5 years to arrive at a more conclusive answer.

On the other hand, the progressively deeper knowledge about the different steps related to collagen synthesis, processing, and degradation has generated a series of experimental works on fibrosis, focused on agents which may favorably alter some of these events. Among them, proline analogues have been used in experimental lung fibrosis with the aim of inhibiting the hydroxylation of prolyl residues, with controversial results.

Thus, Ryley et al.[243] and Kelley et al.[244] have demonstrated that proline analogues partially prevent the structural modifications, the increase of collagen, and the functional changes in bleomycin-induced pulmonary fibrosis in hamsters and rats. In contrast, Kehrer and Witschi,[234] in lung fibrosis developed in mice with BHT and O_2, found no effect at all caused by this compound, using the same dose and form of administration as Ryley.

Several reasons may explain these contradictory findings, including the different species and forms of inducing fibrosis, the "quality" of the lung fibrosis of each experiment, the pool size, and the different degree of incorporation of analogues into noncollagen proteins.

However, with further studies and methodological corrections (for example, to determine and reach therapeutic pulmonary levels), proline analogues should be effective in inhibiting lung fibrosis.

Beta-aminopropionitrile (BAPN), which inhibits lysyl oxidase and therefore decreases collagen cross-linking, has also been assayed in experimental pulmonary fibrosis.

Ryley et al.[245] evaluated this compound in bleomycin-induced lung damage in hamsters and found that BAPN reduces the rate of mortality, prevents excessive collagen accumulation, and reverses the functional alterations. This protective effect of BAPN has been confirmed by other authors.[246] Nevertheless, BAPN is a very potent lathyrogenic substance which may cause severe systemic alterations. We have used it to induce experimental emphysema in rats,[247] and after several months the animals showed skeletal deformities and aortic aneurysm, among other things. Thus, the main problem with this compound is to find subtoxic doses and a selective action over the lungs which allow it to be used for long periods.

Two exciting approaches for the future could arise with the utilization of some molecules which "naturally" participate in collagen metabolism.

One of them could be the amino-terminal extension peptides of procollagen, which seem to exert a feedback inhibition effect on collagen synthesis, at least over fibroblasts in culture.[248,249] The studies performed with this segment of procollagen type I suggest that a certain amount of it is required for the proper control of type I collagen production, and a defect in this action may accunt for excessive collagen synthesis in scleroderma.[249]

On the other hand, gamma interferon, a keystone lymphokine in the immune system, is a potent and selective inhibitor of collagen production and has recently been analyzed for its possible effects on PSS dermal fibroblasts. For this purpose, Rosenbloom et al.[250] studied five cell lines obtained from patients with a rapidly progressive disease which exhibited increased collagen synthesis. His group observed that the recombinant gamma interferon provoked a remarkable inhibition of collagen production in a concentration-dependent manner. In addition, using specific complementary DNA probes, they demonstrated that this effect can be explained by a decrease in specific collagen mRNAs.

Interestingly, nonrecombinant and recombinant gamma interferon have been used in phases I and II clinical trials in different types of neoplasias with an overall response rate from 2 to 9%, without signs of severe toxicity. In addition, there is a recent report of its efficacy in rheumatoid arthritis.[251]

I believe that the use of aminopropeptides of procollagens and gamma interferon represent two very attractive perspectives which must be explored *in vivo*, at least initially in fibrosis experimental models. Since both molecules play a normal role in the complex mechanisms involved in the regulation of collagen metabolism *in vivo*, they can effectively prevent its overproduction without the need for using "exogenous drugs". However, caution is also necessary because if there should be a decrease of cell receptors, these factors could not be useful.

Undoubtedly, all these "antifibrotic substances", D-penicillamine, colchicine, proline analogues, aminoterminal peptide of procollagens, BAPN, gamma interferon, and others, constitute important attempts to render reversible a pathological condition which has usually been considered irreversible and untreatable. All these investigations mentioned, although very incipient yet, suggest that this latter notion is probably an incorrect point of view and that the pharmacological control of fibrosis may be an achievable objective.

Nevertheless, there are several questions that must be answered. How do these drugs really act *in vivo* in the damaged organ? What effects can they have on healthy tissues for which modifications of collagen metabolism could be deleterious? In which patients and with what peculiar lung alterations may they be useful? For how long can these factors and drugs be administered? And the most relevant questions: if we should be able to stop or even reverse the fibrotic process, how could the regeneration of the delicate and complex alveolar-capillary structures which have been destroyed (end-stage lung) be reinitiated? Thus, how would the patient recover lung function in terms of gas exchange?

In this context, the best form of handling pulmonary fibrosis is probably through replacement of the lesioned lung by another lung, that is, through lung transplantation; and several attempts have had encouraging results.

Cooper et al.[252] have recently reported their results with unilateral lung transplantation for pulmonary fibrosis. Currently, they have performed five single lung transplantations for end-stage pulmonary fibrosis, and four patients are alive and well at 3, 6, 17, and 30 months. Moreover, three of them returned to normal employment. The authors attribute its success to a careful patient and donor selection, the use of cyclosporine, and the utilization of an omental pedicle to protect and improve healing of the bronchial anastomosis.

Without question, once the inherent problems related to the lung transplant, such as rejection, infection, and surgical complications, among others, have been overcome, the replacement of a terminal fibrotic lung by a new normal or almost normal lung will in the future be the ideal therapy for these patients.

REFERENCES

1. **Fulmer, J. D.,** An introduction to the interstitial lung diseases, in *Clinics in Chest Medicine. Symposium on Interstitial Lung Diseases,* Vol. 3 (No. 3), Fulmer, J. D., Ed., W. B. Saunders, Philadelphia, 1982, 457.
2. **Fulmer, J. D. and Crystal, R. G.,** Interstitial lung disease, in *Current Pulmonology,* Simmons, D. H., Ed., Houghton Mifflin, Boston, 1979, 1.
3. **Crystal, R. G., Gadek, J. E., Ferrans, V. J., Fulmer, J. D., Line, N. R., and Hunninghake, G. W.,** Interstitial lung disease: current concepts of pathogenesis, staging and therapy, *Am. J. Med.,* 70, 542, 1981.
4. **Selman, M. and Chapela, R.,** Neumopatias intersticiales difusas cronicas, in *Medicina Interna,* Uribe, M., E., Editorial Medica Panamericana, Mexico, D. F., 1987, chap. 360.
5. **Hamman, L. and Rich, A. R.,** Fulminating diffuse interstitial fibrosis of the lungs, *Trans. Am. Clin. Climatol. Assoc.,* 51, 154, 1935.
6. **Liebow, A. A., Steer, A., and Billingsley, J. G.,** Desquamative interstitial pneumonia, *Am. J. Med.,* 39, 369, 1965.
7. **Scadding, J. G.,** Fibrosing alveolitis, *Br. Med. J.,* 2, 686, 1964.
8. **Scadding, J. G. and Hinson, K. W. F.,** Diffuse fibrosing alveolitis (diffuse interstitial fibrosis of the lungs) — correlation of histology at biopsy with prognosis, *Thorax,* 22, 291, 1967.

9. **Rudd, R. M., Haslam, P. L., and Turner-Warwick, M.,** Cryptogenic fibrosing alveolitis. Relationship of pulmonary physiology and bronchoalveolar lavage to response to treatment and prognosis, *Am. Rev. Respir. Dis.*, 124, 1, 1981.
10. **Pinsker, K. L., Schneyer, B., Becker, N., and Kamholz, L. S.,** Usual interstitial pneumonia following Texas A2 influenza infection, *Chest*, 80, 123, 1981.
11. **Vergnon, J. M., De The, G., Weynants, P., Vincent, M., Mornex, J. F., and Brune, J.,** Cryptogenic fibrosing alveolitis and Epstein-Barr virus: an association?, *Lancet*, 2, 768, 1984.
12. **Thet, L. A., Parra, S., and Shelburne, J. D.,** Sequential changes in lung morphology during the repair of acute oxygen induced lung injury in adult rats, *Exp. Lung Res.*, 11, 209, 1986.
13. **Fasske, E. and Morgenroth, K.,** Experimental bleomycin lung in mice. A contribution to the pathogenesis of pulmonary fibrosis, *Lung*, 161, 133, 1983.
14. **Smith, L. J.,** Lung damage induced by butylated hydroxytoluene in mice, *Am. Rev. Respir. Dis.*, 130, 895, 1984.
15. **Bowden, D. H.,** Macrophages, dust, and pulmonary diseases, *Exp. Lung Res.*, 12, 89, 1987.
16. **Salvaggio, J. E. and deShazo, R. D.,** Pathogenesis of hypersensitivity pneumonitis, *Chest*, 89 (Suppl.), 190, 1986.
17. **Thomas, P. D. and Hunninghake, G. W.,** Current concepts of the pathogenesis of sarcoidosis, *Am. Rev. Respir. Dis.*, 135, 747, 1987.
18. **Selman, M., Montano, M., Montfort, I., and Perez-Tamayo, R.,** A new model of diffuse interstitial pulmonary fibrosis in the rat, *Exp. Mol. Pathol.*, 43, 375, 1985.
19. **Davis, W. B., Fells, G. A., Sun, X. H., Gadek, J. E., Venet, A., and Crystal, R. G.,** Eosinophil-mediated injury to lung parenchyma cells and interstitial matrix. A possible role for eosinophils in chronic inflammatory disorders of the lower respiratory tract, *J. Clin. Invest.*, 74, 269, 1984.
20. **Hunninghake, G. W. and Moseley, P. L.,** Immunological abnormalities of chronic noninfectious pulmonary diseases, in *Immunology of the Lung and Upper Respiratory Tract*, Bienenstock, J., Ed., McGraw-Hill, New York, 1984, 345.
21. **Laskin, D. L., Kimura, T., Sakakibara, Sh., Riley, D. J., and Berg, R. A.,** Chemotactic activity of collagen-like polypeptides for human peripheral blood neutrophils, *J. Leukocyte Biol.*, 39, 255, 1986.
22. **Kawanami, O., Ferrans, V. J., Fulmer, J. D., and Crystal, R. G.,** Ultrastructure of pulmonary mast cells in patients with fibrotic lung disorders, *Lab. Invest.*, 40, 717, 1979.
23. **Bitterman, P. B., Saltzman, L. E., Adelberg, S., Ferrans, V. J., and Crystal, R. G.,** Alveolar macrophage replication. One mechanism for the expansion of the mononuclear phagocyte population in the chronically inflamed lung, *J. Clin. Invest.*, 74, 460, 1984.
24. **Cantin, A. M., North, S. L., Fells, G. A., Hubbard, R. C., and Crystal, R. G.,** Oxidant-mediated epithelial cell injury in idiopathic pulmonary fibrosis, *J. Clin. Invest.*, 79, 1665, 1987.
25. **Peterson, M. W., Monick, M., and Hunninghake, G. W.,** Prognostic role of eosinophils in pulmonary fibrosis, *Chest*, 92, 51, 1987.
26. **Dreisin, R. B., Schwarz, M. I., Theofilopoulus, A. N., and Stanford, R. E.,** Circulating immune complexes in the idiopathic interstitial pneumonias, *N. Engl. J. Med.*, 298, 353, 1978.
27. **Kravis, T. C., Ahmed, A., Brown, T. E., Fulmer, J. D., and Crystal, R. G.,** Pathogenic mechanisms in pulmonary fibrosis. Collagen-induced migration inhibition factor production and cytotoxicity mediated by lymphocytes, *J. Clin. Invest.*, 58, 1223, 1976.
28. **Carvajal, R., Gonzalez, R., Vargas, F., and Selman, M.,** Cell-mediated immunity against connective tissue in experimental pulmonary fibrosis, *Lung*, 160, 131, 1982.
29. **Israel, H.,** Sarcoidosis, in *Current Pulmonology*, Vol. 1, Simmons, D. H., Ed., Houghton Mifflin, Boston, 1979, 163.
30. **Carrington, C. B., Gaensler, E. A., Coutu, R. E., Fitzgerald, M. X., and Gupta, R. G.,** Natural history and treated course of usual and desquamative interstitial pneumonia, *N. Engl. J. Med.*, 298, 801, 1978.
31. **Fink, J. N.,** Hypersensitivity pneumonitis, *J. Allergy Clin. Immunol.*, 74, 1, 1984.
32. **Raghu, G., Striker, L. J., Hudson, L. D., and Striker, G. E.,** Extracellular matrix in normal and fibrotic lungs, *Am. Rev. Respir. Dis.*, 131, 281, 1985.
33. **Abrahamson, D. R.,** Recent studies on the structure and pathology of basement membranes, *J. Pathol.*, 149, 257, 1986.
34. **Vracko, R.,** Significance of basal lamina for regeneration of injured lung, *Virchows Arch. A:*, 355, 264, 1972.
35. **Atkins, F. M., Friedman, M. M., Subba Rao, P. V., and Metcalfe, D. D.,** Interactions between mast cells, fibroblasts and connective tissue components, *Int. Arch. Allergy Appl. Immunol.*, 77, 96, 1985.
36. **Uitto, V. J., Schwartz, D., and Veis, A.,** Degradation of basement-membrane collagen by neutral proteases from human leukocytes, *Eur. J. Biochem.*, 105, 409, 1980.
37. **Haschek, W. M. and Witchi, H. P.,** Pulmonary fibrosis — a possible mechanism, *Toxicol. Appl. Pharmacol.*, 51, 475, 1979.

38. **Selman, M., Montano, M., Montfort, I., and Perez-Tamayo, R.,** The duration of pulmonary paraquat toxicity — enhancement effect of O_2 in the rat, *Exp. Mol. Pathol.*, 43, 388, 1985.
39. **Katzanstein, A. L.,** Pathogenesis of "fibrosis" in interstitial pneumonia: an electron microscopic study, *Hum. Pathol.*, 16, 1015, 1985.
40. **Seyer, J. M.,** Mediators of increased collagen synthesis in fibrosing organs, *Funam. Appl. Toxicol.*, 5, 228, 1985.
41. **Postlethwaite, A. E., Smith, G. N., Mainardi, C. L., Seyer, J. M., and Kang, A. H.,** Lymphocyte modulation of fibroblast function in vitro: stimulation and inhibition of collagen production by different effector molecules, *J. Immunol.*, 132, 2470, 1984.
42. **Heppleston, A. G. and Styles, J. A.,** Activity of a macrophage factor in collagen formation by silica, *Nature (London)*, 214, 521, 1967.
43. **Schmidt, J. A., Oliver, C. N., Lepe-Zuniga, J. L., Green, I., and Gery, I.,** Silica-stimulated monocytes release fibroblast proliferation factors identical to interleukin-1, *J. Clin. Invest.*, 73, s1462, 1984.
44. **Gritter, H. L., Adamson, I. Y. R., and King, G. M.,** Modulation of fibroblast activity by normal and silica-exposed alveolar macrophages, *J. Pathol.*, 148, 263, 1986.
45. **Rennard, S. I., Hunninghake, G. W., Bitterman, P. B., and Crystal, R. G.,** Production of fibronectin by the human alveolar macrophage: mechanism for the recruitment of fibroblasts to sites of tissue injury in interstitial lung diseases, *Proc. Natl. Acad. Sci. U.S.A.*, 78, 7147, 1981.
46. **Bitterman, P. B., Rennard, S. I., Hunninghake, G. W., and Crystal, R. G.,** Human alveolar macrophage growth factor for fibroblasts. Regulation and partial characterization, *J. Clin. Invest.*, 70, 806, 1982.
47. **Bitterman, P. B., Adelberg, S., and Crystal, R. G.,** Mechanisms of pulmonary fibrosis. Spontaneous release of the alveolar macrophage-derived growth factor in the interstitial lung disorders, *J. Clin. Invest.*, 72, 1801, 1983.
48. **Kovacs, E. J. and Kelley, J.,** Intra-alveolar release of a competence-type growth factor after lung injury, *Am. Rev. Respir. Dis.*, 133, 68, 1986.
49. **Bitterman, P. B., Wewers, M. D., Rennard, S. I., Adelberg, S., and Crystal, R. G.,** Modulation of alveolar macrophage-driven fibroblast proliteration by alternative macrophage mediators, *J. Clin. Invest.*, 77, 700, 1986.
50. **Duncan, M. R. and Berman, B.,** Interferon is the lymphokine and interferon the monikine responsible for inhibition of fibroblast collagen production and late but not early fibroblast proliferation, *J. Exp. Med.*, 162, 516, 1985.
51. **Schmidt, J. A., Mizel, S. B., Cohen, D., and Green, I.,** Interluekin 1, a potential regulator of fibroblast proliferation, *J. Immunol.*, 128, 2177, 1982.
52. **Shimokado, K., Raines, E. W., Madtes, D. K., Barret, T. B., Benditt, E. P., and Ross, R.,** A significant part of macrophage-derived growth factor consists of at least two forms of PDGF, *Cell*, 43, 277, 1985.
53. **Mornex, J. F., Martinet, Y., Yamauchi, K., Bitterman, P. B., Grotendorst, G. R., Chytil-Weir, A., Martin, G. R., and Crystal, R. G.,** Spontaneous expression of the c-sis gene and release of a platelet-derived growth factorlike molecule by human alveolar macrophages, *J. Clin. Invest.*, 78, 61, 1986.
54. **Lum, H., Tyler, W. S., Hyde, D. M., and Plopper, C. G.,** Morphometry of in situ and lavaged pulmonary alveolar macrophages from control and ozone-exposed rats, *Exp. Lung Res.*, 5, 61, 1983.
55. **Weissler, J. C., Lyons, R. C., Liscomb, M. F., and Towes, G. B.,** Human pulmonary macrophages. Functional comparison of cells obtained from whole lung and by bronchoalveolar lavage, *Am. Rev. Respir. Dis.*, 133, 473, 1986.
56. **Shellito, J. and Kaltreider, H. B.,** Heterogeneity of immunologic function among subfractions of normal rat alveolar macrophages. II. Activation as a determinant of functional activity, *Am. Rev. Respir. Dis.*, 131, 678, 1985.
57. **Bravo, M. A., Avila, R. E., Selman, M., and Antoniades, H. N.,** Platelet-derived growth factor in interstitial lung diseases, submitted for publication.
58. **Welgus, H. G., Campbell, E. J., Bar-Shavit, Z., Senior, R. M., and Teitelbaum, S. L.,** Human alveolar macrophages produce fibroblast-like collagenase and collagenase inhibitor, *J. Clin. Invest.*, 76, 219, 1985.
59. **Schrier, D. J., Phan, S. H., and McGarry, B. M.,** The effects of the nude (nu/nu) mutation on bleomycin-induced pulmonary fibrosis. A biochemical evaluation, *Am. Rev. Respir. Dis.*, 127, 614, 1983.
60. **Thrall, R. S., Lovett, E. J., Barton, R. W., McCormick, J. R., Phan, S. H., and Ward, P. A.,** The effect of T-depletion on the development of bleomycin-induced pulmonary fibrosis in the rat, *Am. Rev. Respir. Dis.*, 121, 99, 1980.
61. **Cathcart, M. C., Edmur, L. I., Ahtiala-Stewart, K., and Ahmad, M.,** Excessive T-helper cell function in patients with IPF: correlation with disease activity, *Clin. Immunol. Immunopathol.*, 43, 382, 1987.
62. **Goto, T., Befus, D., Low, R., and Bienenstock, J.,** Mast cell heterogeneity and hyperplasia in bleomycin-induced pulmonary fibrosis in rats, *Am. Rev. Respir. Dis.*, 130, 797, 1984.

63. **Hawkins, R. A., Claman, H. N., Clark, R. A. F., and Steigerwald, J. C.,** Increased dermal mast cell population in progressive systemic sclerosis: a link in chronic fibrosis?, *Ann. Int. Med.*, 102, 182, 1985.
64. **Takeda, T.,** Phagocytosis of mast cell granules by fibroblasts of the human gingiva, *Virchows Arch. A:*, 406, 197, 1985.
65. **Bjermer, L, Engstrom-Laurent, A., Thunell, M., and Hallgren, R.,** The mast cell and signs of pulmonary fibroblast activation in sarcoidosis, *Int. Arch. Allergy Appl. Immunol.*, 82, 298, 1987.
66. **Hatamochi, A., Fujiwara, K., and Ueki, H.,** Effects of histamine on collagen synthesis by cultured fibroblasts derived from guinea pig skin, *Arch. Dermatol. Res.*, 277, 60, 1985.
67. **Haslam, P. L., Cromwell, O., Dewar, A., and Turner-Warwick, M.,** Evidence of increased histamine levels in lung lavage fluids from patients with cryptogenic fibrosing alveolitis, *Clin. Exp. Immunol.*, 44, 587, 1981.
68. **Ross, R.,** The pathogenesis of atherosclerosis, An update, *N. Engl. J. Med.*, 314, 488, 1986.
69. **Ward, W. F., Molteni, A., Solliday, N. H., and Jones, G. E.,** The relationship between endothelial dysfunction and collagen accumulation in irradiated rat lung, *Int. J. Radiat. Oncol. Biol. Phys.*, 11, 1985.
70. **Diegelmann, R. F. and Lindblad, W. J.,** Cellular sources of fibrotic collagen, *Fundam. Appl. Toxicol.*, 5, 219, 1985.
71. **Crouch, E. D., Moxley, M. A., and Longmore, W.,** Synthesis of collagenous proteins by pulmonary type II epithelial cells, *Am. Rev. Respir. Dis.*, 135, 1118, 1987.
72. **Rojkind, M. and Kershenobich, D.,** Hepatic fibrosis, *Clin. Gastroenterol.*, 10, 737, 1981.
73. **LeRoy, E. C.,** The connective tissue in scleroderma, *Collagen Relat. Res.*, 1, 301, 1981.
74. **Hayakawa, T., Hashimoto, Y., Myokei, Y., Aoyama, H., and Izama, Y.,** Changes in type of collagen during the development of human postburn hypertrophic scars, *Clin. Chim. Acta*, 93, 119, 1979.
75. **Barnes, M. J.,** Collagen in atherosclerosis, *Collagen Relat. Res.*, 5, 65, 1985.
76. **Fulmer, J.D., Bienkowski, R. S., Cowan, M. J., Breul, S. D., Bradley, K. M., Ferrans, V. J., Roberts, W. C., and Crystal, R. G.,** Colagen concentration and rates of synthesis in idiopathic pulmonary fibrosis, *Am. Rev. Respir. Dis.*, 122, 289, 1980.
77. **Seyer, J. M., Hutchenson, E. T., and Kang, A. H.,** Collagen polymorphism in idioapthic pulmonary fibrosis, *J. Clin. Invest.*, 57, 1498, 1976.
78. **Zapol, W. M., Trelstad, R. L., Coffey, J. W., Tsai, I., and Salvador, R. A.,** Pulmonary fibrosis in severe acute respiratory failure, *Am. Rev. Respir. Dis.*, 119, 547, 1979.
79. **Haschek, W. M., Klwin-Szanto, A. J. P., Last, J., Reiser, K. M., and Witschi, H. R.,** Long-term morphologic and biochemical features of experimentally induced lung fibrosis in the mouse, *Lab. Invest.*, 46, 438, 1982.
80. **Chvapil, M. and Peng, Y. M.,** Oxygen and lung fibrosis, *Arch. Environ. Health*, 30, 528, 1975.
81. **Fulmer, J. D., Flint, A., and Law, D. E.,** Experimental granulomatous lung disease in guinea pigs, *Lung*, 161, 337, 1983.
82. **Reiser, K. M., Haschek, W. M., Hesterberg, T. W., and Last, J. A.,** Experimental silicosis. II. Long-term effects of intratracheally instilled quartz on collagen metabolism and morphologic characteristics of rat lungs, *Am. J. Pathol.*, 110, 30, 1983.
83. **Selman, M., Chapela, R., Montano, M., Soto, H., and Diaz, L.,** Changes of collagen content in fibrotic lung disease, *Arch. Invest. Med. (Mexico)*, 13, 93, 1982.
84. **Kirk, J. M. E., Da Costa, P. E., Turner-Warwick, M., Littleton, R. J., and Laurent, G. J.,** Biochemical evidence for an increased and progressive deposition of collagen in lungs of patients with pulmonary fibrosis, *Clin. Sci.*, 70, 39, 1986.
85. **Selman, M., Montano, M., Ramos, C., and Chapela, R.,** Concentration, biosynthesis and degradation of collagen in idiopathic pulmonary fibrosis, *Thorax*, 41, 355, 1986.
86. **Yamaguchi, M., Takahashi, T., Togashi, H., Arai, H., and Motomiya, M.,** The corrected collagen content in paraquat lungs, *Chest*, 90, 251, 1986.
87. **Cheah, K. S.,** Collagen genes and inherited connective tissue disease, *Biochem. J.*, 229, 287, 1985.
88. **Bradley, K. H., McConnell-Breul, S., and Crystal, R. G.,** Collagen in human lung: quantitation of rates of synthesis and partial characterization of composition, *J. Clin. Invest.*, 55, 543, 1975.
89. **Huang, T. W.,** Chemical and histochemical studies of human alveolar collagen fibers, *Am. J. Pathol.*, 86, 81, 1977.
90. **Madri, J. A. and Furthmayr, H.,** Collagen polymorphism in the lung. An immunochemical study of pulmonary fibrosis, *Hum. Pathol.*, 11, 353, 1980.
91. **Batemen, E. D., Turner-Warwick, M., and Adelmann-Grill, B. C.,** Immunohistochemical study of collagen type on human foetal lung and fibrotic lung disease, *Thorax*, 36, 645, 1981.
92. **Bateman, E. D., Turner-Warwick, M., Haslam, P. L., and Adelmann-Grill, B. C.,** Cryptogenic fibrosing alveolitis: prediction of fibrogenic activity from immunohistochemical studies of collagen types in lung biopsy specimens, *Thorax*, 38, 93, 1983.
93. **Seyer, J. M., Kang, A. H., and Rodnan, G.,** Investigation of type I and type III collagens of the lung in progressive systemic sclerosis, *Arthritis Rheum.*, 24, 625, 1981.

94. **Pierce, J. A., Resnich, H., and Henry, P. H.,** Collagen and elastin metabolism in the lung, skin and bones of adult rats, *J. Lab. Clin. Med.*, 69, 485, 1967.
95. **Bienkowski, R. S., Cowan, M. J., McDonald, J. A., and Crystal, R. G.,** Degradation of newly synthesized collagen, *J. Biol. Chem.*, 253, 4356, 1978.
96. **Laurent, G. J.,** Rates of collagen synthesis in lung, skin and muscle obtained in vivo by a simplified method using [^{3}H]-proline, *Biochem. J.*, 206, 535, 1982.
97. **Laurent, G. J. and McAnulty, R. J.,** Protein metabolism during bleomycin-induced pulmonary fibrosis in rabbits. In vivo evidence for collagen accumulation because of increased synthesis and decreased degradation of the newly synthesized collagen, *Am. Rev. Respir. Dis.*, 128, 82, 1983.
98. **Kehrer, J. P., Lee, T. C. C., and Solem, S. M.,** Comparison of in vitro and in vivo rates of collagen synthesis in normal and damaged lung tissue, *Exp. Lung Res.*, 10, 187, 1986.
99. **Phan, S. H., Thrall, R. S., and Ward, P. A.,** Bleomycin-induced pulmonary fibrosis in rats: biochemical demonstration of increased rate of collagen synthesis, *Am. Rev. Respir. Dis.*, 121, 501, 1980.
100. **Lee, Y. C. C. and Kehrer, J. P.,** Increased pulmonary collagen synthesis in mice treated with cyclophosphamide, *Drug Chem. Toxicol.*, 8, 503, 1985.
101. **Morse, C.C., Sigler, C., Lock, S., Hakkinen, P. J., Haschek, W. M., and Witschi, H. P.,** Pulmonary toxicity of cyclophosphamide: a 1-year study, *Exp. Mol. Pathol.*, 42, 251, 1985.
102. **Kehrer, J. P. and Witschi, H. P.,** In vivo collagen accumulation in an experimental model of pulmonary fibrosis, *Exp. Lung Res.*, 1, 259, 1980.
103. **Hesterberg, T. W., Gerriets, J. E., Reiser, K. M., Jackson, A. C., Cross, C. E., and Last, J. A.,** Bleomycin-induced pulmonary fibrosis. Correlation of biochemical, physiological, and histological changes, *Toxicol. Appl. Pharmacol.*, 60, 360, 1981.
104. **Collins, J. F., McCollough, B., Coalson, J. J., and Johanson, W. G.,** Bleomycin-induced diffuse interstitial pulmonary fibrosis in baboons. II. Further studies on connective tissue changes, *Am. Rev. Respir. Dis.*, 123, 305, 1981.
105. **Ehrinpreis, M. N., Giambrone, M. A., and Rojkind, M.,** Liver proline oxidase activity and collagen synthesis in rats with cirrhosis induced by carbon tetrachloride, *Biochim. Biophys. Acta*, 629, 184, 1980.
106. **Krane, S. M.,** The turnover and degradation of collagen, in *Fibrosis*, Evered, D. and Whelan, J., Eds., Pitman Publishing, London, 1985, 97.
107. **Gadek, J. E., Kelman, J. A., Fells, G., Weiberger, S. E., Howitz, A. L., Reynolds, H. Y., Fulmer, J. D., and Crystal, R. G.,** Collagenase in the lower respiratory tract of patients with idiopathic pulmonary fibrosis, *N. Engl. J. Med.*, 301, 737, 1979.
108. **Christner, P., Fein, A., Goldberg, S., Lippman, M., Abrams, W., and Weinbaum, G.,** Collagenase in the lower respiratory tract of patients with adult respiratory distress syndrome, *Am. Rev. Respir. Dis.*, 131, 690, 1985.
109. **Perez-Tamayo, R.,** Cirrhosis of the liver: a reversible disease?, in *Pathology Annual*, Vol. 14 (Part 2), Sommers, S. C. and Rosen, O. O., Eds., Appleton-Century-Crofts, New York, 1979, 183.
110. **Brady, A. H.,** Collagenase in scleroderma, *J. Clin. Invest.*, 56, 1175, 1975.
111. **Zavala, D. C.,** Clinical manifestations and diagnosis of pulmonary fibrosis, *Heart Lung*, 6, 434, 1977.
112. **Crystal, R. G., Fulmer, J. D., Roberst, W. C., Moss, M. L., Line, B. R., and Reynolds, H. Y.,** Idiopathic pulmonary fibrosis. Clinical, histologic, radiographic, physiologic, scintigraphic,, cytologic and biochemical aspects, *Ann. Intern. Med.*, 85, 769, 1976.
113. **Rebuck, A. S., Braude, A. C., and Chamberlain, D. W.,** Arterial $PaCO_2$ as an index of activity in fibrosing alveolitis, *Chest*, 82, 757, 1982.
114. **Galko, B., Grossman, R. F., Day, A., Tenenbaum, J., Kirsh, J., and Rebuck, A. S.,** Hypertrophic pulmonary osteroarthropathy in four patients with interstitial pulmonary disease, *Chest*, 88, 94, 1985.
115. **Rounds, S. and Hill, N. S.,** Pulmonary hypertensive diseases, *Chest*, 85, 397, 1984.
116. **Fraser, R. G.,** The radiology of interstitial lung disease, in *Clinics in Chest Medicine. Symposium on Interstitial Lung Diseases*, Vol. 3 (No. 3), Fulmer, J. D., Ed., W. B. Saunders, Philadelphia, 1982, 475.
117. **Nakata, H., Kimoto, T., Nakayama, T., Kido, M., Miyakazi, N., and Harada, S.,** Diffuse peripheral lung disease: evaluation by high-resolution computed tomography, *Radiology*, 157, 181, 1985.
118. **Wright, P. H., Buxton-Thomas, M., Kreel, L., and Steel, S. J.,** Cryptogenic fibrosing alveolitis: pattern of disease in the lung, *Thorax*, 39, 857, 1984.
119. **Bellamy, E. A., Husband, J. E., Blaquiere, R. M., and Law, M. R.,** Bleomycin-related lung damage: CT evidence, *Radiology*, 156, 155, 1985.
120. **McFadden, R. G., Carr, T. J., and Wood, T. E.,** Proton magnetic resonance imaging to stage activity of interstitial lung disease, *Chest*, 92, 31, 1987.
121. **Vinitski, S., Pearson, M. G., Karlik, S. J., Morgan, W. K. C., Carey, L. S., Perkins, G., Goto, T., and Befus, D.,** Differentiation of parenchymal lung disorders with in vitro proton nuclear magnetic resonance, *Magn. Reson. Med.*, 3, 120, 1986.
122. **Gilbert, R. and Auchincloss, H. J.,** What is a "restrictive" defect?, *Arch. Int. Med.*, 146, 1779, 1986.

123. **Fulmer, J. D., Roberts, W. C., von Gal, E. Ra., and Crystal, R. G.,** Morphologic-physiologic correlates of the severity of fibrosis and degree of cellularity in idiopathic pulmonary fibrosis, *J. Clin. Invest.*, 63, 665, 1979.
124. **Pérez-Neria, J., Selman, M., Rubio, H., Ocana, H., Chapela, R., and Mendoza, A.,** Relationship between lung inflammation or fibrosis and frequency dependence of compliance in interstitial pulmonary diseases, *Respiration*, 52, 254, 1987.
125. **Gibson, G. J., Pride, N. B., Davis, J., and Schroter, R. C.,** Exponential description of the static pressure-volume curve of normal and diseases lungs, *Am. Rev. Respir. Dis.*, 120, 799, 1979.
126. **Fulmer, J. D., Roberts, W. C., von Gal, E. R., and Crystal, R. G.,** Small airways in idiopathic pulmonary fibrosis. Comparison of morphologic and physiologic observations, *J. Clin. Invest.*, 60, 595, 1977.
127. **Allen, D. H., Williams, G. V., and Woolcock, A. J.,** Bird breeder's hypersensitivity pneumonitis: progressive studies of lung function after cessation of exposure to the provoking antigen, *Am. Rev. Respir. Dis.*, 114, 555, 1976.
128. **Martinez, C., Perez, J., Rojas, A., Chapela, R., and Selman, M.,** Via aerea periferica y zona de transicion en fibrosis pulmonary idiopatica y alveolitis alergica extrinseca, *Neumol. Cir. Torax (Mexico)*, 45, 31, 1984.
129. **Collins, J. F., Orozco, C. R., McCullough, B., Coalson, J. J., and Johanson, W. G.,** Pulmonary fibrosis with small-airway disease: a model in non human primates, *Exp. Lung Res.*, 3, 91, 1982.
130. **Weiberger, S. E., Johnson, S. T., and Weiss, S. T.,** Clinical significance of pulmonary function tests. Use and interpretation of the single-breath diffusing capacity, *Chest*, 78, 483, 1980.
131. **Miller, W. C., Heard, J. G., Unger, K. M., and Suich, D. M.,** Anatomical lung shunting in pulmonary fibrosis, *Thorax*, 41, 208, 1986.
132. **Watters, L. C., King, T. E., Schwarz, M. I., Waldron, J. A., Standord, R. E., and Cherniak, R. M.,** A clinical, radiographic, and physiologic scoring system for the longitudinal assessment of patients with idiopathic pulmonary fibrosis, *Am. Rev. Respir. Dis.*, 133, 97, 1986.
133. **Pérez-Padilla, R., West, P., Lertzman, M., and Kryger, M. H.,** Breathing during sleep in patients with interstitial lung disease, *Am. Rev. Respir. Dis.*, 132, 224, 1985.
134. **Hawrylkiewicz, I., Izdebska-Makosa, Z., Grebska, E., and Zielinski, J.,** Pulmonary haemodynamics at rest and on exercise in patients with idiopathic pulmonary fibrosis, *Bull. Eur. Physiopathol. Resp.*, 18, 403, 1982.
135. **Weitzenblum, E., Ehrart, M., Rasaholinjanahary, J., and Hirth, C.,** Pulmonary hemodynamics in idiopathic pulmonary disease and other interstitial pulmonary diseases, *Respiration*, 44, 118, 1983.
136. **Ray, J. F., Lawton, B. R., Myers, W. O., Toyama, W. M., Reyes, C. N., Emanual, D. A., Burns, J. L., Pederson, D. P., Dovenbarger, W. V., Wenzel, F. J., and Sautter, R. D.,** Open pulmonary biopsy. Nineteen-year experience with 416 consecutive operations, *Chest*, 69, 43, 1976.
137. **Gaensler, E. A. and Carrington, C. B.,** Open biopsy for chronic diffuse infiltrative lung disease: clinical, roentgenographic and physiological correlations in 502 patients, *Ann. Thorac. Surg.*, 30, 411, 1980.
138. **Early, G. L., Williams, T. E., and Kilman, J. W.,** Open lung biopsy. Its effects on therapy in the pediatric patient, *Chest*, 87, 467, 1985.
139. **Andersen, H. A. and Fontana, R. S.,** Transbronchoscopic lung biopsy for diffuse pulmonary diseases: technique and results in 450 cases, *Chest*, 62, 125, 1972.
140. **Andersen, H. A.,** Transbronchoscopic lung biopsy for diffuse pulmonary disease, *Chest*, 73 (Suppl.), 734, 1978.
141. **Zavala, D. C.,** Transbronchial biopsy in diffuse lung disease, *Chest*, 73 (Suppl.), 727, 1978.
142. **Poe, R. H., Utell, M. J., Israel, R. H., Hall, W. J., and Eshleman, J. D.,** Sensitivity and specificity of the nonspecific transbronchial lung biopsy, *Am. Rev. Respir. Dis.*, 119, 25, 1979.
143. **Gilman, M. J. and Wang, K. P.,** Transbronchial lung biopsy in sarcoidosis. An approach to determine the optimal number of biopsies, *Am. Rev. Respir. Dis.*, 122, 721, 1980.
144. **Wall, C. P., Gaensler, E. A., Carrington, C. B., and Hayes, J. A.,** Comparison of transbronchial and open biopsies in chronic infiltrative lung diseases, *Am. Rev. Respir. Dis.*, 123, 280, 1981.
145. **Fortoul, T. I., Barrios, R., Chapela, R., and Selman, M.,** comparison of transbronchial and open lung biopsies in interstitial lung diseases, *Arch. Invest. Med. (Mex)*, 19, 7, 1988.
146. **Steel, S. J. and Winstanley, D. P.,** Trephine biopsy for diffuse lung lesions, *Br. Med. J.*, 3, 30, 1967.
147. **Hear, B. E.,** Pathology of interstitial lung diseases with particular reference to terminology, classification and threpine lung biopsy, *Chest*, 69 (Suppl), 252, 1976.
148. **Morales, J., Selman, M., Diaz, X., Villalba, J., Fortoul, T. I., Chapela, R., and Barrios, R.,** Biopsia pulmonar transpleural por toracoscopia en el diagnostico de la neumopatia intersticial difusa, *Arch. Bronconeumol. (Spain)*, 22, 215, 1985.
149. **Newman, S. L., Michel, R. P., and Wang, N. S.,** Longular lung biopsy: is it representative?, *Am. Rev. Respir. Dis.*, 132, 1084, 1985.
150. **Weinstein, L.,** Sensitivity and specificity of lingular segmental biopsies of the lung, *Chest*, 90, 383, 1986.

151. **Reynolds, H. Y.,** Bronchoalveolar lavage, *Am. Rev. Respir. Dis.*, 135, 250, 1987.
152. **Daniele, R. P., Elias, J. A., Epstein, P. E., and Rossman, M. D.,** Bronchoalveolar lavage: role in the pathogenesis, diagnosis and management of interstitial lung disease, *Ann. Intern. Med.*, 102, 93, 1985.
153. **Rennard, S. I., Basset, G., Lecossier, D., O'Donnell, K. M., Pinkston, P., Martin, P. G., and Crystal, R. G.,** Estimation of volume of epithelial lining fluid recovered by lavage using urea as marker of dilution, *J. Appl. Physiol.*, 60, 532, 1986.
154. **Marcy, T. W., Merrill, W. W., Rankin, J. A., and Reynolds, H. Y.,** Limitations of using urea to quantify epithelia lining fluid recovered by bronchoalveolar lavage, *Am. Rev. Respir. Dis.*, 135, 1276, 1987.
155. **Baughman, R. P., Bosken, C. H., London, R. G., Hurtubise, P., and Wessler, T.,** Quantitation of bronchoalveolar lavage with methylene blue, *Am. Rev. Respir. Dis.*, 128, 266, 1983.
156. **Merrill, W., O'Hearn, E., Rankin, J., Naegel, G., Mattay, R. A., and Reynolds, H. Y.,** Kinetics analysis of respiratory tract proteins recovered during a sequential lavage protocol, *Am. Rev. Respir. Dis.*, 126, 617, 1982.
157. **Davis, G., Giancola, M. S., Constanza, M. C., and Low, R. B.,** Analysis of sequential bronchoalveolar lavage samples from healthy human volunteers, *Am. Rev. Respir. Dis.*, 126, 611, 1982.
158. **Dohn, M. N. and Baughman, R. P.,** Effect of changing instilled volume for bronchoalveolar lavage in patients with interstitial lung disease, *Am. Rev. Respir. Dis.*, 132, 390, 1985.
159. **Saltini, C., Hance, A. J., Ferrans, V. J., Basset, R., Bitterman, P. B., and Crystal, R. G.,** Accurate quantification of cells recovered by bronchoalveolar lavage, *Am. Rev. Respir. Dis.*, 130, 650, 1984.
160. **Mordelet-Dambrine, M., Arnoux, A., Stanislas-Leguern, G., Sandron, D., Chretien, J., and Huchon, G.,** Processing of lung lavage fluid causes variability in bronchoalveolar cell count, *Am. Rev. Respir. Dis.*, 130, 305, 1984.
161. **Pingleton, S. K., Harrison, G. F., Stechschulte, D. J., Wesselius, L. J., Kerby, G. R., and Ruth, W. E.,** Effect of location, pH, and temperature of instillate in bronchoalveolar lavage in normal volunteers, *Am. Rev. Respir. Dis.*, 128, 1035, 1983.
162. **García, J. G. N., Wolven, R. G., Garcia, P. L., and Keogh, B. A.,** Assessment of interlobar variations of bronchoalveolar lavage cellular differentials in interstitial lung diseases, *Am. Rev. Respir. Dis.*, 133, 444, 1986.
163. **Nugen, K., Peterson, M., Jolles, H., Monick, M., and Hunninghake, G. W.,** The utility of bilateral bronchoalveolar lavage in patients with interstitial lung diseases, *Am. Rev. Respir. Dis.*, 129, 81A, 1984.
164. **Keogh, B. A., Hunninghake, G. W., Line, B. R., and Crystal, R. G.,** The alveolitis of pulmonary sarcoidosis: evaluation of natural history and alveolitis-dependent changes in lung function, *Am. Rev. Respir. Dis.*, 128, 256, 1983.
165. **Lawrence, E. C., Teague, R. B., Gottlieb, M. D., Jhingran, S. G., and Lieberman, J.,** Serial changes in markers of disease activity with corticosteroids treatment in sarcoidosis, *Am. J. Med.*, 74, 747, 1983.
166. **Ceuppens, J. L., Lacquet, L. M., Marien, G., Demedts, M., Van Den Eeckkhout, A., and Stevens, E.,** Alveolar T-cell subsets in pulmonary sarcoidosis, *Am. Rev. Respir. Dis.*, 129, 563, 1984.
167. **Thrall, R. S. and Barton, R. W.,** A comparison of lymphocyte populations in lung tissue and in bronchoalveolar lavage fluid of rats at various times during the development of bleomycin-induced pulmonary fibrosis, *Am. Rev. Respir. Dis.*, 129, 279, 1984.
168. **Davis, G. S., Brody, A. R., and Craighead, J. E.,** Analysis of airspace and interstitial mononuclear cell populations in human diffuse interstitial lung disease, *Am. Rev. Respir. Dis.*, 118, 7, 1978.
169. **Leatherman, J. W., Michael, A. F., Schwartz, A., and Hoidal, J. R.,** Lung T cells in hypersensitivity pneumonitis, *Ann. Intern. Med.*, 100, 390, 1984.
170. **Barrios, R., Selman, M., Franco, R., Chapela, R., Lopez, J. S., and Fortoul, T. I.,** Subpopulations of T cells in lung biopsies from patients with pigeon breeder's disease, *Lung*, 165, 181, 1987.
171. **Semenzato, G., Chilosi, M., Ossi,E., Trentin, L., Pizzolo, G., Cipriani, A., Agostini, C., Zambella, R., Marcer, G., and Gasparotto, G.,** Bronchoalveolar lavage and lung histology. Comparative analysis of inflammatory and immunocompetent cells in patients with sarcoidosis and hypersensitivity pneumonitis, *Am. Rev. Respir. Dis.*, 132, 400, 1985.
172. **Watters, L. C., Schwarz, M. I., Cherniack, R. M., Waldron, J. A., Thaddens, L.D., Standord, R. E., and King, T. E.,** Idiopathic pulmonary fibrosis. Pretreatment bronchoalveolar lavage cellular constituents and their relationship with lung histopathology and clinical response to therapy, *Am. Rev. Respir. Dis.*, 135, 696, 1987.
173. **Turner-Warwick, M. and Haslam, P. L.,** The value of serial bronchoalveolar lavages in assessing the clinical progress of patients with cryptogenic fibrosing alveolitis, *Am. Rev. Respir. Dis.*, 135, 26, 1987.
174. **Cormier, Y., Belanger, J., and Labiolette, M.,** Prognostic significance of bronchoalveolar lymphocytosis in farmer's lung, *Am. Rev. Respir. Dis.*, 135, 692, 1987.
175. **Cormier, Y., Belanger, J., and Labiolette, M.,** Persistent bronchoalveolar lymphocytosis in asymptomatic farmers, *Am. Rev. Respir. Dis.*, 133, 843, 1986.
176. **Basset, F., Soler, P., Jaurand, M. C., and Bignon, J.,** Ultrastructural examination of broncho-alveolar lavage for diagnosis of pulmonary histiocytosis. X. Preliminary report on 4 cases, *Thorax*, 32, 303, 1977.

177. **Niden, A. H., Mishkin, F. S., and Khurama, M. M. L.,** ^{67}Gallium citrate lung scans in interstitial lung disease, *Chest,* 69 (Suppl.), 266, 1976.
178. **Line, B. R., Fulmer, J. D., Reynolds, H. Y., Roberts, W. C., Jones, E. A., Harris, E. K., and Crystal, R. G.,** Gallium-citrate scanning in the staging of idiopathic pulmonary fibrosis: correlation with physiologic and morphologic features and brochoalveolar lavage, *Am. Rev. Respir. Dis.,* 118, 355, 1978.
179. **Line, B. R., Hunninghake, G. W., Keogh, B. A., Jones, E. A., Johnston, G. S., and Crystal, R. G.,** Gallium-67 scanning to stage the alveolitis of sarcoidosis: correlation with clinical studies, pulmonary function studies and bronchoalveolar lavage, *Am. Rev. Respir. Dis.,* 123, 440, 1981.
180. **Baughman, R. P., Fernandez, M., Bosken, C. H., Mantil, J., and Hurtubisi, P.,** Comparison of gallium-67 scanning, bronchoalveolar lavage, and serum angiotensin-converting enzyme levels in pulmonary sarcoidosis. Predicting response to therapy, *Am. Rev. Respir. Dis.,* 129, 676, 1984.
181. **Gelb, A. F., Dreisin, R. B., Epstein, J. D., Silverthorne, J. D., Bickel, Y., Fields, M., Border, W. A., and Taylor, C. R.,** Immune complexes, gallium lung scans, and bronchoalveolar lavage in idiopathic interstitial pneumonitis-fibrosis. A structure-function clinical study, *Chest,* 84, 148, 1983.
182. **Begin, R., Cantin, A., Drapen, G., Lamoroux, G., Boctor, M., Masse, S., and Rola-Pleszxzynski, M.,** Pulmonary uptake of gallium-67 in asbestos-exposed humans and sheep, *Am. Rev. Respir. Dis.,* 127, 623, 1983.
183. **Low, R. B., Cutroneo, K. R., Davis, G. S., and Giancola, M. S.,** Lavage type III procollagen N-terminal peptides in human pulmonary fibrosis and sarcoidosis, *Lab. Invest.,* 48, 755, 1983.
184. **Kirk, J. M. E., Bateman, E. D., Haslam, P. L., Laurent, G. J., and Turner-Warwick, M.,** Serum type III procollagen peptide concentration in cryptogenic fibrosing aveolitis and its clinical relevance, *Thorax,* 39, 726, 1984.
185. **Watanabe, Y., Yamaki, K., Yamakawa, I., Takagi, K., and Satake, T.,** Type III procollagen N-terminal peptides in experimental pulmonary fibrosis and human respiratory diseases, *Eur. J. Respir. Dis.,* 67, 10, 1985.
186. **Begin, R., Martel, M., Desmarais, Y., Drapeau, G., Boileau, R., Rola-Pleszcyzynski, M., and Masse, S.,** Fibronectin and procollagen 3 levels in bronchoalveolar lavage of asbestos-exposed human subjects and sheep, *Chest,* 89, 237, 1986.
187. **Anttinen, H., Terho, E. O., Jarvensivu, P. M., and Savolainen, E. R.,** Elevated serum galactosylhydroxylysyl glucosyltransferase, a collagen synthesis marker, in fibrosing lung diseases, *Clin. Chim. Acta,* 3, 3, 1985.
188. **Anttinen, H., Terho, E. O., Myllyla, R., and Savolainen, E. R.,** Two serum markers of collagen biosynthesis as possible indicators of irreversible pulmonary impairment in farmer's lung, *Am. Rev. Respir. Dis.,* 133, 88, 1986.
189. **Pirie, H. M. and Selman, I. E.,** A bovine pulmonary disease resembling human diffuse fibrosing alveolitis, *Proc. R. Soc. Med.,* 65, 987, 1972.
190. **Johnson, B., Burton, M., and Qualls, C. W.,** Interstitial pulmonary fibrosis in an African elephant, *J. Am. Vet. Med. Assoc.,* 189, 1188, 1986.
191. **Rossi, G. A., Hunninghake, G. W., Kawanami, O., Ferrans, V. J., Hansen, C. T., and Crystal, R. G.,** Motheaten mice — an animal model with an inherited form of interstitial lung disease, *Am. Rev. Respir. Dis.,* 131, 150, 1985.
192. **Galli, S. J. and Kitamura, Y.,** Genetically mast-cell-deficient W/Wv and Sl/Sld mice. Their value for the analysis of the roles of mast cells in biologic responses in vivo, *Am. J. Pathol.,* 127, 191, 1987.
193. **Carrington, C. B.,** Organizing interstitial pneumonia. Definition of the lesion and attempts to devise an experimental model, *Yale J. Biol. Med.,* 40, 352, 1968.
194. **Saunier, C.,** Fibroses pulmonaires interstitielles experimentales, *Bull. Eur. Physiopathol. Resp.,* 18, 515, 1982.
195. **Tryka, F. A., Skornik, W. A., Godleski, J. J., and Brain, J. D.,** Potentiation of bleomycin-induced lung injury by exposure to 70% oxygen, *Am. Rev. Respir. Dis.,* 126, 1074, 1982.
196. **Bowden, D. H.,** Unraveling pulmonary fibrosis: the bleomycin model, *Lab. Invest.,* 50, 487, 1984.
197. **Adler, K. B., Callahan, L. M., and Evans, J. N.,** Cellular alterations in the alveolar wall in bleomycin-induced pulmonary fibrosis in rats. An ultrastructural morphometric study, *Am. Rev. Respir. Dis.,* 133, 1043, 1986.
198. **Marom, Z., Weinberg, K. S., and Fanburg, B. L.,** Effect of bleomycin on collagenolytic activity of the rat alveolar macrophage, *Am. Rev. Respir. Dis.,* 121, 859, 1980.
199. **Adamson, I. Y. R. and Bowden, D. H.,** The pathogenesis of bleomycin-induced pulmonary fibrosis in mice, *Am. J. Pathol.,* 77, 185, 1974.
200. **Sikic, B. I., Young, D. M., Mimnaugh, E. G., and Gram, T. E.,** Quantification of bleomycin pulmonary toxicity in mice by changes in lung hydroxyproline content and morphometric histopathology, *Cancer Res.,* 38, 787, 1978.
201. **Snider, G. L., Hayes, J. A., and Korthy, A. L.,** Chronic interstitial pulmonary fibrosis produced in hamster by endotracheal bleomycin, *Am. Rev. Respir. Dis.,* 117, 1099, 1978.

202. **Schnrier, D. J., Kinkel, R. G., and Phan, S. H.,** The role of strain variation in murine bleomycin-induced pulmonary fibrosis, *Am. Rev. Respir. Dis.*, 127, 63 1983.
203. **Smith, P., Heath, D., and Kay, J. M.,** The pathogenesis and structure of paraquat-induced pulmonary fibrosis in rats, *J. Pathol.*, 114, 57, 1974.
204. **Butler, C.,** Pulmonary interstitial fibrosis from paraquat in the hamster, *Arch. Pathol.*, 99, 503, 1975.
205. **Popenoe, D.,** Effects of paraquat aerosol on mouse lung, *Arch. Pathol. Lab. Med.*, 103, 331, 1979.
206. **Fukuda, Y., Ferrans, V. J., Schoenberger, C. I. and Rennard, S. I., and Crystal, R. G.,** Patterns of pulmonary structural remodeling after experimental paraquat toxicity. The morphogenesis of intra-alveolar fibrosis, *Am. J. Pathol.*, 118, 452, 1985.
207. **Selman, M.,** unpublished data.
208. **Smith, L. J.,** Hyperoxic lung injury: biochemical, cellular and morphologic characterization in the mouse, *J. Lab. Clin. Med.*, 106, 269, 1985.
209. **Coursin, D. B., Cihla, H. P., Will, J. A., and McCreary, J. L.,** Adaptation to chronic hyperoxia. Biochemical effects and the response to subsequent lethal hyperoxia, *Am. Rev. Respir. Dis.*, 135, 1002, 1987.
210. **Adamson, I. Y. R. and Bowden, D. H.,** Endothelial injury and repair in radiation-induced pulmonary fibrosis, *Am. J. Pathol.*, 112, 224, 1983.
211. **Shrivastava, P. N., Hans, L., and Concannon, J. P.,** Changes in pulmonary compliance and production of fibrosis in x-irriadiated lungs in rats, *Radiology*, 112, 439, 1974.
212. **Ward, W. F., Shih-Hoellwarth, A., Port, C. D., and Kim, Y. Y.,** Modification of radiation-induced pulmonary fibrosis in rats, *Radiology*, 131, 751, 1979.
213. **Vergara, J. A., Raymund, U., and Thet, L. A.,** Changes in lung morphology and cell number in radiation pneumonitis and fibrosis: a quantitative ultrastructural study, *Int. J. Radiat. Oncol. Biol. Phys.*, 13, 723, 1987.
214. **Lugano, E. M., Dauber, J. H., and Daniele, R. P.,** Acute pulmonary silicosis. Lung morphology, histology, and macrophage chemotoxin secretion, *Am. J. Pathol.*, 109, 27, 1982.
215. **Callis, A. H., Sohnle, P. G., Mandel, G. S., Wiessner, J., and Mandel, N. S.,** Kinetics of inflammatory and fibrotic pulmonary changes in a murine model of silicosis, *J. Lab. Clin. Med.*, 105, 547, 1985.
216. **Selman, M., Ramos, C., Montano, M., Montfort, I., and Perez-Tamayo, R.,** Increase of biosynthesis and degradation of collagen in normal lungs induced by soluble factors obtained from experimental pulmonary silicosis, *Arch. Biol. Med. Exp.*, 20, 311, 1987.
217. **Selman, M., Montano, M., Montfort, I., and Perez-Tamayo, R.,** En busca de un modelo de fibrosis pulmonar intersticial cronica, *Patologia*, 19, 231, 1981.
218. **Rinaldo, J., Goldstein, R. H., and Snider, G. L.,** Modification of oxygen toxicity after lung injury by bleomycin in hamsters, *Am. Rev. Respir. Dis.*, 126, 1030, 1982.
219. **Witschi, H. R., Haschek, W. M., Klein-Szanto, A. J. P., and Hakkinen, P. J.,** Potentiation of diffuse lung damage by oxygen: determining variables, *Am. Rev. Respir. Dis.*, 123, 98, 1981.
220. **Witschi, H. R., Haschek, W. M., Meyer, K. R., Ullrich, R. L., and Dolbey, W. E.,** A pathogenetic mechanism in lung fibrosis, *Chest*, 78 (Suppl.), 395, 1980.
221. **Brody, A. R., Soler, P., Basset, F., Haschek, W. M., and Witschi, H.,** Epithelial-mesenchymal association of cells in human pulmonary fibrosis and in BHT-oxygen-induced fibrosis in mice, *Exp. Lung Res.*, 2, 207, 1981.
222. **Turner-Warwick, M., Burrows, B., and Johnson, A.,** Cryptogenic fibrosing alveolitis: clinical features and their influence on survival, *Thorax*, 35, 171, 1980.
223. **Cutroneo, K. R., Rokowski, R., and Counts, D. F.,** Clucocorticoids and collagen syntheisi: comparison of in vivo and cell culture studies, *Collagen Relat. Res.*, 1, 557, 1981.
224. **Sterling, K. M., Di Petrillo, T., Cutroneo, K. R., and Prestyko, A.,** Inhibition of collagen accumulation by glucocorticoids in rat lung after intratracheal bleomycin instillation, *Cancer Res.*, 42, 405, 1982.
225. **Rokowski, R. J., Sheehy, J., and Cutroneo, K. R.,** Glucocorticoid-mediated selective reduction of functionating collagen messenger ribonucleic acid, *Arch. Biochem. Biophys.*, 210, 74, 1981.
226. **Raghu, G.,** Idiopathic pulmonary fibrosis. A rational clinical approach, *Chest*, 92, 148, 1987.
227. **Ozaki, T., Nakayama, T., Ishimi, H., Kawano, T., Yasuoka, S., and Tusbura, E.,** Glucocorticoid receptors in bronchoalveolar cells from patients with idioathic pulmonary fibrosis, *Am. Rev. Respir. Dis.*, 126, 968, 1982.
228. **Lacronique, J. G., Rennard, S. I., Bitterman, P. B., Ozaki, T., and Crystal, R. G.,** Alveolar macrophages in idiopathic pulmonary fibrosis have glucocorticoid receptors, but glucocorticoid therapy does not suppress alveolar macrophage release of fibronectin and alveolar macrophage derived growth factor, *Am. Rev. Respir. Dis.*, 130, 450, 1984.
229. **Keogh, B. A., Bernardo, J., Hunninghake, G. W., Line, B. R., Price, D. L., and Crystal, R. G.,** Effect of intermittent high dose parenteral corticosteroids on the alveolitis of idiopathic pulmonary fibrosis, *Am. Rev. Respir. Dis.*, 127, 18, 1983.
230. **Brown, C. H. and Turner-Warwick, M.,** The treatment of cryptogenic fibrosing alveolitis with immunosuppressant drugs, *Q. J. Med.*, 40, 289, 1971.

231. **Winterbauer, R. H., Hammar, S. P. Hallman, K. O., Hays, J. E., Pardee, N. E., Morgan, E. H., Allen, J. D., Moores, K. D., Bush, W., and Walker, J. H.,** Diffuse interstitial pneumonitis. Clinicopathological correlations in 20 patients treated with prednisone/azathioprine, *Am. J. Med.*, 65, 661, 1978.
232. **Weiland, J., Dorinsky, P., Davis, W. B., Patch, E., Thomson, D., and Gadek, J.,** Use of cyclophosphamide in the therapy of idiopathic pulmonary fibrosis, *Chest*, 89 (Suppl), 150, 1986.
233. **Eisenberg, H.,** The interstitial lung diseases associated with the collagen-vascular disorders, in *Clinics in Chest Medicine. Symposium on Interstitial Lung Disease*, Vol. 3 (No. 3), Fulmer, J.D., Ed., W. B. Saunders, Philadelphia, 1982, 565..
234. **Kehrer, J. P. and Witschi, H.,** The effect of indomethacin, prednisolone and cis-4-hydroxyproline on pulmonary fibrosis produced by butylated hydroxytoluene and oxygen, *Toxicology*, 20, 281, 1981.
235. **Chapela, R., Zuniga, G., and Selman, M.,** D-Penicillamine in the therapy of fibrotic lung diseases, *Int. J. Clin. Pharmacol. Ther. Toxicol.*, 24, 16, 1986.
236. **Steen, V. D., Owens, G. R., Redmon, C., Rodnan, G. P., and Medsger, T. A.,** The effect of D-penicillamine on pulmonary fingings in systemic sclerosis, *Arthritis Rheum.*, 28, 882, 1985.
237. **Ward, W. F., Shih-Hoellwarth, A., and Tuttle, R. D.,** Collagen accumulation in irradiated rat lung. Modification by D-penicillamine, *Radiology*, 146, 533, 1983.
238. **Molteni, A., Ward, W. F., Ts'ao, C., Solliday, N. H., and Dunne, M.,** Monocrotaline-induced pulmonary fibrosis in rats: amelioration by captopril and penicillamine, *Proc. Soc. Exp. Biol. Med.*, 180, 112, 1985.
239. **Murphy, K. C., Atkins, C. J., Offer, R. C., Hogg, J. C., and Stein, H. B.,** Obliterative bronchiolitis in two rheumatoid arthritis patients treated with penicillamine, *Arthritis Rheum.*, 24, 557, 1981.
240. **Bauer, E. A and Valle, K. J.,** Colchicine-induced modulation of collagenase in human skin fibroblast-cultures. I. Stimulation of enzyme synthesis in normal cells, *J. Invest. Dermatol.*, 79, 398, 1982.
241. **Kershenobich, D., Uribe, M., Suarez, G. I., Mata, J. M., Perez-Tamayo, R., and Rojkind, M.,** Treatment of cirrhosis with colchicine. A double-blind randomized trial, *Gastroenterology*, 77, 532, 1979.
242. **Dubrawsky, C., Dubravisky, N. B., and Withers, H. R.,** The effect of colchicine on the accumulation of hydroxyproline and on the lung compliance after irradiation, *Radiat. Res.*, 73, 111, 1978.
243. **Ryley, D. J., Kerr, J. S., Berg, R. A., Ianni, B. D., Pietra, G. G., Edelman, N. H., and Prockop, D. J.,** Prevention of bleomycin-induced pulmonary fibrosis in the hamster by cis-4-hydroxy-l-proline, *Am. Rev. Respir. Dis.*, 123, 388, 1981.
244. **Kelley, J., Newman, R. A., and Evans, J. N.,** Bleomycin-induced pulmonary fibrosis. Prevention with an inhibitor of collagen synthesis, *J. Lab. Clin. Med.*, 96, 954, 1980.
245. **Ryley, D. J., Kerr, J. S., Berg, R. A., Ianni, B. D., Pietra, G. G., Edelman, N. H., and Prockop, D. J.,** Beta-aminopropionitrile prevents bleomycin-induced pulmonary fibrosis in the hamster, *Am. Rev. Respir. Dis.*, 125, 67, 1982.
246. **Fuller, G. C.,** Perspectives for the use of collagen synthesis inhibitors as antifibrotic agents, *J. Med. Chem.*, 24, 651, 1981.
247. **Selman, M., Soto, H., Barrios, R., and Diaz, L.,** Variations in collagen content and biosynthesis in experimental lung emphysema, *Am. Rev. Respir. Dis.*, 125 (Suppl.), 265, 1982.
248. **Krieg, T., Horlein, D., Wiestner, M., and Muller, P. K.,** Aminoterminal extension peptides from type I procollagen normalize excessive collagen synthesis of scleroderma fibroblasts, *Arch. Dermatol. Res.*, 263, 171, 1978.
249. **Perlish, J. S., Timpl, R., and Fleischmajer, R.,** Collagen synthesis regulation by the aminopropeptide of procollagen I in normal and scleroderma fibroblasts, *Arthritis Rheum.*, 28, 647, 1985.
250. **Rosenbloom, J., Feldman, G., Freundlich, B., and Jimenez, S.,** Inhibition of excessive scleroderma fibroblast collagen production by recombinant gamma-interferon. Association with a coordinate decrease in types I and III procollagen messenger RNA levels, *Arthritis Rheum.*, 29, 851, 1986.
251. **Bonnem, E. M. and Oldham, R. K.,** Gamma-interferon: physiology and speculation on its role in medicine, *J. Biol. Response Mod.*, 6, 275, 1987.
252. **Cooper, J. D., Pearson, F. G., Patterson, G. A., Todd, T. R. J., Ginsberg, R. J., Goldberg, M., and De Majo, W. A. P.,** Technique of successful lung transplantation in humans, *Thorac. Cardiovasc. Surg.*, 93, 173, 1987.

INDEX

A

B

C

D

M

N

O

P

For Product Safety Concerns and Information please contact our EU
representative GPSR@taylorandfrancis.com
Taylor & Francis Verlag GmbH, Kaufingerstraße 24, 80331 München, Germany

www.ingramcontent.com/pod-product-compliance
Ingram Content Group UK Ltd.
Pitfield, Milton Keynes, MK11 3LW, UK
UKHW052221190926
475702UK00006B/1143
* 9 7 8 1 1 3 8 5 0 6 7 8 7 *